From Euclidean to Hilbert Spaces

To my mentors, Sissa Abbati and Renzo Cirelli, who taught me the importance of rigor in mathematics, and to Brunella, Paola, Clara and Tommo, whose passion for their work has both helped and brought joy to many

From Euclidean to Hilbert Spaces

Introduction to Functional Analysis and its Applications

Edoardo Provenzi

iSTE

WILEY

First published 2021 in Great Britain and the United States by ISTE Ltd and John Wiley & Sons, Inc.

ISTE Ltd
27-37 St George's Road
London SW19 4EU
UK

www.iste.co.uk

John Wiley & Sons, Inc.
111 River Street
Hoboken, NJ 07030
USA

www.wiley.com

Library of Congress Control Number: 2021937006

British Library Cataloguing-in-Publication Data
A CIP record for this book is available from the British Library
ISBN 978-1-78630-682-1

MIX
Paper from responsible sources
FSC® C013604

Contents

Preface

This book provides an introduction to the key theoretical concepts associated with Hilbert spaces and with operators defined over these spaces.

Our decision to dedicate a whole book to the subject of Hilbert spaces stems from a simple observation: of all the infinite dimensional vector spaces, Hilbert spaces bear the closest resemblance to finite dimensional Euclidean spaces, that is, $\mathbb{R}^n$ or $\mathbb{C}^n$, which provide the framework for classical analysis and linear algebra.

The topological subtleties which come into play when using infinite dimensions mean that certain conditions (which are always verified in finite dimensions) must be posed in order to maintain the validity of known results from Euclidian spaces. For Hilbert spaces, one of these topological conditions is completeness, that is, any Cauchy sequence must converge in the space in which it is defined.

From this perspective, the theory of Hilbert spaces may be seen as an elegant conjunction of algebra, analysis and topology. It draws on the work of some of the great mathematicians of the early 20th century, including Riesz, Banach and, evidently, Hilbert, who established the conditions needed to extend classical algebra and analysis into infinite dimensions.

One particularly important linear operator, the Fourier transform, appears on multiple occasions throughout this book. We start by examining the properties of this transform in finite dimensions, with the discrete Fourier transform, before extending it to infinite dimensions, considering the use of this operator in a range of different domains, including signal and image processing.

A clear understanding of the concepts introduced in this book is essential for mathematicians, physicists or engineers hoping to progress in any field, whether applied or theoretical. These concepts provide access to tools and techniques developed over a particularly rich, creative period in the history of mathematics, which remain relevant for both pure and applied forms of the subject.

The author would like to thank Olivier Husson for his assistance in producing the majority of the figures included in this book.

April 2021

1

Inner Product Spaces (Pre-Hilbert)

This chapter will focus on inner product spaces, that is, vector spaces with a scalar product, specifically those of *finite dimension*.

1.1. Real and complex inner products

In real Euclidean spaces $\mathbb{R}^2$ and $\mathbb{R}^3$, the inner product of two vectors v, w is defined as the real number:

$$v \bullet w = \langle v, w \rangle = \|v\| \|w\| \cos(\vartheta)$$

where ϑ is the smallest angle between v and w and $\| \ \|$ represents the norm (or the magnitude) of the vectors.

Using the inner product, it is possible to define the orthogonal projection of vector v in the direction defined by vector w. A distinction must be made between:

– the *scalar projection* of v in the direction of w: $\|v\| \cos(\theta) = \frac{\langle v, w \rangle}{\|w\|}$; and

– the *vector projection* of v in the direction of w: $\|v\| \cos(\theta) \frac{w}{\|w\|} = \frac{\langle v, w \rangle}{\|w\|^2} w$;

where $\frac{w}{\|w\|}$ is the unit vector in the direction of w. Evidently, the roles of v and w can be reversed.

The absolute value of the scalar projection measures the "similarity" of the directions of two vectors. To understand this concept, consider two remarkable relative positions between v and w:

– if v and w possess the same direction, then the angle between them ϑ is either null or π, hence $\cos(\vartheta) = \pm 1$, that is, the absolute value of the scalar projection of v in direction w is $\|v\|$;

– however, if v and w are perpendicular, then $\vartheta = \frac{\pi}{2}$ and hence $\cos(\vartheta) = 0$, showing that the scalar projection of v in direction w is null.

When the position of v relative to w falls somewhere in the interval between the two vectors described above, the absolute value of the scalar projection of v in the direction of w falls between 0 and $\|v\|$; this explains its use to measure the similarity of the direction of vectors.

In this book, we shall consider vector spaces which are far more complex than $\mathbb{R}^2$ and $\mathbb{R}^3$, and the measure of vector similarity obtained through projection supplies crucial information concerning the coherence of directions.

Before we can obtain this information, we must begin by moving from Euclidean spaces $\mathbb{R}^2$ and $\mathbb{R}^3$ to abstract vector spaces. The general definition of an inner product and an orthogonal projection in these spaces may be seen as an extension of the previous definitions, permitting their application to spaces in which our representation of vectors is no longer applicable.

Geometric properties, which can only be apprehended and, notably, visualized in two or three dimensions, must be replaced by a set of algebraic properties which can be used in any dimension.

Evidently, these algebraic properties must be necessary and sufficient to characterize the inner product of vectors in a plane or in real space. This approach, in which we generalize concepts which are "intuitive" in two or three dimensions, is a classic approach in mathematics.

In this chapter, the symbol V will be used to describe a vector space defined over the field $\mathbb{K}$, where $\mathbb{K}$ is either $\mathbb{R}$ or $\mathbb{C}$ and is of finite dimension $n < +\infty$. Field $\mathbb{K}$ contains the *scalars* used to construct linear combinations between vectors in V. Note that two finite dimensional vector spaces are isomorphic if and only if they are of the same dimension. Furthermore, if we establish a basis $B = (b_1, \ldots, b_n)$ for V, an *isomorphism* between V and $\mathbb{K}^n$ can be constructed as follows:

$$\begin{array}{llll} I: & V & \longrightarrow & \mathbb{K}^n \\ & v = [v]_B = \sum\limits_{i=1}^{n} v_i b_i & \longmapsto & \begin{pmatrix} v_1 \\ \vdots \\ v_n \end{pmatrix} \end{array}$$

that is, I associates each $v \in V$ with the vector of $\mathbb{K}^n$ given by the scalar components of v in relation to the established basis B. Since I is an isomorphism, it follows that $\mathbb{K}^n$ *is the prototype of all vector spaces of dimension* n *over a field* $\mathbb{K}$.

DEFINITION 1.1.– *Let* V *be a vector space defined over a field* $\mathbb{K}$.

A $\mathbb{K}$-form over V is an application defined over $V \times V$ with values in $\mathbb{K}$, that is:

$$\begin{aligned} \phi : V \times V &\longrightarrow \mathbb{K} \\ (v, w) &\longmapsto \phi(v, w). \end{aligned}$$

DEFINITION 1.2.– *Let V be a real vector space. A couple $(V, \langle , \rangle)$ is said to be a real inner product space (or a real pre-Hilbert space) if the form $\langle , \rangle$ is:*

1) bilinear, i.e.[1] *linear in relation to each argument (the other being fixed):*

$$\langle v_1 + v_2, w_1 + w_2 \rangle = \langle v_1, w_1 \rangle + \langle v_1, w_2 \rangle + \langle v_2, w_1 \rangle + \langle v_2, w_2 \rangle, \\ \forall\, v_1, v_2, w_1, w_2 \in V$$

and:

$$\langle \alpha v, \beta w \rangle = \alpha \langle v, \beta w \rangle = \beta \langle \alpha v, w \rangle = \alpha\beta \langle v, w \rangle, \quad \forall \alpha, \beta \in \mathbb{R},\ v, w \in V$$

2) symmetrical: $\langle v, w \rangle = \langle w, v \rangle$, $\forall v, w \in V$;

3) defined: $\langle v, v \rangle = 0 \iff v = 0_V$, the null vector of the vector space V;

4) positive: $\langle v, v \rangle > 0\ \forall v \in V,\ v \neq 0_V$.

Upon reflection, we see that, for a real form over V, *the symmetry and bilinearity requirements are equivalent to requiring symmetry and linearity on the left-hand side*, that is:

$$\langle v_1 + v_2, w \rangle = \langle v_1, w \rangle + \langle v_2, w \rangle, \quad \langle \alpha v, w \rangle = \alpha \langle v, w \rangle, \quad \forall v, w \in V,\ \forall \alpha \in \mathbb{R}$$

The simplest and most important example of a real inner product is the *canonical inner product*, defined as follows: let $v = (v_1, v_2, \ldots, v_n)$, $w = (w_1, w_2, \ldots, w_n)$ be two vectors in $\mathbb{R}^n$ written with their components in relation to any given, but fixed, basis $(b_i)_{i=1}^n$ in $\mathbb{R}^n$. The canonical inner product of v and w is:

$$\langle v, w \rangle \equiv \sum_{i=1}^{n} v_i w_i = v^t \cdot w = v \cdot w^t,$$

where v^t and w^t in the final equations are the transposed vectors of v and w, giving us the matrix product of a line vector (treated as a $1 \times n$ matrix) and a column vector (treated as an $n \times 1$ matrix).

The extension of these definitions to complex vector spaces is not particularly straightforward. First, note that if V is a *complex* vector space, then *there is no bilinear and definite-positive transformation* over $V \times V$. In this case, any vector $v \in V$ would give the following:

$$\langle iv, iv \rangle = i^2 \langle v, v \rangle = -\langle v, v \rangle \leqslant 0 \text{ since } \langle v, v \rangle \geqslant 0 \text{ by positivity.}$$

1 i.e. is the abbreviation of the Latin expression "id est", meaning "that is". This term is often used in mathematical literature.

As we shall see, the property of positivity is essential in order to define a norm (and thus a distance, and by extension, a topology) from a complex inner product. To obtain an algebraic structure for complex scalar products which remains compatible with a topological structure, we are therefore forced to abandon the notion of bilinearity, and to search for an alternative.

We could consider antilinearity[2], i.e.

$\langle \alpha v, \beta w\rangle = \bar{\alpha}\bar{\beta}\langle v, w\rangle$

But it has the same problem as bilinearity, $\langle iv, iv\rangle = (-i)(-i)\langle v, v\rangle = i^2\langle v, v\rangle = -\langle v, v\rangle^2 \leqslant 0$.

A simple analysis shows that, in order to avoid losing the positivity, it is sufficient to request the linearity with respect to one variable and the antilinearity with respect to the other. This property is called *sesquilinearity*[3].

The choice of the linear and antilinear variable is entirely arbitrary.

By convention, the antilinear component is placed on the right-hand side in mathematics, but on the left-hand side in physics.

We have chosen to adopt the mathematical convention here, i.e. $\langle \alpha v, \beta w\rangle = \alpha\bar{\beta}\langle v, w\rangle$.

Next, it is important to note that *sesquilinearity and symmetry are incompatible*: if both properties were verified, then $\langle v, \alpha w\rangle = \bar{\alpha}\langle v, w\rangle$, and also $\langle v, \alpha w\rangle = \langle \alpha w, v\rangle = \alpha\langle w, v\rangle = \alpha\langle v, w\rangle$. Thus, $\langle v, \alpha w\rangle = \bar{\alpha}\langle v, w\rangle = \alpha\langle v, w\rangle$ which can only be verified if $\alpha \in \mathbb{R}$.

Thus $\langle , \rangle$ cannot be both sesquilinear and symmetrical when working with vectors belonging to a complex vector space.

The example shown above demonstrates that, instead of symmetry, the property which must be verified for every vector pair v, w is $\langle v, w\rangle = \overline{\langle w, v\rangle}$, that is, changing the order of the vectors in $\langle , \rangle$ must be equivalent to complex conjugation.

A transform which verifies this property is said to be *Hermitian*[4].

---

2 The symbols z^* and $\bar{z}$ represent the *complex conjugation*, i.e. if $z \in \mathbb{C}$, $z = a + ib$, $a, b \in \mathbb{R}$, then $z^* = \bar{z} = a - ib$. We recall that $\sum_{k=1}^{n} \overline{z_k} = \overline{\sum_{k=1}^{n} z_k}$, $\prod_{k=1}^{n} \overline{z_k} = \overline{\prod_{k=1}^{n} z_k}$, $|z|^2 = z\bar{z}$ and $z = \bar{z}$ if and only if $z \in \mathbb{R}$.

3 Sesqui comes from the Latin *semisque*, meaning *one and a half times*. This term is used to highlight the fact that there are not two instances of linearity, but one "and a half", due to the presence of the complex conjugation.

4 For the French mathematician Charles Hermite (1822, Dieuze-1901, Paris).

These observations provide full justification for Definition 1.3.

DEFINITION 1.3.– *Let V be a complex vector space. The pair $(V, \langle , \rangle)$ is said to be a complex inner product space (or a complex pre-Hilbert space) if $\langle , \rangle$ is a complex form which is:*

1) sesquilinear:

$$\langle v_1 + v_2, w_1 + w_2 \rangle = \langle v_1, w_1 \rangle + \langle v_1, w_2 \rangle + \langle v_2, w_1 \rangle + \langle v_2, w_2 \rangle$$

$\forall\, v_1, v_2, w_1, w_2 \in V$, *and:*

$$\xrightarrow[\text{Linearity on the left}]{\text{Antilinearity on the right}} \langle \alpha v, \beta w \rangle = \alpha \langle v, \beta w \rangle = \bar{\beta} \langle \alpha v, w \rangle = \alpha \bar{\beta} \langle v, w \rangle$$

$\forall\, \alpha, \beta \in \mathbb{C}, \forall\, v, w \in V$;

2) Hermitian: $\langle v, w \rangle = \overline{\langle w, v \rangle}$, $\forall v, w \in V$;

3) definite: $\langle v, v \rangle = 0 \iff v = 0_V$, *the null vector of the vector space* V;

4) positive: $\langle v, v \rangle > 0\ \forall v \in V,\ v \neq 0_V$.

As in the case of the canonical inner product, for a complex form over V, *the symmetry and sesquilinearity requirement is equivalent to requiring the Hermitian property and linearity on the left-hand side*; if these properties are verified, then:

$$\langle v, \alpha w \rangle = \overline{\langle \alpha w, v \rangle} = \overline{\alpha \langle w, v \rangle} = \bar{\alpha} \overline{\langle w, v \rangle} = \bar{\alpha} \overline{\overline{\langle v, w \rangle}} = \bar{\alpha} \langle v, w \rangle, \quad \forall \alpha \in \mathbb{C}.$$

Considering the sum of n, rather than two, vectors, sesquilinearity is represented by the following formulae:

$$\boxed{\langle \sum_{i=1}^{n} \alpha_i v_i\, ,\, w \rangle = \sum_{i=1}^{n} \alpha_i \langle v_i, w \rangle} \qquad [1.1]$$

$$\boxed{\langle v\, ,\, \sum_{i=1}^{n} \alpha_i w_i \rangle = \sum_{i=1}^{n} \overline{\alpha_i} \langle v, w_i \rangle} \qquad [1.2]$$

In $\mathbb{C}^n$, the *complex Euclidean inner product* is defined by:

$$\langle v, w \rangle \equiv \sum_{i=1}^{n} v_i \overline{w_i} = v \cdot (\overline{w})^t = v^t \cdot \overline{w}$$

where $v = (v_1, v_2, \ldots, v_n), w = (w_1, w_2, \ldots, w_n) \in \mathbb{C}^n$ are written with their components in relation to any given, but fixed, basis $(b_i)_{i=1}^n$ in $\mathbb{C}^n$.

The symbol $\mathbb{K}$ will be used throughout to represent either $\mathbb{R}$ or $\mathbb{C}$ in the context of properties which are valid independently of the reality or complexity of the inner product.

THEOREM 1.1.– Let $(V, \langle\ ,\ \rangle)$ be an inner product space. We have:

1) $\langle v, 0_V \rangle = 0\ \forall v \in V$;

2) if $\langle u, w \rangle = \langle v, w \rangle\ \forall w \in V$, then u and v must coincide;

3) $\langle v, w \rangle = 0\ \forall v \in V \iff w = 0_V$, i.e. the null vector is the only vector which is orthogonal to all of the other vectors.

PROOF.–

1) $\langle v, 0_V \rangle = \langle v, 0_V + 0_V \rangle = \langle v, 0_V \rangle + \langle v, 0_V \rangle$ by linearity, i.e. $\langle v, 0_V \rangle - \langle v, 0_V \rangle = 0 = \langle v, 0_V \rangle$.

2) $\langle u, w \rangle = \langle v, w \rangle\ \forall w \in V$ implies, by linearity, that $\langle u - v, w \rangle = 0\ \forall w \in V$ and thus, notably, considering $w = u - v$, we obtain $\langle u - v, u - v \rangle = 0$, implying, due to the definite positiveness of the inner product, that $u - v = 0_V$, i.e. $u = v$.

3) If $w = 0_V$, then $\langle v, w \rangle = 0\ \forall v \in V$ using property (1). Inversely, by hypothesis, it holds that $\langle v, w \rangle = 0 = \langle v, 0_V \rangle\ \forall v \in V$, but then property (2) implies that $w = 0_V$. □

Finally, let us consider a typical property of the complex inner product, which results directly from a property of complex numbers.

THEOREM 1.2.– Let $(V, \langle\ ,\ \rangle)$ be a complex inner product space. Thus:

$$\Im(\langle v, w \rangle) = \Re(\langle v, iw \rangle) \qquad \forall v, w \in V$$

PROOF.– Consider any complex number $z = a + ib$, so $-iz = b - ia$, hence $b = \Im(z) = \Re(-iz)$. Taking $z = \langle v, w \rangle$, we obtain $\Im(\langle v, w \rangle) = \Re(-i\langle v, w \rangle) = \Re(\langle v, iw \rangle)$ by sesquilinearity. □

1.2. The norm associated with an inner product and normed vector spaces

If $(V, \langle, \rangle)$ is an inner product space over $\mathbb{K}$, then a norm on V can be defined as follows:

$$\begin{aligned} \|\ \|: V &\to \mathbb{R}_0^+ = [0, +\infty) \\ v &\to \|v\| = \sqrt{\langle v, v \rangle} \end{aligned}$$

Note that $\|v\|$ is well defined since $\langle v, v \rangle \geqslant 0\ \forall v \in V$. Once a norm has been established, it is always possible to define a *distance* between two vectors v, w in V: $d(v, w) = \|v - w\|$.

The vector $v \in V$ such that $\|v\| = 1$ is known as a *unit vector*. Every vector $v \in V$ can be *normalized* to produce a unit vector, simply by dividing it by its norm.

NOTABLE EXAMPLES.–

$$(\mathbb{R}^n, \langle , \rangle) : \|v\| = \sqrt{\sum_{i=1}^{n} v_i^2}$$

$$(\mathbb{C}^n, \langle , \rangle) : \|v\| = \sqrt{\sum_{i=1}^{n} v_i \bar{v}_i} = \sqrt{\sum_{i=1}^{n} |v_i|^2}$$

Three properties of the norm, which should already be known, are listed below. Taking any $v, w \in V$, and any $\alpha \in \mathbb{K}$:

1) $\|v\| \geqslant 0$, $\|v\| = 0 \iff v = 0_V$;

2) $\|\alpha v\| = |\alpha| \|v\|$ (homogeneity);

3) $\|v + w\| \leqslant \|v\| + \|w\|$ (triangle inequality).

DEFINITION 1.4 (normed vector space).– *A normed vector space is a pair* $(V, \| \ \|)$ *given by a vector space* V *and a function, called a norm,* $\| \ \| : V \to \mathbb{R}_0^+$, *satisfying the three properties listed above.*

A norm $\| \ \|$ *is Hilbertian if there exists an inner product* $\langle \ , \ \rangle$ *on* V *such that* $\|v\| = \sqrt{\langle v, v \rangle}$ $\forall v \in V$.

Canonically, an inner product space is therefore a normed vector space. Counter-examples can be used to show that the reverse is not generally true.

Note that, by definition, $\langle v, v \rangle = \|v\| \, \|v\|$, but, in general, the magnitude of the inner product between two different vectors is dominated by the product of their norms. This is the result of the well-known inequality shown below.

THEOREM 1.3 (Cauchy-Schwarz inequality).– For all $v, w \in (V, \langle \ , \ \rangle)$ we have:

$$\boxed{|\langle v, w \rangle| \leqslant \|v\| \|w\|}$$

PROOF.– Dozens of proofs of the Cauchy-Schwarz inequality have been produced. One of the most elegant proofs is shown below, followed by the simplest one:

– *first proof*: if $w = 0_V$, then the inequality is verified trivially with $0 = 0$. If $w \neq 0_V$, then we can define $z = v - \frac{\langle v,w\rangle}{\|w\|^2}w$, i.e. $v = \frac{\langle v,w\rangle}{\|w\|^2}w + z$, and we note that:

$$\langle z, w\rangle = \langle v - \frac{\langle v,w\rangle}{\|w\|^2}w, w\rangle = \langle v,w\rangle - \frac{\langle v,w\rangle}{\cancel{\|w\|^2}}\cancel{\langle w,w\rangle} = 0$$

thus:

$$\begin{aligned}
\|v\|^2 &= \langle v,v\rangle = \left\langle \frac{\langle v,w\rangle}{\|w\|^2}w + z, \frac{\langle v,w\rangle}{\|w\|^2}w + z \right\rangle \\
&= \frac{\langle v,w\rangle}{\|w\|^2}\left\langle w, \frac{\langle v,w\rangle}{\|w\|^2}w + z \right\rangle + \left\langle z, \frac{\langle v,w\rangle}{\|w\|^2}w + z \right\rangle \\
&= \frac{\langle v,w\rangle}{\|w\|^2}\frac{\overline{\langle v,w\rangle}}{\cancel{\|w\|^2}}\cancel{\langle w,w\rangle} + \frac{\langle v,w\rangle}{\|w\|^2}\langle w,z\rangle + \frac{\overline{\langle v,w\rangle}}{\|w\|^2}\langle z,w\rangle + \langle z,z\rangle \\
&= \frac{|\langle v,w\rangle|^2}{\|w\|^2} + \|z\|^2
\end{aligned}$$

as the two intermediate terms in the penultimate step are zero, since $\langle z,w\rangle = \langle w,z\rangle = 0$.

As $\|z\|^2 \geqslant 0$, we have seen that:

$$\|v\|^2 = \frac{|\langle v,w\rangle|^2}{\|w\|^2} + \|z\|^2 \geqslant \frac{|\langle v,w\rangle|^2}{\|w\|^2}$$

i.e. $|\langle v,w\rangle|^2 \leqslant \|v\|^2\|w\|^2$, hence $|\langle v,w\rangle| \leqslant \|v\|\|w\|$;

– *second proof* (in one line!): $\forall t \in \mathbb{R}$ we have:

$$\begin{aligned}
&0 \leqslant \|tv - w\|^2 = \langle tv - w, tv - w\rangle = t^2\|v\|^2 - 2t\langle v,w\rangle + \|w\|^2 \\
&\iff \Delta = 4\langle v,w\rangle^2 - 4\|v\|^2\|w\|^2 \leqslant 0
\end{aligned}$$ □

The Cauchy-Schwarz inequality allows the concept of the angle between two vectors to be generalized for abstract vector spaces. In fact, it implies the existence of a coefficient k between -1 and $+1$ such that $\langle v,w\rangle = \|v\|\|w\|k$, but, given that the restriction of cos to $[0,\pi]$ creates a bijection with $[-1,1]$, this means that there is only one $\vartheta \in [0,\pi]$ such that $\langle v,w\rangle = \|v\|\|w\|\cos\vartheta$. $\vartheta \in [0,\pi]$ is known as the angle between the two vectors v and w.

Another very important property of the norm is as follows.

THEOREM 1.4.– Let $(V, \| \ \|)$ be an arbitrary normed vector space and $v, w \in V$. We have:

$$\boxed{\big|\|v\| - \|w\|\big| \leqslant \|v - w\|} \qquad [1.3]$$

PROOF.– On one side:

$$\|v\| = \|v - w + w\| = \|(v - w) + w\| \leqslant \|v - w\| + \|w\|$$

by the triangle inequality, thus $\|v\| - \|w\| \leqslant \|v - w\|$. On the other side:

$$\|w\| = \|w - v + v\| = \|(w - v) + v\| \leqslant \|w - v\| + \|w\|$$

thus $\|w\| - \|v\| \leqslant \|v - w\|$, i.e. $\|v\| - \|w\| \geqslant -\|v - w\|$.

Hence, $-\|v - w\| \leqslant \|v\| - \|w\| \leqslant \|v - w\|$, i.e. $|\|v\| - \|w\|| \leqslant \|v - w\|$. □

The following formula is also extremely useful.

THEOREM 1.5 (Carnot's theorem).– Taking $v, w \in (V, \langle\ ,\ \rangle)$:

$$\boxed{\|v \pm w\|^2 = \|v\|^2 + \|w\|^2 \pm 2\langle v, w\rangle, \quad (\mathbb{K} = \mathbb{R})} \qquad [1.4]$$

and

$$\boxed{\|v \pm w\|^2 = \|v\|^2 + \|w\|^2 \pm \langle v, w\rangle \pm \langle w, v\rangle, \quad (\mathbb{K} = \mathbb{C})} \qquad [1.5]$$

PROOF.– Direct calculation:

$$\begin{aligned}\|v \pm w\|^2 &= \langle v \pm w, v \pm w\rangle = \langle v, v\rangle \pm \langle v, w\rangle \pm \langle w, v\rangle + \langle w, w\rangle \\ &= \|v\|^2 + \|w\|^2 \pm \langle v, w\rangle \pm \langle w, v\rangle\end{aligned}$$ □

If $\mathbb{K} = \mathbb{C}$, then $\langle w, v\rangle = \overline{\langle v, w\rangle}$, and since, if $z = a + ib = \Re(z) + i\Im(z)$, then $z + \bar{z} = 2a = 2\Re(z)$, we can rewrite [1.5] as:

$$\boxed{\|v \pm w\|^2 = \|v\|^2 + \|w\|^2 \pm 2\Re(\langle v, w\rangle)} \qquad [1.6]$$

The laws presented in this section have immediate consequences which will be highlighted in section 1.2.1.

1.2.1. *The parallelogram law and the polarization formula*

The parallelogram law in $\mathbb{R}^2$ is shown in Figure 1.1. This law can be generalized on a vector space with an arbitrary inner product.

THEOREM 1.6 (Parallelogram law).– Let $(V, \langle , \rangle)$ be an inner product space on $\mathbb{K}$. Thus, $\forall v, w \in V$:

$$\|v + w\|^2 + \|v - w\|^2 = 2(\|v\|^2 + \|w\|^2)$$

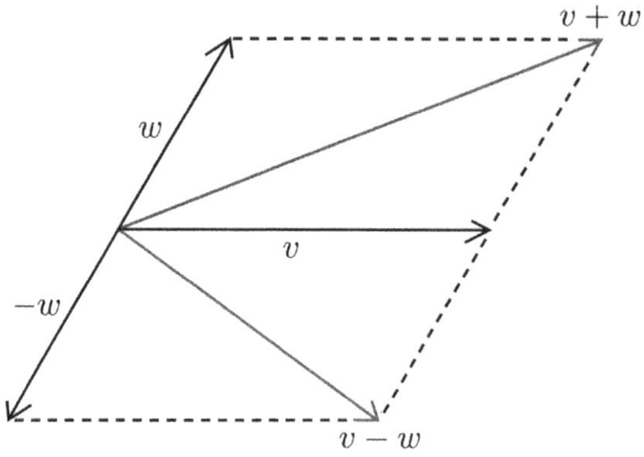

Figure 1.1. *Parallelogram law in $\mathbb{R}^2$: The sum of the squares of the two diagonal lines is equal to two times the sum of the squares of the edges v and w. For a color version of this figure, see www.iste.co.uk/provenzi/spaces.zip*

PROOF.– A direct consequence of law [1.4] or law [1.5] taking $\|v+w\|^2$ then $\|v-w\|^2$. □

As we have seen, an inner product induces a norm. The polarization formula can be used to "reverse" roles and write the inner product using the norm.

THEOREM 1.7 (Polarization formula).– Let $(V, \langle , \rangle)$ be an inner product space on $\mathbb{K}$. In this case, $\forall v, w \in V$:

$$\langle v, w\rangle = \frac{1}{4}\left(\|v+w\|^2 - \|v-w\|^2\right), \quad (\mathbb{K}=\mathbb{R})$$

and:

$$\langle v, w\rangle = \frac{1}{4}\left[\|v+w\|^2 - \|v-w\|^2 + i\left(\|v+iw\|^2 - \|v-iw\|^2\right)\right], \quad (\mathbb{K}=\mathbb{C})$$

PROOF.– This law is a direct consequence of law [1.4], in the real case. For the complex case, w is replaced by iw in law [1.5], and by sesquilinearity, we obtain:

$$\|v \pm iw\|^2 = \|v\|^2 + \|w\|^2 \mp i\langle v, w\rangle \pm i\langle w, v\rangle$$

By direct calculation, we can then verify that $\|v+w\|^2 - \|v-w\|^2 + i\,\|v+iw\|^2 - i\,\|v-iw\|^2 = 4\langle v, w\rangle$. □

It may seem surprising that something as simple as the parallelogram law may be used to establish a necessary and sufficient condition to guarantee that a norm over a vector space will be induced by an inner product, that is, the norm is Hilbertian. This notion will be formalized in Chapter 4.

1.3. Orthogonal and orthonormal families in inner product spaces

The "geometric" definition of an inner product in $\mathbb{R}^2$ and $\mathbb{R}^3$ indicates that this product is zero if and only if ϑ, the angle between the vectors, is $\pi/2$, which implies $\cos(\vartheta) = 0$.

In more complicated vector spaces (e.g. polynomial spaces), or even Euclidean vector spaces of more than three dimensions, it is no longer possible to visualize vectors; their orthogonality must therefore be "axiomatized" *via* the nullity of their scalar product.

DEFINITION 1.5.– *Let* $(V, \langle , \rangle)$ *be a real or complex inner product space of finite dimension* n. *Let* $F = \{v_1, \cdots, v_n\}$ *be a family of vectors in* V. *Thus:*

– *F is an* orthogonal family *of vectors if each different vector pair has an inner product of 0:* $\langle v_i, v_j \rangle = 0$;

– *F is an* orthonormal family *if it is orthogonal and, furthermore,* $\|v_i\| = 1\ \forall i$. *Thus, if* $\{v_i\}_{i=1}^n$ *is an orthogonal family,* $\{u_i = \|v_i\|^{-1} v_i\}_{i=1}^n$ *is an orthonormal family.*

An orthonormal family (unit and orthogonal vectors) may be characterized as follows:

$$\langle v_i, v_j \rangle = \delta_{i,j} = \begin{cases} 1 & \text{if } i = j \\ 0 & \text{if } i \neq j \end{cases} \qquad \textit{Orthonormal family}$$

$\delta_{i,j}$ is the *Kronecker delta*[5].

1.4. Generalized Pythagorean theorem

The Pythagorean theorem can be generalized to abstract inner product spaces. The general formulation of this theorem is obtained using a lemma.

LEMMA 1.1.– Let $(V, \langle , \rangle)$ be a real or complex inner product space. Let $u \in V$ be orthogonal to all vectors $v_1, \ldots, v_n \in V$. Hence, u is also orthogonal to all vectors in V obtained as a linear combination of $v_1, \ldots, v_n$.

PROOF.– Let $w = \sum_{i=1}^{n} \alpha_i v_i$, $\alpha_i \in \mathbb{K}\ \forall i = 1, \ldots, n$, be an arbitrary linear combination of vectors $v_1, \ldots, v_n$. By direct calculation:

$$\langle u, w \rangle = \langle u, \sum_{i=1}^{n} \alpha_i v_i \rangle \underset{\text{(sesquilinearity)}}{=} \sum_{i=1}^{n} \overline{\alpha_i} \langle u, v_i \rangle \underset{u \perp v_i}{=} \sum_{i=1}^{n} \overline{\alpha_i} 0 = 0 \qquad \square$$

5 Leopold Kronecker (1823, Liegnitz-1891, Berlin).

THEOREM 1.8 (Generalized Pythagorean theorem).– Let $(V, \langle , \rangle)$ be an inner product space on $\mathbb{K}$. Let $u, v \in V$ be orthogonal to each other. Hence:

$$\|u + v\|^2 = \|u\|^2 + \|v\|^2$$

More generally, if the vectors $v_1, \ldots, v_n \in V$ are orthogonal, then:

$$\left\| \sum_{i=1}^{n} v_i \right\|^2 = \sum_{i=1}^{n} \|v_i\|^2 \quad \Longleftrightarrow \quad \|v_1 + \ldots + v_n\|^2 = \|v_1\|^2 + \ldots + \|v_n\|^2$$

PROOF.– The two-vector case can be proven thanks to Carnot's formula:

$$\begin{aligned} \|u + v\|^2 &= \langle u + v, u + v \rangle \\ &= \langle u, u \rangle + \cancelto{0}{\langle u, v \rangle} + \cancelto{0}{\langle v, u \rangle} + \langle v, v \rangle \\ &= \|u\|^2 + \|v\|^2 \end{aligned}$$

Proof for cases with n vectors is obtained by recursion:

– the case where $n = 2$ is demonstrated above;

– we suppose that $\left\| \sum_{i=1}^{n-1} v_i \right\|^2 = \sum_{i=1}^{n-1} \|v_i\|^2$ (recursion hypothesis);

– now, we write $u = v_n$ and $z = \sum_{i=1}^{n-1} v_i$, so $u \perp z$ using Lemma 1.1. Hence, using case $n = 2$: $\|u + z\|^2 = \|u\|^2 + \|z\|^2$, but:

$$u + z = v_n + \sum_{i=1}^{n-1} v_i = \sum_{i=1}^{n} v_i$$

so:

$$\|u + z\|^2 = \left\| \sum_{i=1}^{n} v_i \right\|^2$$

and:

$$\|u\|^2 + \|z\|^2 = \|v_n\|^2 + \left\| \sum_{i=1}^{n-1} v_i \right\|^2 \underset{\text{(Recursion hypothesis)}}{=} \|v_n\|^2 + \sum_{i=1}^{n-1} \|v_i\|^2 = \sum_{i=1}^{n} \|v_i\|^2$$

giving us the desired thesis. □

Note that *the Pythagorean theorem thesis is a double implication if and only if V is real*, in fact, using law [1.6] we have that $\|u + v\|^2 = \|u\|^2 + \|v\|^2$ holds true if and only if $\Re(\langle u, v \rangle) = 0$, which is equivalent to orthogonality if and only if V is real.

The following result gives information concerning the distance between any two vectors within an orthonormal family.

THEOREM 1.9.– Let $(V, \langle , \rangle)$ be an inner product space on $\mathbb{K}$ and let F be an orthonormal family in V. The distance between any two elements of F is constant and equal to $\sqrt{2}$.

PROOF.– Using the Pythagorean theorem: $\|u + (-v)\|^2 = \|u\|^2 + \|v\|^2 = 2$, from the fact that $u \perp v$. □

1.5. Orthogonality and linear independence

The orthogonality condition is more restrictive than that of linear independence: *all orthogonal families are free.*

THEOREM 1.10.– Let F be an orthogonal family in $(V, \langle , \rangle)$, $F = \{v_1, \cdots, v_n\}$, $v_i \neq 0 \; \forall i$, then F is free.

PROOF.– We need to prove the linear independence of the elements v_i, that is, $\sum_{i=1}^{n} a_i v_i = 0 \implies a_i = 0 \; \forall i$. To this end, we calculate the inner product of the linear combination $\sum_{i=1}^{n} a_i v_i$ and an arbitrary vector v_j with $j \in \{1, \ldots, n\}$:

$$\langle \sum_{i=1}^{n} a_i v_i, v_j \rangle \underset{[1.1]}{=} \sum_{i=1}^{n} a_i \langle v_i, v_j \rangle \underset{(\langle v_i, v_j \rangle \neq 0 \Leftrightarrow i=j)}{=} a_j \langle v_j, v_j \rangle = a_j \|v_j\|^2$$

By hypothesis, none of the vectors in F are zero; the hypothesis that $\sum_{i=1}^{n} a_i v_i = 0$ therefore implies that:

$$\underbrace{\langle 0, v_j \rangle}_{= 0} = a_j \underbrace{\|v_j\|^2}_{\neq 0} \Rightarrow a_j = 0.$$

This holds for any $j \in \{1, \ldots, n\}$, so the orthogonal family F is free. □

Using the general theory of vector spaces in finite dimensions, an immediate corollary can be derived from Theorem 1.10.

COROLLARY 1.1.– An orthogonal family of n non-null vectors in a space $(V, \langle , \rangle)$ of dimension n is a basis of V.

DEFINITION 1.6.– *A family of n non-null orthogonal vectors in a vector space $(V, \langle , \rangle)$ of dimension n is said to be an* orthogonal basis *of V. If this family is also orthonormal, it is said to be an* orthonormal basis *of V.*

The extension of the orthogonal basis concept to inner product spaces of infinite dimensions will be discussed in Chapter 5. For the moment, it is important to note that *an orthogonal basis is made up of the maximum number of mutually orthogonal vectors in a vector space.* Taking n to represent the dimension of the space V and proceeding by reductio ad absurdum, imagine the existence of another vector $u^* \in V$, $u \neq 0$, orthogonal to all of the vectors in an orthogonal basis $(u_i)_{i=1}^n$; in this case, the set $(u^*, u_i)_{i=1}^n$ would be free as orthogonal vectors are linearly independent, and the dimension of V would be $n + 1$ instead of n! This property is usually expressed by saying that an orthogonal family is a basis if it is not a subset of another orthogonal family of vectors in V.

Note that in order to determine the components of a vector in relation to an arbitrary basis, we must solve a linear system of n equations with n unknown variables. In fact, if $v \in V$ is any vector and (u_i) $i = 1, \ldots, n$ is a basis of V, then the components of v in (u_i) are the scalars $\alpha_1, \ldots, \alpha_n$ such that:

$$v = \sum_{i=1}^{n} \alpha_i u_i \qquad \Longleftrightarrow \qquad \begin{cases} v_1 = \sum\limits_{i=1}^{n} \alpha_i u_{i,1} \\ \vdots \\ v_n = \sum\limits_{i=1}^{n} \alpha_i u_{i,n}, \end{cases}$$

where $u_{i,j}$ is the j-th component of vector u_i.

However, in the presence of an orthogonal or orthonormal basis, components are determined by inner products, as seen in Theorem 1.11.

Note, too, that solving a linear system of n equations with n unknown variables generally involves far more operations than the calculation of inner products; this highlights one advantage of having an orthogonal basis for a vector space.

THEOREM 1.11.– Let $B = \{u_1, \ldots, u_n\}$ be an *orthogonal* basis of $(V, \langle , \rangle)$. Then:

$$\boxed{v = \sum_{i=1}^{n} \frac{\langle v, u_i \rangle}{\|u_i\|^2} u_i}$$

Notably, if B is an *orthonormal* basis, then:

$$\boxed{v = \sum_{i=1}^{n} \langle v, u_i \rangle u_i}$$

PROOF.– B is a basis, so there exists a set of scalars $\alpha_1, \dots, \alpha_n$ such that $v = \sum_{j=1}^{n} \alpha_j u_j$. Consider the inner product of this expression of v with a fixed vector u_i, $i \in \{1, \dots, n\}$:

$$\langle v, u_i \rangle = \langle \sum_{j=1}^{n} \alpha_j u_j, u_i \rangle = \sum_{j=1}^{n} \alpha_j \langle u_j, u_i \rangle \underset{(u_i \perp u_j \ \forall i \neq j)}{=} \alpha_i \langle u_i, u_i \rangle = \alpha_i \|u_i\|^2$$

so $\alpha_i = \frac{\langle v, u_i \rangle}{\|u_i\|^2} \ \forall i = 1, \cdots, n$, and thus $v = \sum_{i=1}^{n} \frac{\langle v, u_i \rangle}{\|u_i\|^2} u_i$. If B is an orthonormal basis, $\|u_i\| = 1$ giving the second law in the theorem. □

Geometric interpretation of the theorem: The theorem that we are about to demonstrate is the generalization of the decomposition theorem of a vector in plane $\mathbb{R}^2$ or in space $\mathbb{R}^3$ on a canonical basis of unit vectors on axes. To simplify this, consider the case of $\mathbb{R}^2$.

If $\hat{\imath}$ and $\hat{\jmath}$ are, respectively, the unit vectors of axes x and y, then the decomposition theorem says that:

$$v = \underbrace{\|v\| \cos \alpha}_{\langle v, \hat{\imath} \rangle} \hat{\imath} + \underbrace{\|v\| \cos \beta}_{\langle v, \hat{\jmath} \rangle} \hat{\jmath} = \langle v, \hat{\imath} \rangle \hat{\imath} + \langle v, \hat{\jmath} \rangle \hat{\jmath}$$

which is a particular case of the theorem above.

We will see that the *Fourier series* can be viewed as a further generalization of the decomposition theorem on an orthogonal or orthonormal basis.

1.6. Orthogonal projection in inner product spaces

The definition of orthogonal projection can be extended by examining the geometric and algebraic properties of this operation in $\mathbb{R}^2$ and $\mathbb{R}^3$. Let us begin with $\mathbb{R}^2$.

In the Euclidean space $\mathbb{R}^2$, the inner product of a vector v and a *unit* vector evidently gives us the orthogonal projection of v in the direction defined by this vector, as shown in Figure 1.2 with an orthogonal projection along the x axis.

The properties verified by this projection are as follows:

1) projecting onto the x axis a second time, vector $P_x v$ obviously remains unchanged given that it is already on the x axis, i.e. $P_x^2(v) := P_x(P_x v) = P_x v$ $\forall v \in V$. Put differently, the operator P_x bound to the x axis is the identity of this axis;

2) the difference vector between v and its projection $v - P_x v$ is orthogonal to the x axis, as we see from Figure 1.3;

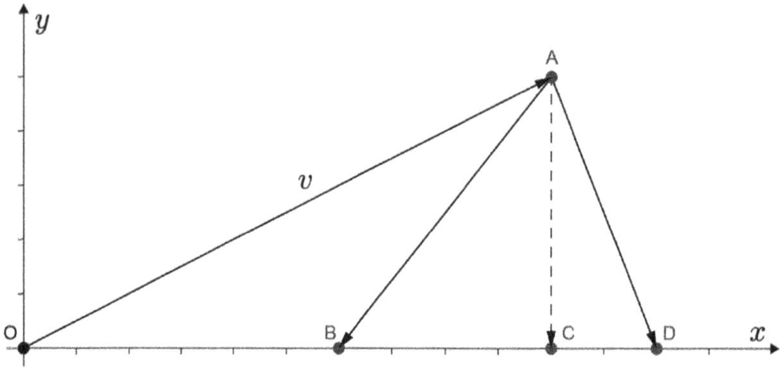

Figure 1.2. *Orthogonal projection $P_x v = \vec{OC}$ and diagonal projections $\vec{OB}$ and $\vec{OD}$ of a vector in $v \in \mathbb{R}^2$ onto the x axis. For a color version of this figure, see www.iste.co.uk/provenzi/spaces.zip*

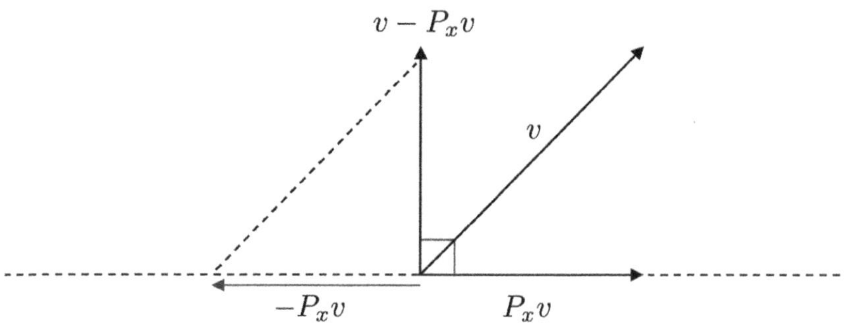

Figure 1.3. *Visualization of property 2 in $\mathbb{R}^2$. For a color version of this figure, see www.iste.co.uk/provenzi/spaces.zip*

3) $P_x v$ minimizes the distance between the terminal point of v and the x axis. In Figure 1.2, $\vec{AB}$ and $\vec{AD}$ are, in fact, the hypotenuses of right-angled triangles ABC and ACD; on the other hand, $\vec{AC}$ is another side of these triangles, and is therefore smaller than $\vec{AB}$ and $\vec{AD}$. $\vec{AC}$ is the distance between the terminal point of v and the terminal point of $P_x v$, while $\vec{AB}$ and $\vec{AD}$ are the distances between the terminal point of v and the diagonal projections of v onto x rooted at B and D, respectively.

We wish to define an orthogonal projection operation for an abstract inner product space of dimension n which retains these same geometric properties.

Analyzing orthogonal projections in $\mathbb{R}^3$ helps us to establish an idea of the algebraic definition of this operation. Figure 1.4 shows a vector $v \in \mathbb{R}^3$ and the plane produced by the *orthogonal* vectors u_1 and u_2. We see that the projection p of v onto this plane is the vector sum of the orthogonal projections $p_1 = \frac{\langle v,u_1\rangle}{\|u_1\|^2} u_1$ and $p_2 = \frac{\langle v,u_2\rangle}{\|u_2\|^2} u_2$ onto the two vectors u_1 and u_2 taken separately, i.e. $p = p_1 + p_2 = \sum_{i=1}^{2} \frac{\langle v,u_i\rangle}{\|u_i\|^2} u_i$.

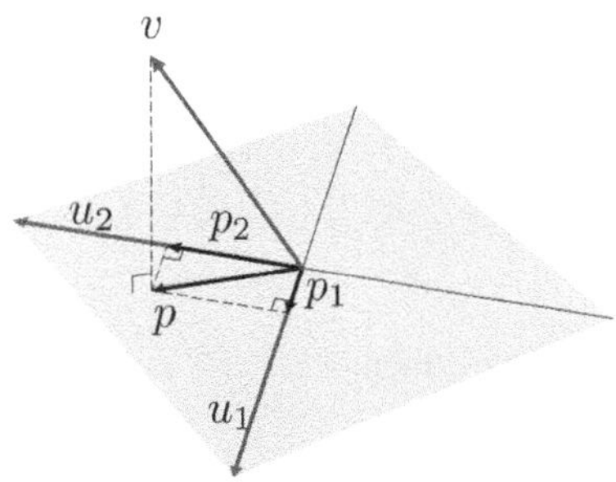

Figure 1.4. *Orthogonal projection* p *of a vector in* $\mathbb{R}^3$ *onto the plane produced by two unit vectors. For a color version of this figure, see www.iste.co.uk/provenzi/spaces.zip*

Generalization should now be straightforward: consider an inner product space $(V, \langle,\rangle)$ of dimension n and an *orthogonal family* of non-zero vectors $F = \{u_1, \ldots, u_m\}$, $m \leqslant n$, $u_i \neq 0_V$ $\forall i = 1, \ldots, m$.

The vector subspace of V produced by all linear combinations of the vectors of F shall be written Span(F):

$$\text{span}(F) \equiv S = \left\{ s \in V \ : \ \exists \alpha_1, \ldots, \alpha_m \in \mathbb{K} \text{ such that } s = \sum_{j=1}^{m} \alpha_j u_j \right\}$$

The *orthogonal projection operator* or *orthogonal projector* of a vector $v \in V$ onto S is defined as the following application, which is obviously linear:

$$\begin{aligned} P_S : V &\longrightarrow S \subseteq V \\ v &\longmapsto \boxed{P_S(v) = \sum_{i=1}^{m} \frac{\langle v, u_i\rangle}{\|u_i\|^2} u_i} \end{aligned}$$

Theorem 1.12 shows that the orthogonal projection defined above retains all of the properties of the orthogonal projection demonstrated for $\mathbb{R}^2$.

THEOREM 1.12.– Using the same notation as before, we have:

1) if $s \in S$ then $P_S(s) = s$, i.e. the action of P_S on the vectors in S is the identity;

2) $\forall v \in V$ and $s \in S$, the *residual* vector of the projection, i.e. $v - P_S(v)$, is $\perp$ to S:

$$\langle v - P_S(v), s\rangle = 0 \quad \Longleftrightarrow \quad v - P_S(v) \perp s$$

3) $\forall v \in V$ et $s \in S$: $\|v - P_S(v)\| \leqslant \|v - s\|$ and the equality holds if and only if $s = P_S(v)$. We write:

$$\boxed{P_S(v) = \underset{s \in S}{argmin}\, \|v - s\|}$$

PROOF.–

1) Let $s \in S$, i.e. $s = \sum_{j=1}^{m} \alpha_j u_j$, then:

$$P_S(s) = \sum_{i=1}^{m} \frac{\langle \sum_{j=1}^{m} \alpha_j u_j, u_i\rangle}{\|u_i\|^2} u_i = \sum_{i=1}^{m} \frac{\sum_{j=1}^{m} \alpha_j \langle u_j, u_i\rangle}{\|u_i\|^2} u_i$$

$$\underset{(u_i \perp u_j\ \forall i \neq j)}{=} \sum_{i=1}^{m} \frac{\alpha_i \langle u_i, u_i\rangle}{\|u_i\|^2} u_i = \sum_{i=1}^{m} \alpha_i u_i = s$$

2) Consider the inner product of $P_S(v)$ and a fixed vector u_j, $j \in \{1, \ldots, m\}$:

$$\langle P_S(v), u_j\rangle = \langle \sum_{i=1}^{m} \frac{\langle v, u_i\rangle}{\|u_i\|^2} u_i, u_j\rangle \underset{\text{(linearity)}}{=} \sum_{i=1}^{m} \frac{\langle v, u_i\rangle}{\|u_i\|^2} \langle u_i, u_j\rangle$$

$$\underset{(u_i \perp u_j\ \forall i \neq j)}{=} \frac{\langle v, u_j\rangle}{\|u_j\|^2} \langle u_j, u_j\rangle = \langle v, u_j\rangle$$

hence:

$$\langle v, u_j\rangle - \langle P_S(v), u_j\rangle = 0 \quad \underset{\text{linearity of } \langle\ ,\ \rangle}{\Longleftrightarrow} \quad \langle v - P_S(v), u_j\rangle = 0 \quad \forall j \in \{1, ..., m\}$$

Lemma 1.1 guarantees that $\langle v - P_S(v), s\rangle = 0\ \forall s = \sum_{j=1}^{m} \alpha_j u_j$.

3) It is helpful to rewrite the difference $v - s$ as $v - P_S(v) + P_S(v) - s$. From property 2, $v - P_S(v) \perp S$, however $P_S(v), s \in S$ so $P_S(v) - s \in S$. Hence $(v - P_S(v)) \perp (P_S(v) - s)$. The generalized Pythagorean theorem implies that:

$$\|v - s\|^2 = \|v - P_S(v) + P_S(v) - s\|^2 = \|v - P_S(v)\|^2 + \underbrace{\|P_S(v) - s\|^2}_{\geqslant 0} \geqslant \|v - P_S(v)\|^2$$

hence $\|v - s\| \geqslant \|v - P_S(v)\|\ \forall v \in V,\ s \in S$.

Evidently, $\|P_S(v) - s\|^2 = 0$ if and only if $s = P_S(v)$, and in this case $\|v - s\|^2 = \|v - P_S(v)\|^2$. □

The theorem demonstrated above tells us that the vector in the vector subspace $S \subseteq V$ which is the most "similar" to $v \in V$ (in the sense of the norm induced by the inner product) is given by the orthogonal projection. The generalization of this result to infinite-dimensional Hilbert spaces will be discussed in Chapter 5.

As already seen for the projection operator in $\mathbb{R}^2$ and $\mathbb{R}^3$, the non-negative scalar quantity $\frac{|\langle v,u_i \rangle|}{\|u_i\|}$ gives a measure of the importance of $\frac{u_i}{\|u_i\|}$ in the reconstruction of the best approximation of v in S *via* the formula $P_S(v) = \sum_{i=1}^{m} \frac{\langle v,u_i \rangle}{\|u_i\|^2} u_i$: if this quantity is large, then $\frac{u_i}{\|u_i\|}$ is very important to reconstruct $P_S(v)$, otherwise, in some circumstances, it may be ignored. In the applications to signal compression, a usual strategy consists of reordering the summation that defines $P_S(v)$ in descent order of the quantities $\frac{|\langle v,u_i \rangle|}{\|u_i\|}$ and trying to eliminate as many small terms as possible without degrading the signal quality.

This observation is crucial to understanding the significance of the Fourier decomposition, which will be examined in both discrete and continuous contexts in the following chapters.

Finally, note that the seemingly trivial equation $v = v - s + s$ is, in fact, far more meaningful than it first appears when we know that $s \in S$: in this case, we know that $v - s$ and s are orthogonal.

The decomposition of a vector as the sum of a component belonging to a subspace S and a component belonging to its orthogonal is known as the *orthogonal projection theorem*.

This decomposition is unique, and its generalization for infinite dimensions, alongside its consequences for the geometric structure of Hilbert spaces, will be examine in detail in Chapter 5.

1.7. Existence of an orthonormal basis: the Gram-Schmidt process

As we have seen, projection and decomposition laws are much simpler when an orthonormal basis is available.

Theorem 1.13 states that in a finite-dimensional inner product space, an orthonormal basis can always be constructed from a free family of generators.

THEOREM 1.13.– (The iterative Gram-Schmidt process[6]) If $(v_1, \ldots, v_n)$, $n \leqslant \infty$ is a basis of $(V, \langle,\rangle)$, then an orthonormal basis of $(V, \langle,\rangle)$ can be obtained from $(v_1, \ldots, v_n)$.

PROOF.– This proof is constructive in that it provides the method used to construct an orthonormal basis from any arbitrary basis.

– Step 1: normalization of v_1:

$$u_1 = \frac{v_1}{\|v_1\|}$$

– Step 2, illustrated in Figure 1.5: v_2 is projected in the direction of u_1, that is, we consider $\langle v_2, u_1\rangle u_1$. We know from Theorem 1.12 that the vector difference $v_2 - \langle v_2, u_1\rangle u_1$ is orthogonal to u_1. The result is then normalized:

$$u_2 = \frac{v_2 - \langle v_2, u_1\rangle u_1}{\|v_2 - \langle v_2, u_1\rangle u_1\|}$$

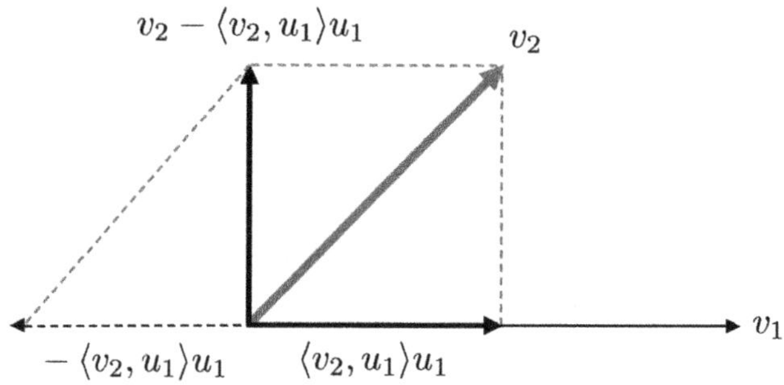

Figure 1.5. *Illustration of the second step in the Gram-Schmidt orthonormalization process. For a color version of this figure, see www.iste.co.uk/provenzi/spaces.zip*

– Step n, by iteration:

$$u_n = \frac{v_n - (\langle v_n, u_{n-1}\rangle u_{n-1} + \ldots + \langle v_n, u_1\rangle u_1)}{\|v_n - (\langle v_n, u_{n-1}\rangle u_{n-1} + \ldots + \langle v_n, u_1\rangle u_1)\|} \qquad \square$$

1.8. Fundamental properties of orthonormal and orthogonal bases

The most important properties of an orthonormal basis are listed in Theorem 1.14.

6 Jørgen Pedersen Gram (1850, Nustrup-1916, Copenhagen), Erhard Schmidt (1876, Tatu-1959, Berlin).

THEOREM 1.14.– *Let $(u_1, \dots, u_n)$ be an orthonormal basis of $(V, \langle , \rangle)$, $dim(V) = n$. Then, $\forall v, w \in V$:*

1) Decomposition theorem on an orthonormal basis*:*

$$\boxed{v = \sum_{i=1}^{n} \langle v, u_i \rangle u_i} \qquad [1.7]$$

2) Parseval's identity[7]*:*

$$\boxed{\langle v, w \rangle = \sum_{i=1}^{n} \langle v, u_i \rangle \langle u_i, w \rangle} \qquad [1.8]$$

3) Plancherel's theorem[8]*:*

$$\boxed{\|v\|^2 = \sum_{i=1}^{n} |\langle v, u_i \rangle|^2} \qquad [1.9]$$

Proof of 1: an immediate consequence of Theorem 1.12. Given that $(u_1, \dots, u_n)$ is a basis, $v \in \text{span}(u_1, \dots, u_n)$; furthermore, $(u_1, \dots, u_n)$ is orthonormal, so $v = P_S(v) = \sum_{i=1}^{n} \langle v, u_i \rangle u_i$. It is not necessary to divide by $\|u_i\|^2$ when summing since $\|u_i\| = 1 \; \forall i$.

Proof of 2: using point 1 it is possible to write $v = \sum_{i=1}^{n} \langle v, u_i \rangle u_i$, and calculating the inner product of v, written in this way, and w, using equation [1.1], we obtain:

$$\langle v, w \rangle = \langle \sum_{i=1}^{n} \langle v, u_i \rangle u_i, w \rangle = \sum_{i=1}^{n} \langle v, u_i \rangle \langle u_i, w \rangle$$

Proof of 3: writing $w = v$ on the left-hand side of Parseval's identity gives us $\langle v, v \rangle = \|v\|^2$. On the right-hand side, we have:

$$\sum_{i=1}^{n} \langle v, u_i \rangle \langle u_i, v \rangle = \sum_{i=1}^{n} \langle v, u_i \rangle \overline{\langle v, u_i \rangle} = \sum_{i=1}^{n} |\langle v, u_i \rangle|^2$$

hence $\|v\|^2 = \sum_{i=1}^{n} |\langle v, u_i \rangle|^2$. □

7 Marc-Antoine de Parseval des Chêsnes (1755, Rosières-aux-Salines-1836, Paris).

8 Michel Plancherel (1885, Bussy-1967, Zurich).

NOTE.–

1) The *physical interpretation of Plancherel's theorem* is as follows: the energy of v, measured as the square of the norm, can be decomposed using the sum of the squared moduli of each projection of v on the n directions of the orthonormal basis $(u_1, ..., u_n)$.

In Fourier theory, the directions of the orthonormal basis are fundamental harmonics (sines and cosines with defined frequencies): this is why Fourier analysis may be referred to as *harmonic analysis*.

2) If $(u_1, \ldots, u_n)$ is an orthogonal, rather than an orthonormal, basis, then using the projector formula and Theorem 1.12, the results of Theorem 1.14 can be written as:

a) *decomposition of $v \in V$ on an orthogonal basis*:

$$v = \sum_{i=1}^{n} \frac{\langle v, u_i \rangle}{\|u_i\|^2} u_i \qquad [1.10]$$

b) *Parseval's identity for an orthogonal basis*:

$$\langle v, w \rangle = \sum_{i=1}^{n} \frac{\langle v, u_i \rangle \langle u_i, w \rangle}{\|u_i\|^2} \qquad [1.11]$$

c) *Plancherel's theorem for an orthogonal basis*:

$$\|v\|^2 = \sum_{i=1}^{n} \frac{|\langle v, u_i \rangle|^2}{\|u_i\|^2} \qquad [1.12]$$

The following exercise is designed to test the reader's knowledge of the theory of finite-dimensional inner product spaces. The two subsequent exercises explicitly include inner products which are non-Euclidean.

Exercise 1.1

Consider the complex Euclidean inner product space $\mathbb{C}^3$ and the following three vectors:

$$u = (0, i, 2i), \quad v = (2i, 0, -i), \quad w = \left(0, i, \frac{1}{2}e^{-\frac{\pi}{2}i}\right)$$

1) Determine the orthogonality relationships between vectors u, v, w.

2) Calculate the norm of u, v, w and the Euclidean distances between them.

3) Verify that (u, v, w) is a (non-orthogonal) basis of $\mathbb{C}^3$.

4) Let S be the vector subspace of $\mathbb{C}^3$ generated by u and w. Calculate $P_S v$, the orthogonal projection of v onto S. Calculate $d(v, P_S v)$, that is, the Euclidean distance between v and its projection onto S, and verify that this minimizes the distance between v and the vectors of S (*hint*: look at the square of the distance).

5) Using the results of the previous questions, determine an orthogonal basis and an orthonormal basis for $\mathbb{C}^3$ without using the Gram-Schmidt orthonormalization process (*hint*: remember the geometric relationship between the residual vector r and the subspace S).

6) Given a vector $a = (2i, -1, 0)$, write the decomposition of a and Plancherel's theorem in relation to the orthonormal basis identified in point 5. Use these results to identify the vector from the orthonormal basis which has the heaviest weight in the decomposition of a (and which gives the best "rough approximation" of a). Use a graphics program to draw the progressive vector sum of a, beginning with the rough approximation and adding finer details supplied by the other vectors.

Solution to Exercise 1.1

1) Evidently, $e^{-\frac{\pi}{2}i} = -i$, so by directly calculating the inner products: $\langle u, v\rangle = -2$, $\langle u, w\rangle = 0$ et $\langle v, w\rangle = \frac{1}{2}$.

2) By direct calculation: $\|u\|^2 = 5$, $\|v\|^2 = 5$, $\|w\|^2 = \frac{5}{4}$. After calculating the difference vectors, we obtain: $d(u, v) = \|u - v\| = \sqrt{14}$, $d(u, w) = \|u - w\| = \frac{5}{2}$, $d(v, w) = \|v - w\| = \frac{\sqrt{21}}{2}$.

3) The three vectors u, v, w are linearly independent, so they form a basis in $\mathbb{C}^3$. This basis is not orthogonal since only vectors u and w are orthogonal.

4) $S = \text{span}(u, w)$. Since (u, w) is an orthogonal basis in S, we can write:

$$P_S(v) = \frac{\langle v, u\rangle}{\|u\|^2} u + \frac{\langle v, w\rangle}{\|w\|^2} w = (0, 0, -i)$$

The residual vector of the projection of v on S is $r = v - P_S v = (2i, 0, 0)$ and thus $d(v, P_S v)^2 = \|r\|^2 = 4$. The most general vector in S is $s = \alpha u + \beta w = (0, (\alpha+\beta)i, (2\alpha - \frac{\beta}{2})i)$ and $d(v, s)^2 = \|v - s\|^2 = 4 + (\alpha+\beta)^2 + (2\alpha - \frac{\beta}{2} + 1)^2 \geqslant 4 = d(v, P_S v)^2$. This confirms that $P_S v$ is the vector in S with the minimum distance from v in relation to the Euclidean norm.

5) r is orthogonal to S, which is generated by u and w, hence (u, w, r) is a set of orthogonal vectors in $\mathbb{C}^3$, that is, an orthogonal basis of $\mathbb{C}^3$. To obtain an orthonormal basis, we then simply divide each vector by its norm:

$$(\hat{u}, \hat{w}, \hat{r}) \equiv \left(\frac{u}{\|u\|}, \frac{w}{\|w\|}, \frac{r}{\|r\|}\right) = \left(\left(0, \frac{i}{\sqrt{5}}, \frac{2i}{\sqrt{5}}\right), \left(0, \frac{2i}{\sqrt{5}}, -\frac{i}{\sqrt{5}}\right), (i, 0, 0)\right)$$

6) Decomposition: $a = \langle a, \hat{u}\rangle\hat{u} + \langle a, \hat{w}\rangle\hat{w} + \langle a, \hat{r}\rangle\hat{r} = \frac{i}{\sqrt{5}}\hat{u} + \frac{2i}{\sqrt{5}}\hat{w} + 2\hat{r}$.

Plancherel's theorem: $\|a\|^2 = 5 = |\langle a, \hat{u}\rangle|^2 + |\langle a, \hat{w}\rangle|^2 + |\langle a, \hat{r}\rangle|^2 (= \frac{1}{5} + \frac{4}{5} + 4 = 5)$.

The vector with the heaviest weight in the reconstruction of a is thus $\hat{r}$: this vector gives the best rough approximation of a. By calculating the vector sum of this rough representation and the other two vectors, we can reconstruct the "fine details" of a, first with $\hat{w}$ and then with $\hat{u}$. □

Exercise 1.2

Let $M(n, \mathbb{C})$ be the space of $n \times n$ complex matrices. The application $\phi : M(n, \mathbb{C}) \times M(n, \mathbb{C}) \to \mathbb{C}$ is defined by:

$$\phi(A, B) = \text{tr}(B^\dagger A)$$

where $B^\dagger := \overline{B}^t$ denotes the adjoint matrix of B and tr is the matrix trace. Prove that ϕ is an inner product.

Solution to Exercise 1.2

The distributive property of matrix multiplication for addition and the linearity of the trace establishes the linearity of ϕ in relation to the first variable.

Now, let us prove that ϕ is Hermitian. Let $A = (a_{i,j})_{1 \leqslant i,j \leqslant n}$ and $B = (b_{i,j})_{1 \leqslant i,j \leqslant n}$ be two matrices in $M(n, \mathbb{C})$. Let $(c_{i,j})_{1 \leqslant i,j \leqslant n} = (\overline{b_{j,i}})_{1 \leqslant i,j \leqslant n}$ be the coefficients of the matrix $B^\dagger$ and let $(d_{i,j})_{1 \leqslant i,j \leqslant n} = (\overline{a_{j,i}})_{1 \leqslant i,j \leqslant n}$ be the coefficients of $A^\dagger$.

This gives us:

$$\begin{aligned}\overline{\phi(A, B)} = \overline{\text{tr}(B^\dagger A)} &= \overline{\text{tr}\left[\left(\sum_{k=1}^{n} c_{i,k} a_{k,j}\right)_{i,j}\right]} = \overline{\text{tr}\left[\left(\sum_{k=1}^{n} \overline{b_{k,i}} a_{k,j}\right)_{i,j}\right]} \\ &= \sum_{i,k=0}^{n} b_{k,i} \overline{a_{k,i}} = \sum_{i,k=0}^{n} d_{i,k} b_{k,i} = \text{tr}(A^\dagger B) \\ &= \phi(B, A)\end{aligned}$$

Thus, ϕ is a sesquilinear Hermitian form. Furthermore, ϕ is positive:

$$\forall A \in M(n, \mathbb{C}),\ \phi(A, A) = \sum_{i,k=0}^{n} |a_{k,i}|^2 \geqslant 0$$

It is also definite:

$$\begin{aligned}\phi(A, A) = 0 &\iff \sum_{i,k=0}^{n} |a_{k,i}|^2 = 0 \\ &\iff \forall 1 \leqslant k, i \leqslant n,\ a_{k,i} = 0 \\ &\iff A = 0\end{aligned}$$

Thus, ϕ is an inner product. □

Exercise 1.3

Let $E = \mathbb{R}[X]$ be the vector space of single variable polynomials with real coefficients. For $P, Q \in E$, take:

$$\Phi(P, Q) = \int_{-1}^{1} \frac{P(t)Q(t)}{\sqrt{1-t^2}} dt$$

1) Remember that $f(t) \underset{t \to t_0}{=} O(g(t))$ means that $\exists\, a, C > 0$ such that $|t - t_0| < a \implies |f(t)| \leqslant C\,|g(t)|$. Prove that for all $P, Q \in E$, this is equal to:

$$\frac{P(t)Q(t)}{\sqrt{1-t^2}} \underset{t \to 1}{=} O\left(\frac{1}{\sqrt{1-t}}\right)$$

and:

$$\frac{P(t)Q(t)}{\sqrt{1-t^2}} \underset{t \to -1}{=} O\left(\frac{1}{\sqrt{1+t}}\right)$$

Use this result to deduce that Φ is definite over $E \times E$.

2) Prove that Φ is an inner product over E, which we shall note $\langle\ ,\ \rangle$.

3) For $n \in \mathbb{N}$, let T_n be the n-th *Chebyshev polynomial*, that is, the only polynomial such that $\forall \theta \in \mathbb{R}$, $T_n(\cos\theta) = \cos(n\theta)$. Applying the substitution $t = \cos\theta$, show that $(T_n)_{n \in \mathbb{N}}$ is an orthogonal family in E. Hint: use the trigonometric formula [1.13]:

$$\frac{1}{2}(\cos((n+m)\theta)+\cos((n-m)\theta)) = \cos(n\theta)\cos(m\theta) \qquad \forall n, m \in \mathbb{N}. \quad [1.13]$$

4) Prove that for all $n \in \mathbb{N}$, $(T_0, \ldots, T_n)$ is an orthogonal basis of $\mathbb{R}_n[X]$, the vector space of polynomials in $\mathbb{R}[X]$ of degree less than or equal to n. Deduce that $(T_n)_{n \in \mathbb{N}}$ is an orthogonal basis in the algebraic sense: every element in E is a *finite* linear combination of elements in the basis of E.

5) Calculate the norm of T_n for all n and deduce an orthonormal basis (in the algebraic sense) of E using this result.

Solution to Exercise 1.3

1) We write $f(t) = \frac{P(t)Q(t)}{\sqrt{1-t^2}} = \frac{P(t)Q(t)}{\sqrt{1-t}\sqrt{1+t}}$. Since P and Q are polynomials, the function $t \mapsto \frac{P(t)Q(t)}{\sqrt{1+t}}$ is continuous in a neighborhood $V_1(1)$ and thus, according to the Weierstrass theorem, it is bounded in this neighborhood, that is, $\exists\, C_1 > 0$ such that $t \in V_1(1) \implies \left|\frac{P(t)Q(t)}{\sqrt{1+t}}\right| \leqslant C_1$. Similarly, the function $t \mapsto \frac{P(t)Q(t)}{\sqrt{1-t}}$ is continuous in a neighborhood $V_2(-1)$, thus $\exists\, C_2 > 0$ such that $t \in V_2(-1) \implies \left|\frac{P(t)Q(t)}{\sqrt{1-t}}\right| \leqslant C_2$. This gives us:

$$t \in V_1(1) \implies |f(t)| \leqslant C_1 \frac{1}{\sqrt{1-t}} \iff f(t) \underset{t\to 1}{=} O\left(\frac{1}{\sqrt{1-t}}\right)$$

and:

$$t \in V_2(-1) \implies |f(t)| \leqslant C_2 \frac{1}{\sqrt{1+t}} \iff f(t) \underset{t\to -1}{=} O\left(\frac{1}{\sqrt{1+t}}\right)$$

This implies that the integral defining Φ is definite; $f(t)$ is continuous over $(-1, 1)$ and therefore can be integrated. The result which we have just proved shows that $f(t)$ is integrable in a right neighborhood of –1 and a left neighborhood of 1, as the integral of its absolute value is incremented by an integrable function in both cases.

2) The bilinearity of Φ is obtained from the linearity of the integral using direct calculation. Its symmetry is a consequence of that of the dot product between functions. The only property which is not immediately evident is definite positiveness. Let us start by proving positiveness:

$$\Phi(P, P) = \int_{-1}^{1} \frac{P^2(t)}{\sqrt{1-t^2}} dt \geqslant 0$$

and[9]:

$$\Phi(P, P) = 0 \iff \frac{P^2(t)}{\sqrt{1-t^2}} dt = 0 \text{ a.e. on } (-1, 1) \iff P(t) = 0 \text{ a.e. on } (-1, 1)$$

but the only polynomial with an infinite number of roots is the null polynomial $\mathbf{0}(t) \equiv 0$, so $P = \mathbf{0}$. Φ is therefore an inner product on E.

9 a.e.: almost everywhere (see Chapter 3).

3) For all $n, m \in \mathbb{N}$:

$\langle T_n, T_m \rangle = \int_{-1}^{1} \frac{T_n(t)T_m(t)}{\sqrt{1-t^2}} dt \qquad (t = \cos\theta,\ dt = -\sin\theta d\theta)$

$$t = \cos\theta = -1 \iff \theta = \pi,\ t = \cos\theta = 1 \iff \theta = 0$$

$$\begin{aligned} &= \int_{\pi}^{0} -\sin\theta \frac{T_n(\cos\theta)T_m(\cos\theta)}{\sqrt{1-\cos^2(\theta)}} d\theta \\ &= \int_{0}^{\pi} \sin\theta \frac{\cos(n\theta)\cos(m\theta)}{|\sin\theta|} d\theta \\ &= \int_{0}^{\pi} \cancel{\sin\theta} \frac{\cos(n\theta)\cos(m\theta)}{\cancel{\sin\theta}} d\theta \quad (\sin\theta \geqslant 0 \text{ on } [0,\pi]) \\ &= \int_{0}^{\pi} \cos(n\theta)\cos(m\theta) d\theta \\ &= \frac{1}{2}\left(\int_{0}^{\pi} \cos((n+m)\theta)d\theta + \int_{0}^{\pi} \cos((n-m)\theta)d\theta\right) \quad \text{(from [1.13])} \end{aligned}$$

So, for all $n \neq m$, we have:

$$\langle T_n, T_m \rangle = \frac{1}{2}\left(\left[\frac{\sin((n+m)\theta)}{n+m}\right]_0^{\pi} + \left[\frac{\sin((n-m)\theta)}{n-m}\right]_0^{\pi}\right) = 0$$

that is, Chebyshev polynomials form an orthogonal family of polynomials in relation to the inner product defined above.

4) The family $(T_0, T_1, \ldots, T_n)$ is an orthogonal (and thus free) of $n+1$ elements of $\mathbb{R}[X]$, which is of dimension $n+1$, meaning that it is an orthogonal basis of $\mathbb{R}_n[X]$. To show that $(T_n)_{n\in\mathbb{N}}$ is a basis in the algebraic sense of E, consider a polynomial $P \in E$ of an arbitrary degree $d \in \mathbb{N}$, i.e. $P \in \mathbb{R}_d[X]$, and note that $(T_0, T_1, \ldots, T_d)$ is an orthogonal (free) family of generators of $\mathbb{R}_d[X]$, that is, a basis in the algebraic sense of the term.

5) The norm of T_n is calculated using the following equality:

$$\langle T_n, T_m \rangle = \frac{1}{2}\left(\int_{0}^{\pi} \cos((n+m)\theta)d\theta + \int_{0}^{\pi} \cos((n-m)\theta)d\theta\right)$$

which was demonstrated in point 3. Taking $n = m$, we have:

$$\|T_n\|^2 = \langle T_n, T_n \rangle = \frac{1}{2}\left(\int_{0}^{\pi} \cos(2n\theta)d\theta + \int_{0}^{\pi} d\theta\right) = \frac{1}{2}\left(\left[\frac{\sin 2n\theta}{2n}\right]_0^{\pi} + \pi\right) =$$

$$\|T_n\|^2 = \langle T_n, T_n \rangle = \tfrac{1}{2}\left(\int_0^{\pi} \cos(2n\theta)d\theta + \int_0^{\pi} d\theta\right) =$$

$$= \begin{cases} \frac{1}{2}\left(\int_0^{\pi} d\theta + \int_0^{\pi} d\theta\right) = \pi & \text{if } n = 0 \\ \frac{1}{2}\left(\left[\frac{\sin 2n\theta}{2n}\right]_0^{\pi} + \pi\right) = \frac{\pi}{2} & \text{if } n \geqslant 1, \end{cases}$$

hence $\|T_0\| = \sqrt{\pi}$ and $\|T_n\| = \sqrt{\pi/2}$ for $n \geqslant 1$. Finally, the family:

$$\left\{\frac{T_0}{\sqrt{\pi}}\right\} \cup \left\{\sqrt{\frac{2}{\pi}}T_n\right\}_{n\geqslant 1}$$

is an orthonormal basis of the vector space of first-order polynomials with real coefficients E. □

1.9. Summary

In this chapter, we have examined the properties of real and complex inner products, highlighting their differences. We noted that the symmetrical and bilinear properties of the real inner product must be replaced by conjugate symmetry and sesquilinearity in order to obtain a set of properties which are compatible with definite positivity. This final property is essential in order to produce a norm from a scalar product.

We noted that the prototype for all inner product spaces, or pre-Hilbert spaces, of finite dimension n is the Euclidean space $\mathbb{K}^n$, where $\mathbb{K} = \mathbb{R}$ or $\mathbb{K} = \mathbb{C}$.

Using the inner product, the concept of orthogonality between vectors can be extended to any inner product space. Two vectors are orthogonal if their inner product is null. The null vector is the only vector which is orthogonal to all other vectors, and the property of definite positiveness means that it is the only vector to be orthogonal to itself. If two vectors have the same inner product with all other vectors, that is, the same projection in every direction, then these vectors coincide.

A norm on a vector space is said to be a Hilbert norm if an inner product can be defined which generates the norm in a canonical manner. Remarkably, a norm is a Hilbert norm if and only if it satisfies the parallelogram law; this holds true for both finite and infinite dimensions. The polarization law can be used to define an inner product which is compatible with a Hilbert norm.

Vector orthogonality implies linear independence, guaranteeing that a set of n orthogonal vectors in a vector space of dimension n will constitute a basis. The expansion of a vector on an orthonormal basis is trivial: the components in relation to this basis are the inner products of the vector with the basis vectors. It is therefore much simpler to calculate components in such cases because, if the basis is not orthonormal, then a linear system of equations must be solved.

The concept of orthogonal projection on a vector subspace S was also presented. Given an orthogonal basis of this space, the projection can be represented as an expansion over the vectors of the basis, with coefficients given by the inner products (which are normalized if the basis is not orthonormalized). We have seen that the difference between a vector and its orthogonal projection, known as the residual vector, is orthogonal to the projection subspace S. We also demonstrated that the orthogonal projection is the vector in S which minimizes the distance (in relation to the norm of the vector space) between the vector and the vectors of S.

Given an inner product space, of finite or infinite dimensions, an orthonormal basis can always be defined using the Gram-Schmidt orthonormalization algorithm.

Finally, we proved the important Parseval identity and Plancherel's theorem in relation to an orthonormal or orthogonal basis. The extension of these properties to infinite dimensions is presented in Chapter 5.

2

The Discrete Fourier Transform and its Applications to Signal and Image Processing

The information presented in the previous chapter (Chapter 1) concerning complex inner product spaces and their properties lays the foundations for a very simple introduction to the discrete Fourier transform (DFT).

We simply need to prove that certain functions of complex exponentials constitute an orthogonal basis for a complex inner product space of finite dimension.

From a mathematical standpoint, the DFT is a simple change of basis in a vector space; however, its interpretation is of crucial importance and is extremely useful in the context of applications, notably in signal theory, as we shall see in section 2.6.

This section draws on the excellent work of M. Frazier (2001).

2.1. The space $\ell^2(\mathbb{Z}_N)$ and its canonical basis

In order to introduce the vector space in which the DFT is to be constructed, we need to make a few adjustments to the notation used thus far.

We shall continue to work with complex vectors with a number of components N, $1 < N < +\infty$, but a vector in $\mathbb{C}^N$ will be considered as a finite sequence.

Our first task is to define $\mathbb{Z}_N$.

DEFINITION 2.1.– *Two integers $i, j \in \mathbb{Z}$ are said to be* congruent modulo N *if their difference is divisible by N, that is:*

$$\frac{a-b}{N} = m \in \mathbb{Z},$$

meaning that we can write $a = b + mN$. The (Gaussian) notation for two integers which are congruent modulo N is:

$$a \equiv b \pmod N$$

Congruence modulo N can be shown to be an equivalence relationship in $\mathbb{Z}$. Like all equivalence relationships, it creates a partition of $\mathbb{Z}$ into distinct equivalence classes. The set of these equivalence classes is written as:

$$\mathbb{Z}_N = \mathbb{Z}\,(\text{mod } N)$$

A (finite) sequence of complex values on $\mathbb{Z}_N$ is a function:

$$\begin{aligned} z : \mathbb{Z}_N &\to \mathbb{C} \\ j &\mapsto z(j) \end{aligned}$$

In practice, circular or "clock" arithmetic is applied: this consists of identifying a sequence defined on $\mathbb{Z}_N$ as a sequence defined on $\{0, 1, \ldots, N-1\}$ and extended to $\mathbb{Z}$ by N-periodicity:

$$\boxed{z(j + mN) = z(j)} \quad \forall j, m \in \mathbb{Z}$$

that is, given the definition of $z(j)$ when $j \in \{0, 1, \ldots, N-1\}$, in order to determine $z(j)$ when $j \notin \{0, \ldots, N-1\}$, we must add an integer multiple of N to j. This is written as mN, $m \in \mathbb{Z}$ such that $\bar{j} = j + mN \in \{0, 1, \ldots, N-1\}$. We then define $z(j + mN) = z(\bar{j})$. An example is shown below.

EXAMPLE.–

$N = 12$, $z = (1, i, i, \sqrt{2}i, 0, 0, 0, -1, 0, 0, 0, 2)$, that is:

$$\begin{cases} z(0) = 1 \\ z(1) = z(2) = i \\ z(3) = \sqrt{2}i \\ z(4) = z(5) = z(6) = 0 \\ z(7) = -1 \\ z(8) = z(9) = z(10) = 0 \\ z(11) = 2 \end{cases}$$

extended by 12-periodicity to $\mathbb{Z}$. Determine $z(-21)$. Since $N = 12$, we must find the integer $m \neq 0$ such that $-21 + 12m \in \{0, 1, \dots, 11\}$:

$$-21 + 12m = \begin{cases} -21 + 12 = -9 & m = 1 \\ -21 + 24 = 3 & m = 2 \\ -21 + 36 = 15 & m = 3 \\ \dots \end{cases}$$

The value of m for which $-21 + 12m$ falls within $\{0, \dots, 11\}$ is $m = 2$, and in this case we have $-21 + 2 \cdot 12 = 3$, which implies $z(-21) = z(3) = \sqrt{2}i$.

Despite the fact that $\mathbb{Z}_N$ is often considered to represent the set of canonical representatives $\{0, 1, \dots, N-1\}$, we can, in fact, *consider z to be defined over any sub-set of $\mathbb{Z}$ given by N consecutive integers, and not necessarily over* $\{0, \dots, N-1\}$. This convention will be used throughout this book.

The complex vector space that will be used in this section is the set of all sequences of complex values over $\mathbb{Z}_N$:

$$\boxed{\ell^2(\mathbb{Z}_N) = \{z : \mathbb{Z}_N \to \mathbb{C}\}}$$

The reason for using this particular notation will become clear later.

$\ell^2(\mathbb{Z}_N)$ is a complex vector space with the usual scalar summation and multiplication operators, that is, given $z, w \in \ell^2(\mathbb{Z}_N)$, $\alpha \in \mathbb{C}$, the sum and multiplication by a complex vector are defined as follows:

$$\begin{aligned} z + w : \mathbb{Z}_N &\to \mathbb{C} \\ j &\mapsto (z+w)(j) = z(j) + w(j) \end{aligned}$$

$$\begin{aligned} \alpha z : \mathbb{Z}_N &\to \mathbb{C} \\ j &\mapsto (\alpha z)(j) = \alpha z(j) \end{aligned}$$

$\ell^2(\mathbb{Z}_N)$ is of dimension N: the application which associates each sequence $z \in \ell^2(\mathbb{Z}_N)$ with its images $(z(0), z(1), \dots, z(N-1))$:

$$\begin{aligned} \ell^2(\mathbb{Z}_N) &\longleftrightarrow \mathbb{C}^N \\ z &\longleftrightarrow (z(0), z(1), \dots, z(N-1)) = \begin{pmatrix} z(0) \\ z(1) \\ \vdots \\ z(N-1) \end{pmatrix} \end{aligned}$$

is a *linear isomorphism* (the proof is left to the reader). z will be represented as a row vector or as a column vector as the case requires.

The isomorphism above allows us to define the *canonical basis B of $\ell^2(\mathbb{Z}_N)$* as the set of the following N sequences:

$$B = (e_0, e_1, \dots, e_{N-1}), \quad e_j(k) = \delta_{j,k} = \begin{cases} 1 & k = j \\ 0 & k \neq j \end{cases}$$

We can also introduce an *inner product into $\ell^2(\mathbb{Z}_N)$* using:

$$\langle z, w \rangle = \sum_{k=0}^{N-1} z(k)\overline{w(k)}$$

so $z, w \in \ell^2(\mathbb{Z}_N)$ are orthogonal if and only if $\langle z, w \rangle = \sum_{k=0}^{N-1} z(k)\overline{w(k)} = 0$.

The norm induced by this inner product is:

$$\|z\| = \left(\sum_{k=0}^{N-1} |z(k)|^2 \right)^{\frac{1}{2}}$$

which will be referred to as the *$\ell^2(\mathbb{Z}_N)$ norm*.

2.1.1. *The orthogonal basis of complex exponentials in $\ell^2(\mathbb{Z}_N)$*

In this section, we are going to define the function system that will be essential to the development of the DFT.

First, we recall these basic facts:

1) for an arbitrary $z \in \mathbb{C}$, $z = \rho[\cos\alpha + i \sin\alpha] = \rho e^{i\alpha}$, $\rho, \alpha \in \mathbb{R}$, $\rho \geqslant 0$;

2) Euler's formulas: $\cos\alpha = \frac{1}{2}(e^{i\alpha} + e^{-i\alpha})$, $\sin\alpha = \frac{1}{2i}(e^{i\alpha} - e^{-i\alpha})$;

3) $|z| = 1 \Leftrightarrow z = e^{i\alpha}$;

4) $e^{i\alpha} = e^{i(\alpha+2\pi k)}$, $k \in \mathbb{Z}$;

5) as a specific instance of the previous point, if $\alpha = 0$, we obtain:

$$\boxed{e^{2\pi i k} = 1} \qquad \forall k \in \mathbb{Z} \, ;$$

6) $e^{i\alpha} e^{i\beta} = e^{i(\alpha+\beta)}$;

7) $(e^{i\alpha})^n = e^{in\alpha}$;

8) $\overline{e^{i\alpha}} = e^{-i\alpha}$;

9) given $z = \rho e^{i\alpha}$, the solutions to the equation $w^N = z$ are the N complex roots given by the equation: $w_m = \sqrt[N]{\rho} e^{i\frac{2\pi m+\alpha}{N}}$, $m = 0, \dots, N-1$;

10) specifically:

$$\text{roots } N\text{-ths of the unit} : \boxed{\omega_m = e^{2\pi i \frac{m}{N}}}, \qquad m = 0, \ldots, N-1.$$

We also need to recall the geometric summation formula, defined by:

$$S_k = 1 + z + z^2 + \ldots + z^{k-1} + z^k = \sum_{j=0}^{k} z^j$$

If $z = 1$, then $S_k = k + 1$. If $z \neq 1$, we observe that:

$$(1 - z)S_k = 1 + z + z^2 + \ldots + z^k - (z + z^2 + \ldots + z^k + z^{k+1}) = 1 - z^{k+1}$$

hence:

$$\sum_{j=0}^{k} z^j = \begin{cases} \frac{1-z^{k+1}}{1-z} & \text{if } z \in \mathbb{C}\backslash\{1\} \\ k+1 & \text{if } z = 1 \end{cases}$$

Now, consider the sequences in $\ell^2(\mathbb{Z}_N)$ defined by the following complex exponentials:

$$\begin{aligned} \mathcal{E}_m : \mathbb{Z}_N &\longrightarrow \mathbb{C} \\ n &\longmapsto \mathcal{E}_m(n) \end{aligned}$$

where:

$$\begin{cases} \mathcal{E}_0(n) = 1 \\ \mathcal{E}_1(n) = e^{2\pi i \frac{n}{N}} \\ \mathcal{E}_2(n) = e^{2\pi i \frac{2n}{N}} \\ \vdots \\ \mathcal{E}_{N-1}(n) = e^{2\pi i \frac{(N-1)n}{N}} \end{cases}$$

Hence:

– $\mathcal{E}_0$ is the constant sequence $\mathcal{E}_0(n) \equiv 1 \ \forall n \in \mathbb{Z}_N$;

– $\mathcal{E}_1$ is the sequence $\mathcal{E}_1 = \left(1, e^{2\pi i \frac{1}{N}}, e^{2\pi i \frac{2}{N}}, \ldots, e^{2\pi i \frac{(N-1)}{N}}\right)$;

– $\mathcal{E}_2$ is the sequence $\mathcal{E}_2 = \left(1, e^{2\pi i \frac{2}{N}}, e^{2\pi i \frac{4}{N}}, \ldots, e^{2\pi i \frac{2(N-1)}{N}}\right)$;

– $\mathcal{E}_{N-1}$ is the sequence $\mathcal{E}_{N-1} = \left(1, e^{2\pi i \frac{N-1}{N}}, e^{2\pi i \frac{2(N-1)}{N}}, \ldots, e^{2\pi i \frac{(N-1)^2}{N}}\right)$.

The general sequence is:

$$\boxed{\mathcal{E}_m(n) = e^{2\pi i \frac{mn}{N}} = (\omega_m)^n} \qquad \forall m, n = 0, \ldots, N-1$$

where $(\omega_m)^n$ is the n-th power of the N-th roots of the unit, $\forall n \in \{0, ..., N-1\}$, so:

$$(\omega_m)^n = \left(e^{2\pi i \frac{m}{N}}\right)^n = e^{2\pi i \frac{mn}{N}}$$

From formula $z = e^{i\alpha} = [\cos\alpha + i\sin\alpha]$, we know that the system defined above is *a set of sequences of values which oscillate at different frequencies*, since the arguments of the cos and sin functions change with the coefficients m and n. As we shall see, the signification of these frequencies is crucial to Fourier analysis.

For now, let us focus on proving that the exponential system defined above is an orthogonal basis of $\ell^2(\mathbb{Z}_N)$. This proof relies on a preliminary lemma.

LEMMA 2.1.– For all $j, k \in \{0, 1, \ldots, N-1\}$, we have:

$$\sum_{n=0}^{N-1} e^{2\pi i n \frac{j-k}{N}} = \sum_{n=0}^{N-1} e^{-2\pi i n \frac{j-k}{N}} = N\delta_{j,k} = \begin{cases} N & j = k \\ 0 & j \neq k \end{cases} \qquad [2.1]$$

The physical interpretation of this key formula will be discussed later. Before going further with the proof, note that in the case where $j, k \in \mathbb{Z}_N$, $j \neq k$, we have $j - k \in \{1, 2, \ldots, N-1\}$, so $\frac{j-k}{N} = -\frac{k-j}{N} \notin \mathbb{Z}$.

PROOF.– This proof covers the first summation, but it is evident that this demonstration also holds for the second summation. We start by using the properties of complex exponentials to rewrite the formula as follows:

$$\sum_{n=0}^{N-1} e^{2\pi i n \frac{j-k}{N}} = \sum_{n=0}^{N-1} \left(e^{2\pi i \frac{j-k}{N}}\right)^n$$

Let us analyze the following two cases:

– if $j = k$, the exponentials in the sum are equal to 1, since $e^{2\pi i \frac{0}{N}} = 1$, and thus:

$$\sum_{n=0}^{N-1} e^{2\pi i n \frac{j-j}{N}} = \sum_{n=0}^{N-1} 1 = N$$

– if $j \neq k$, the exponentials are $\neq 1$, so, using the geometric summation formula:

$$\begin{aligned}\sum_{n=0}^{N-1} \left(e^{2\pi i \frac{j-k}{N}}\right)^n &= \frac{1 - \left(e^{2\pi i \frac{j-k}{N}}\right)^{N-1+1}}{1 - e^{2\pi i \frac{j-k}{N}}} \\ &= \frac{1 - e^{2\pi i \frac{(j-k)N}{N}}}{1 - e^{2\pi i \frac{j-k}{N}}} \\ &= \frac{1 - e^{2\pi i (j-k)}}{1 - e^{2\pi i \frac{j-k}{N}}}\end{aligned}$$

Since $j - k = m \in \mathbb{Z}$, $e^{2\pi i(j-k)} = 1$, the numerator of the final formula is 0 when $j \neq k$. The denominator, on the other hand, never cancels out; as we saw in the remark before the proof, if $j \neq k$, then $\frac{j-k}{N} \notin \mathbb{Z}$. In this case, the summation is equal to 0. □

The demonstration that $\mathcal{E}$ is an orthogonal basis of $\ell^2(\mathbb{Z}_N)$ is now trivial.

THEOREM 2.1.– $\mathcal{E} = (\mathcal{E}_0, \dots, \mathcal{E}_{N-1})$ is an orthogonal basis of $\ell^2(\mathbb{Z}_N)$.

PROOF.– $\mathcal{E}$ is given by N elements of an N –dimensional inner product space, so if we can prove that $\mathcal{E}$ is an orthogonal family, then the theorem is also proved. We know that an orthogonal family is free, and a free family of N vectors in an N–dimensional vector space is a basis.

We thus calculate the inner products $\langle \mathcal{E}_j, \mathcal{E}_k \rangle$, $\forall j, k \in \{0, \dots, N-1\}$:

$$\langle \mathcal{E}_j, \mathcal{E}_k \rangle = \sum_{n=0}^{N-1} \mathcal{E}_j(n)\overline{\mathcal{E}_k(n)} = \sum_{n=0}^{N-1} e^{2\pi i \frac{jn}{N}} e^{-2\pi i \frac{kn}{N}} = \sum_{n=0}^{N-1} e^{2\pi i \frac{(j-k)n}{N}} = N\delta_{j,k}$$

using Lemma 2.1 to give us the final equality, which proves that $\langle \mathcal{E}_j, \mathcal{E}_k \rangle = N\delta_{j,k}$, that is, the elements in the basis are mutually orthogonal. □

If we consider that $j = k = m$ in equation $\langle \mathcal{E}_j, \mathcal{E}_k \rangle = N\delta_{j,k}$, then $\langle \mathcal{E}_m, \mathcal{E}_m \rangle = N\delta_{m,m} = N$, hence:

$$\boxed{\|\mathcal{E}_m\|^2 = N}, \quad \boxed{\|\mathcal{E}_m\| = \sqrt{N}}, \qquad \forall m \in \{0, 1, \dots, N-1\}$$

Now, let us consider two examples in which the expression of the complex exponentials is particularly simple: $N = 2$ and $N = 4$ (the expression using $N = 3$ is not quite so simple):

1) $N = 2$. $\ell^2(\mathbb{Z}_2) = \{z = (z(0), z(1)) \in \mathbb{C}^2\}$, in this case $\mathcal{E}_m(n) = e^{2\pi i \frac{mn}{2}} = e^{\pi imn}$ and thus:

$$m = 0: \qquad \mathcal{E}_0 = \left(e^{\pi i 0\cdot 0}, e^{\pi i 0\cdot 1}\right) = (1, 1)$$

$$m = 1: \qquad \mathcal{E}_1 = \left(e^{\pi i 1\cdot 0}, e^{\pi i 1\cdot 1}\right) = \left(1, e^{\pi i}\right)$$

However, $e^{\pi i} = \cos(\pi) + i\sin(\pi) = -1$, so $\mathcal{E}_1 = (1, -1)$. Thus:

$$\mathcal{E} = ((1, 1), (1, -1)) \qquad [2.2]$$

is the basis of complex exponentials in $\ell^2(\mathbb{Z}_2)$. Note the presence of a constant sequence (the first) and an oscillating sequence (the second). This particular feature of the basis will be discussed in greater detail later.

2) $N = 4$. $\ell^2(\mathbb{Z}_4) = \{z = (z(0), z(1), z(2), z(3)) \in \mathbb{C}^4\}$: the Fourier basis is obtained from four complex sequences, each with four components. Verification that the basis of complex exponentials of $\ell^2(\mathbb{Z}_4)$ is:

$$\mathcal{E} = ((1,1,1,1),(1,i,-1,-i),(1,-1,1,-1),(1,-i,-1,i)) \quad [2.3]$$

is left to the reader. Results [1.10], [1.11] and [1.12] from section 1.8 may be used to write the following formulas, which are valid for any two elements $z, w \in \ell^2(\mathbb{Z}_N)$:

– decomposition on the orthogonal basis $\mathcal{E}$:

$$z = \sum_{m=0}^{N-1} \frac{\langle z, \mathcal{E}_m \rangle}{N} \mathcal{E}_m \quad [2.4]$$

– Parseval's identity for the orthogonal basis $\mathcal{E}$:

$$\langle z, w \rangle = \sum_{m=0}^{N-1} \frac{\langle z, \mathcal{E}_m \rangle \langle \mathcal{E}_m, w \rangle}{N} \quad [2.5]$$

– Plancherel's theorem for $\mathcal{E}$:

$$\|z\|^2 = \sum_{m=0}^{N-1} \frac{|\langle z, \mathcal{E}_m \rangle|^2}{N} \quad [2.6]$$

The expressions above are calculated explicitly in section 2.3.

There are several ways of renormalizing the basis $\mathcal{E}$. Two of the most widespread approaches, which can also be used to define the DFT, are discussed in the next two sections.

2.2. The orthonormal Fourier basis of $\ell^2(\mathbb{Z}_N)$

As we saw in section 2.1.1, the norm $\ell^2(\mathbb{Z}_N)$ of all sequences $\mathcal{E}_m$ is $\sqrt{N}$; evidently, an orthonormal basis can therefore be obtained by dividing by $\sqrt{N}$. This justifies Definition 2.2.

DEFINITION 2.2.– *The orthonormal Fourier basis of $\ell^2(\mathbb{Z}_N)$ is the set:*

$$E = (E_0, E_1, E_2, \ldots, E_{N-1})$$

of the N sequences $E_m \in \ell^2(\mathbb{Z}_N)$:

$$\begin{aligned} E_m : \mathbb{Z}_N &\longrightarrow \mathbb{C} \\ n &\longmapsto E_m(n) \end{aligned}$$

where:

$$\begin{cases} E_0(n) = \frac{1}{\sqrt{N}} \\ E_1(n) = \frac{1}{\sqrt{N}} e^{2\pi i \frac{n}{N}} \\ E_2(n) = \frac{1}{\sqrt{N}} e^{2\pi i \frac{2n}{N}} \\ \vdots \\ E_{N-1}(n) = \frac{1}{\sqrt{N}} e^{2\pi i \frac{(N-1)n}{N}} \end{cases}$$

The general sequence of the orthonormal Fourier basis is:

$$\boxed{E_m(n) = \frac{1}{\sqrt{N}} e^{2\pi i \frac{mn}{N}} = \frac{1}{\sqrt{N}} (\omega_m)^n} \qquad \forall m, n = 0, \dots, N-1$$

and the orthonormality formula $\langle E_j, E_k \rangle = \delta_{j,k}$ holds true.

Using formulas [2.2] and [2.3], we can say that:

$$E = \frac{1}{\sqrt{2}} ((1, 1), (1, -1)) \qquad [2.7]$$

is the orthonormal Fourier basis of $\ell^2(\mathbb{Z}_2)$ and:

$$E = \frac{1}{2} ((1, 1, 1, 1), (1, i, -1, -i), (1, -1, 1, -1), (1, -i, -1, i)) \qquad [2.8]$$

is the orthonormal Fourier basis of $\ell^2(\mathbb{Z}_4)$.

The translation of theorem 1.14 for $\ell^2(\mathbb{Z}_N)$ equipped with the orthonormal Fourier basis is as follows. Given arbitrary elements $z, w \in \ell^2(\mathbb{Z}_N)$, we have:

– a decomposition on the orthonormal Fourier basis:

$$z = \sum_{m=0}^{N-1} \langle z, E_m \rangle E_m \qquad [2.9]$$

– Parseval's identity:

$$\langle z, w \rangle = \sum_{m=0}^{N-1} \langle z, E_m \rangle \langle E_m, w \rangle \qquad [2.10]$$

– Plancherel's theorem:

$$\|z\|^2 = \sum_{m=0}^{N-1} |\langle z, E_m \rangle|^2 \qquad [2.11]$$

2.3. The orthogonal Fourier basis of $\ell^2(\mathbb{Z}_N)$

Although the normalization constant $1/\sqrt{N}$, which appears in the definition of the orthonormal Fourier basis, might appear to be the most logical choice for normalizing the basis $\mathcal{E}$ in $\ell^2(\mathbb{Z}_N)$, another normalization is more commonly used in practical applications. The reason for this choice, shown below, is that it simplifies the writing of several other formulas.

DEFINITION 2.3.– *The orthogonal Fourier basis of $\ell^2(\mathbb{Z}_N)$ is the set:*

$$F = (F_0, F_1, F_2, \ldots, F_{N-1})$$

of N sequences $F_m \in \ell^2(\mathbb{Z}_N)$:

$$\begin{aligned} F_m : \mathbb{Z}_N &\longrightarrow \mathbb{C} \\ n &\longmapsto F_m(n) \end{aligned}$$

where:

$$\begin{cases} F_0(n) = \frac{1}{N} \\ F_1(n) = \frac{1}{N}e^{2\pi i \frac{n}{N}} \\ F_2(n) = \frac{1}{N}e^{2\pi i \frac{2n}{N}} \\ \vdots \\ F_{N-1}(n) = \frac{1}{N}e^{2\pi i \frac{(N-1)n}{N}} \end{cases}$$

The general sequence of the orthogonal Fourier basis is:

$$\boxed{F_m(n) = \frac{1}{N}e^{2\pi i \frac{mn}{N}} = \frac{1}{N}(\omega_m)^n} \qquad \forall m, n = 0, \ldots, N-1$$

The relationships between the three bases $\mathcal{E}$, E and F are:

$$E_m = \frac{\mathcal{E}_m}{\sqrt{N}}, \quad F_m = \frac{\mathcal{E}_m}{N}, \quad F_m = \frac{E_m}{\sqrt{N}} \qquad \forall m \in \{0, 1, \ldots, N\} \qquad [2.12]$$

Using the formulas above, the orthogonal Fourier bases of $\ell^2(\mathbb{Z}_2)$ and $\ell^2(\mathbb{Z}_4)$ are easy to calculate:

– orthogonal Fourier basis of $\ell^2(\mathbb{Z}_2)$:

$$F = \frac{1}{2}((1,1),(1,-1)) \qquad [2.13]$$

– orthogonal Fourier basis of $\ell^2(\mathbb{Z}_4)$:

$$F = \frac{1}{4}((1,1,1,1),(1,i,-1,-i),(1,-1,1,-1),(1,-i,-1,i)) \qquad [2.14]$$

Again, using relationship [2.12], we can determine the equivalents of formulas [2.9] or [2.4], [2.10] or [2.5] and [2.11] or [2.6] for two arbitrary elements $z, w \in \ell^2(\mathbb{Z}_N)$:

– decomposition on the orthogonal Fourier basis:

$$z = N \sum_{m=0}^{N-1} \langle z, F_m \rangle F_m$$

– Parseval's identity for the orthogonal Fourier basis:

$$\langle z, w \rangle = N \sum_{m=0}^{N-1} \langle z, F_m \rangle \langle F_m, w \rangle$$

– Plancherel's theorem for the orthogonal Fourier basis:

$$\|z\|^2 = N \sum_{m=0}^{N-1} |\langle z, F_m \rangle|^2$$

Table 2.1 supplies a helpful summary of the differences between these bases and formulas:

$$\mathcal{E}_m(n) = e^{2\pi i \frac{mn}{N}}, \; E_m(n) = \frac{1}{\sqrt{N}} e^{2\pi i \frac{mn}{N}}, \; F_m(n) = \frac{1}{N} e^{2\pi i \frac{mn}{N}}$$

Basis	Decomposition	Parseval's identity	Plancherel 's theorem
$\mathcal{E}$	$z = \sum_{m=0}^{N-1} \frac{\langle z, \mathcal{E}_m \rangle}{N} \mathcal{E}_m$	$\langle z, w \rangle = \sum_{m=0}^{N-1} \frac{\langle z, \mathcal{E}_m \rangle \langle \mathcal{E}_m, w \rangle}{N}$	$\|z\|^2 = \sum_{m=0}^{N-1} \frac{|\langle z, \mathcal{E}_m \rangle|^2}{N}$
E	$z = \sum_{m=0}^{N-1} \langle z, E_m \rangle E_m$	$\langle z, w \rangle = \sum_{m=0}^{N-1} \langle z, E_m \rangle \langle E_m, w \rangle$	$\|z\|^2 = \sum_{m=0}^{N-1} |\langle z, E_m \rangle|^2$
F	$z = N \sum_{m=0}^{N-1} \langle z, F_m \rangle F_m$	$\langle z, w \rangle = N \sum_{m=0}^{N-1} \langle z, F_m \rangle \langle F_m, w \rangle$	$\|z\|^2 = N \sum_{m=0}^{N-1} |\langle z, F_m \rangle|^2.$

Table 2.1. *Different normalizations of Fourier bases and relative formulas*

2.4. Fourier coefficients and the discrete Fourier transform

The definition of the DFT varies from author to author and from application to application. The two most widespread definitions use the orthonormal basis E and a blend of the orthogonal bases $\mathcal{E}$ and F.

These two versions are useful for different reasons:

– using the orthonormal basis E allows us to obtain unitary operators;

– using a blend of the orthogonal bases $\mathcal{E}$ and F makes it possible to simplify many formulas, including the convolution formula, widely used in applications, which will be discussed later.

For the purposes of this book, we shall use formulas obtained by a blend of the orthogonal bases $\mathcal{E}$ and F. This decision was made for reasons of coherency with various mathematical programs, notably MATLAB.

First, let us reconsider the following decomposition:

$$z = \sum_{m=0}^{N-1} \frac{\langle z, \mathcal{E}_m \rangle}{N} \mathcal{E}_m = \sum_{m=0}^{N-1} \langle z, \mathcal{E}_m \rangle \frac{\mathcal{E}_m}{N}$$

However, $\mathcal{E}_m/N = F_m$, so:

$$z = \sum_{m=0}^{N-1} \langle z, \mathcal{E}_m \rangle F_m$$

that is, any given element $z \in \ell^2(\mathbb{Z}_N)$ can be decomposed over the orthogonal Fourier basis F with the components given by the inner products of z with elements of the basis $\mathcal{E}$.

Using the definition of the inner product of $\ell^2(\mathbb{Z}_N)$, we can write:

$$\langle z, \mathcal{E}_m \rangle = \sum_{n=0}^{N-1} z(n)\overline{\mathcal{E}_m(n)} = \sum_{n=0}^{N-1} z(n)\overline{e^{2\pi i \frac{mn}{N}}} = \sum_{n=0}^{N-1} z(n) e^{-2\pi i \frac{mn}{N}}$$

DEFINITION 2.4.– *Given any $z \in \ell^2(\mathbb{Z}_N)$, the complex vectors $\langle z, \mathcal{E}_m \rangle$, $m \in \{0, 1, \ldots, N-1\}$ are known as the Fourier coefficients of z, noted $\hat{z}(m)$. Explicitly:*

$$\boxed{\hat{z}(m) = \sum_{n=0}^{N-1} z(n) e^{-2\pi i \frac{mn}{N}}} \quad \text{Fourier coefficients of z} \qquad [2.15]$$

The sequence of Fourier coefficients of z is written using $\hat{z} \in \ell^2(\mathbb{Z}_N)$:

$$\hat{z} = (\hat{z}(0), \hat{z}(1), \hat{z}(2), \ldots, \hat{z}(N-1)) \qquad [2.16]$$

The linear operator which transforms $z \in \ell^2(\mathbb{Z}_N)$ into the sequence $\hat{z} \in \ell^2(\mathbb{Z}_N)$ of its Fourier coefficients, that is:

$$\begin{aligned} \text{DFT} \equiv \hat{} \equiv \mathcal{F} : \ell^2(\mathbb{Z}_N) &\longrightarrow \ell^2(\mathbb{Z}_N) \\ z &\longmapsto \text{DFT}(z) \equiv \hat{z} \equiv \mathcal{F}(z) \end{aligned}$$

$$\hat{z}(m) = \sum_{n=0}^{N-1} z(n)e^{-2\pi i \frac{mn}{N}} \qquad \forall m \in \{0, 1, \ldots, N-1\}$$

is known as the discrete Fourier transform, or DFT.

It is important to note that *the variable of* z *is* n*, while the variable of* $\hat{z}$ *is* m. The interpretation of n and m in the context of signal theory will be given in section 2.6; for now, note simply that n is the discrete value of an instant in time (or a position in space) at which a signal z is measured, whereas m is proportional to the oscillation frequency of a wave (harmonic) and is a multiple of a fundamental frequency. The DFT is used to translate a description of a signal in terms of temporal (or spatial) samples into a description in terms of signal frequencies. This notion will be formalized in section 2.6.

Using the definitions given above, the decomposition of z may be written as follows:

$$z = \sum_{m=0}^{N-1} \hat{z}(m)F_m \qquad [2.17]$$

that is, the Fourier coefficients of z are the components of z in the orthogonal Fourier basis F:

$$\hat{z} = [z]_F \qquad [2.18]$$

Using the notation introduced above, the theorem of decomposition on the orthonormal Fourier basis, Parseval's identity and Plancherel's theorem may be rewritten as:

– *decomposition of* z *on the orthogonal Fourier basis*:

$$\boxed{z(n) = \frac{1}{N}\sum_{m=0}^{N-1} \hat{z}(m)e^{2\pi i \frac{mn}{N}}} \qquad \forall n = 0, 1, \ldots, N-1 \qquad [2.19]$$

– *Parseval's identity*:

$$\boxed{\langle z, w\rangle = \frac{1}{N}\sum_{m=0}^{N-1} \hat{z}(m)\overline{\hat{w}(m)} = \frac{1}{N}\langle \hat{z}, \hat{w}\rangle} \qquad [2.20]$$

– *Plancherel's theorem* :

$$\boxed{\|z\|^2 = \frac{1}{N}\sum_{m=0}^{N-1} |\hat{z}(m)|^2 = \frac{1}{N}\|\hat{z}\|^2} \qquad [2.21]$$

2.4.1. *The inverse discrete Fourier transform*

It is interesting to compare formulas [2.15] and [2.19]:

$$\hat{z}(m) = \sum_{n=0}^{N-1} z(n) e^{-2\pi i \frac{mn}{N}}, \quad z(n) = \frac{1}{N} \sum_{m=0}^{N-1} \hat{z}(m) e^{2\pi i \frac{mn}{N}}$$
$$\forall n, m \in \{0, 1, \ldots, N-1\}$$

The first relationship states that given the values of $z(n)$, the values of $\hat{z}(m)$ can be reconstructed using formula [2.15].

The second relationship states that given the values of $\hat{z}(m)$, the values of $z(n)$ can be reconstructed using formula [2.19].

There is thus a "duality" between the two formulas: it is possible to obtain sequence z from sequence $\hat{z}$ and *vice versa* using relationships [2.15] and [2.19].

This duality is formalized in Definition 2.5 and Theorem 2.2.

DEFINITION 2.5.– *The linear operator:*

$$\begin{array}{rcl} \text{IDFT} \equiv \check{} \equiv \mathcal{F}^{-1} : \ell^2(\mathbb{Z}_N) & \longrightarrow & \ell^2(\mathbb{Z}_N) \\ z & \longmapsto & \text{IDFT}(z) \equiv \check{z} \equiv \mathcal{F}^{-1}(z) \end{array}$$

$$\boxed{\check{z}(n) = \frac{1}{N} \sum_{m=0}^{N-1} z(m) e^{2\pi i \frac{mn}{N}}} \qquad \forall n \in \{0, 1, \ldots, N-1\}$$

is known as the inverse discrete Fourier transform, *or* IDFT.

THEOREM 2.2.– The IDFT is the inverse linear operator of the DFT and *vice versa*:

$$\text{IDFT} = \text{DFT}^{-1}, \quad \text{DFT} = \text{IDFT}^{-1}$$

or, in other terms,

$$\check{\hat{z}} = z, \quad \hat{\check{z}} = z \qquad \forall z \in \ell^2(\mathbb{Z}_N)$$

PROOF.– We wish to prove that the composition between the DFT and the IDFT and between the IDFT and the DFT gives the identity operator id: the DFT∘IDFT=IDFT∘DFT= id, $id(z) = z$, $\forall z \in \ell^2(\mathbb{Z}_N)$.

We start by verifying that, given an arbitrary sequence $z \in \ell^2(\mathbb{Z}_N)$ and applying the DFT to obtain the sequence of Fourier coefficients $\hat{z} \in \ell^2(\mathbb{Z}_N)$, it is possible to obtain the original sequence by applying the IDFT:

$$\begin{array}{ccccc} \ell^2(\mathbb{Z}_N) & \underset{\text{DFT}}{\longrightarrow} & \ell^2(\mathbb{Z}_N) & \underset{\text{IDFT}}{\longrightarrow} & \ell^2(\mathbb{Z}_N) \\ z & \longmapsto & \hat{z} & \longmapsto & \check{\hat{z}} = z \end{array}$$

Before writing the composition, it is important to note that the summation index – the symbol of which is unimportant – should not be confused with the fixed variables n, m in $\check{z}(n)$ and $\hat{z}(m)$. To avoid this problem we will use the neutral symbol j.

$$\begin{aligned}
\check{\hat{z}}(n) &= \frac{1}{N}\sum_{m=0}^{N-1} \hat{z}(m)e^{2\pi i\frac{mn}{N}} = \frac{1}{N}\sum_{m=0}^{N-1}\left(\sum_{j=0}^{N-1} z(j)e^{-2\pi i\frac{mj}{N}}\right)e^{2\pi i\frac{mn}{N}} \\
&= \frac{1}{N}\sum_{m=0}^{N-1}\sum_{j=0}^{N-1} z(j)e^{2\pi i m\frac{n-j}{N}} \\
&= \frac{1}{N}\sum_{j=0}^{N-1} z(j)\left(\sum_{m=0}^{N-1} e^{2\pi i m\frac{n-j}{N}}\right) \\
&\underset{(\text{Lemma } 2.1)}{=} \frac{1}{N}\sum_{j=0}^{N-1} z(j)N\delta_{j,n} \\
&= z(n) \quad \forall n \in \{0, 1, \ldots, N-1\}
\end{aligned}$$

Now, let us verify that the inverse composition produces the same identity:

$$\begin{array}{ccccc}
\ell^2(\mathbb{Z}_N) & \underset{\text{IDFT}}{\longrightarrow} & \ell^2(\mathbb{Z}_N) & \underset{\text{DFT}}{\longrightarrow} & \ell^2(\mathbb{Z}_N) \\
z & \longmapsto & \check{z} & \longmapsto & \hat{\check{z}} = z.
\end{array}$$

$$\begin{aligned}
\hat{\check{z}}(m) &= \sum_{n=0}^{N-1} \check{z}(n)e^{-2\pi i\frac{mn}{N}} = \sum_{n=0}^{N-1}\left(\frac{1}{N}\sum_{j=0}^{N-1} z(j)e^{2\pi i\frac{jn}{N}}\right)e^{-2\pi i\frac{mn}{N}} \\
&= \frac{1}{N}\sum_{n=0}^{N-1}\sum_{j=0}^{N-1} z(j)e^{2\pi i n\frac{j-m}{N}} \\
&= \frac{1}{N}\sum_{j=0}^{N-1} z(j)\left(\sum_{n=0}^{N-1} e^{2\pi i n\frac{j-m}{N}}\right) \\
&\underset{(\text{Lemma} 2.1)}{=} \frac{1}{N}\sum_{j=0}^{N-1} z(j)N\delta_{j,m} \\
&= z(m) \quad \forall m \in \{0, 1, \ldots, N-1\}
\end{aligned}$$

Thus, $\check{\hat{z}}(n) = z(n)$ and $\hat{\check{z}}(m) = z(m)$, $\forall n, m \in \{0, 1, \ldots, N-1\}$ which concludes our proof. □

Note the similarity between the DFT and the IDFT: the only differences are the coefficient $1/N$ and the sign of the complex exponential.

We wish to draw the reader's attention to the formulas demonstrated above:

$$\boxed{\check{\hat{z}}(n) = \frac{1}{N}\sum_{m=0}^{N-1} \hat{z}(m)e^{2\pi i\frac{mn}{N}} = z(n)} \qquad \forall n \in \mathbb{Z}_N$$

$$\boxed{\hat{\check{z}}(m) = \sum_{n=0}^{N-1} \check{z}(n) e^{-2\pi i \frac{mn}{N}} = z(m)} \qquad \forall m \in \mathbb{Z}_N$$

DEFINITION 2.6.– *The pair* $(z, \hat{z}) \in \ell^2(\mathbb{Z}_N) \times \ell^2(\mathbb{Z}_N)$ *is known as a* Fourier pair.

2.4.2. *Definition of the DFT and the IDFT with the orthonormal Fourier basis*

An alternative definition of Fourier coefficients, the DFT and the IDFT, more commonly found in a theoretical mathematical context, uses the orthonormal Fourier basis E:

– $z, w \in \ell^2(\mathbb{Z}_N)$;

– Fourier coefficients:

$$\hat{z}(m) = \frac{1}{\sqrt{N}} \sum_{n=0}^{N-1} z(n) e^{-2\pi i \frac{mn}{N}} \qquad [2.22]$$

The notation $\hat{z}$ *in the following formulas in this list (and only these formulas) refers to the Fourier coefficients above.*

– decomposition on the orthonormal Fourier basis:

$$z(m) = \frac{1}{\sqrt{N}} \sum_{n=0}^{N-1} \hat{z}(n) e^{2\pi i \frac{mn}{N}}$$

– DFT :

$$\hat{z}(m) = \frac{1}{\sqrt{N}} \sum_{n=0}^{N-1} z(n) e^{-2\pi i \frac{mn}{N}} \qquad \forall m \in \{0, 1, \ldots, N-1\}$$

– IDFT :

$$\check{z}(n) = \frac{1}{\sqrt{N}} \sum_{m=0}^{N-1} z(m) e^{2\pi i \frac{mn}{N}} \qquad \forall n \in \{0, 1, \ldots, N-1\}$$

– Parseval's identity:

$$\langle z, w \rangle = \sum_{m=0}^{N-1} \hat{z}(m) \overline{\hat{w}(m)} = \langle \hat{z}, \hat{w} \rangle$$

– Plancherel's theorem:

$$\|z\|^2 = \sum_{m=0}^{N-1} |\hat{z}(m)|^2 = \|\hat{z}\|^2$$

Box 2.1. *Discrete orthonormal Fourier analysis*

As we can see, the greatest advantage of using the orthonormal Fourier basis in defining the objects used in Fourier analysis is that the DFT and the IDFT are operators which conserve the inner product, and consequently the norm; they are therefore represented using unitary matrices.

We also see that, independently of the definition used, *the product of the coefficients of* $\hat{z}$ *and* $\check{z}$ *must always be equal to* $1/N$ *to guarantee that* IDFT = DFT^{-1}.

2.4.3. *The real (orthonormal) Fourier basis*

The Fourier basis and DFT can be written using real notation. The advantage of a real DFT is that, if z is real, we can avoid the need to introduce imaginary components. For simplicity's sake, we shall focus on the orthonormal Fourier basis.

First, we must determine whether N is even or odd. Let us begin with the case where N is even: $N = 2M$, $M \in \mathbb{N}$, $M \geqslant 1$. In this case, $\forall n = 0, 1, \ldots, N-1$, we write:

$$\begin{cases} c_0(n) = \frac{1}{\sqrt{N}} \\ c_m(n) = \sqrt{\frac{2}{N}} \cos\left(\frac{2\pi mn}{N}\right) & m = 1, 2, ..., M-1 \\ c_M(n) = \frac{1}{\sqrt{N}} \cos\left(\frac{2\pi \frac{N}{2} n}{N}\right) = \frac{(-1)^n}{\sqrt{N}} \\ s_m(n) = \sqrt{\frac{2}{N}} \sin\left(\frac{2\pi mn}{N}\right) & m = 1, 2, \ldots, M-1 \end{cases}$$

If $N = 2M + 1$ is odd, c_0, c_m and s_m are defined in the same way as above, but $m = N/2$ should not be considered as in this case $N/2$ is not an integer.

THEOREM 2.3.– The set $\{c_0, c_1, \ldots, c_{M-1}, c_M, s_1, \ldots, s_{M-1}\}$, when $N = 2M$, or the set $\{c_0, c_1, \ldots, c_{M-1}, s_1, \ldots, s_{M-1}\}$, when $N = 2M+1$, is an orthonormal basis of $\ell^2(\mathbb{Z}_N)$. Thus, for all $z \in \ell^2(\mathbb{Z}_N)$:

$$z = \sum_{m=0}^{M} \langle z, c_m \rangle c_m + \sum_{m=1}^{M-1} \langle z, s_m \rangle s_m \quad (N = 2M)$$

$$z = \sum_{m=0}^{M-1} \langle z, c_m \rangle c_m + \sum_{m=1}^{M-1} \langle z, s_m \rangle s_m \quad (N = 2M + 1)$$

DEFINITION 2.7.– *The real orthonormal Fourier basis of* $\ell^2(\mathbb{Z}_N)$ *is the set of sequences of* $\ell^2(\mathbb{Z}_N)$ $\{c_0, c_1, \ldots, c_{M-1}, c_M, s_1, \ldots, s_{M-1}\}$ *when* $N = 2M$, *our the set of sequences of* $\ell^2(\mathbb{Z}_N)$ $\{c_0, c_1, \ldots, c_{M-1}, s_1, \ldots, s_{M-1}\}$ *when* $N = 2M + 1$.

The relationship with the Fourier coefficients is obtained using the following formulas:

$$\begin{cases} \langle z, c_0 \rangle = \frac{\hat{z}(0)}{\sqrt{N}} \\ \langle z, c_M \rangle = \frac{\hat{z}(M)}{\sqrt{N}} \\ \langle z, c_m \rangle = \frac{1}{\sqrt{2N}}(\hat{z}(m) + \hat{z}(N-m)), \quad m = 1, 2, \ldots, M-1 \\ \langle z, s_m \rangle = \frac{-i}{\sqrt{2N}}(\hat{z}(m) - \hat{z}(N-m)), \quad m = 1, 2, \ldots, M-1 \\ \hat{z}(0) = \sqrt{N}\langle z, c_0 \rangle \\ \hat{z}(M) = \sqrt{N}\langle z, c_M \rangle \\ \hat{z}(m) = \sqrt{N/2}(\langle z, c_m \rangle - i\langle z, s_m \rangle), \quad m = 1, 2, \ldots, M-1 \\ \hat{z}(m) = \sqrt{N/2}(\langle z, c_{N-m} \rangle + i\langle z, s_{N-m} \rangle), \quad m = M+1, M+2, \ldots, N-1 \end{cases}$$

2.5. Matrix interpretation of the DFT and the IDFT

By definition, the DFT transforms sequences of $\ell^2(\mathbb{Z}_N)$ represented in the canonical basis B of $\ell^2(\mathbb{Z}_N)$ into sequences of $\ell^2(\mathbb{Z}_N)$ represented in the orthogonal Fourier basis F of $\ell^2(\mathbb{Z}_N)$ [2.17]:

$$\begin{aligned} \text{DFT}: \ \ell^2(\mathbb{Z}_N) &\longrightarrow \ell^2(\mathbb{Z}_N) \\ z = [z]_B &\longmapsto \text{DFT}(z) = \hat{z} = [z]_F \end{aligned}$$

The DFT is thus the operator used to operate the change from the canonical basis B of $\ell^2(\mathbb{Z}_N)$ to the Fourier basis F of $\ell^2(\mathbb{Z}_N)$, and, consequently, the IDFT is the opposite operator.

We wish to establish a matrix representation of these two linear operators DFT and IDFT. To do this, we shall use a notation which is widely used in literature concerning the DFT: $\omega_N = e^{-\frac{2\pi i}{N}}$. Using the properties of complex exponentials, we can write:

$$\omega_N^{mn} = e^{-2\pi i \frac{mn}{N}}$$

and the Fourier coefficients can thus be written as:

$$\hat{z}(m) = \sum_{n=0}^{N-1} z(n) e^{-2\pi i \frac{mn}{N}} = \sum_{n=0}^{N-1} z(n) \omega_N^{mn}$$

We define the matrix W_N containing the elements ω_N^{mn} :

$$w_{mn} = \omega_N^{mn}$$

that is, explicitly:

$$W_N = \begin{pmatrix} 1 & 1 & 1 & 1 & \dots & 1 \\ 1 & \omega_N & \omega_N^2 & \omega_N^3 & \dots & \omega_N^{N-1} \\ 1 & \omega_N^2 & \omega_N^4 & \omega_N^6 & \dots & \omega_N^{2(N-1)} \\ 1 & \omega_N^3 & \omega_N^6 & \omega_N^9 & \dots & \omega_N^{3(N-1)} \\ \vdots & \vdots & \vdots & \vdots & \ddots & \vdots \\ 1 & \omega_N^{N-1} & \omega_N^{2(N-1)} & \omega_N^{3(N-1)} & \dots & \omega_N^{(N-1)(N-1)} \end{pmatrix} \qquad [2.23]$$

This $N \times N$ matrix is called *Sylvester matrix*[1]. It is *symmetrical*: $W_N = W_N^t$, i.e. $w_{mn} = w_{nm}$ (an obvious consequence of the definition of w_{mn} , since $\omega_N^{mn} = \omega_N^{nm}$) and each line or column is obtained by the geometric progression[2] of a power of ω_N. A matrix of this type is known as a Vandermonde matrix[3].

By convention, when considering W_N, we examine the variability of the indices of the lines and columns between 0 and $N - 1$ (in place of canonical variability, from 1 to N). This convention is the reason why all elements in the first line ($m = 0$) and all elements in the first column ($n = 0$) are equal to 1.

If we apply W_N to z *considered as a column vector* in $\mathbb{C}^N$, then, by the definition of matrix product, we obtain a vector $W_N z$ whose m-th component $(W_N z)(m)$[4] is given by:

$$(W_N z)(m) = \sum_{n=0}^{N-1} w_{mn} z(n) = \sum_{n=0}^{N-1} z(n) e^{-2\pi i \frac{mn}{N}} = \hat{z}(m), \qquad \forall m \in \mathbb{Z}_N,$$

thus:

$$\boxed{\hat{z} = W_N z} \qquad \forall z \in \ell^2(\mathbb{Z}_N)$$

Using the same approach, we can verify that the IDFT is implemented *via* the conjugate matrix of W_N normalized by the coefficient $1/N$ (transposition is not required as W_N is symmetrical):

$$\boxed{W_N^{-1} = \frac{1}{N}\overline{W_N}}, \qquad \boxed{\check{z} = W_N^{-1} z} \qquad \forall z \in \ell^2(\mathbb{Z}_N)$$

1 James Joseph Sylvester (1814, London-1897, London).

2 A geometric progression of reason r is the sequence of powers $1 = r^0, r = r^1, r^2, r^3, \dots, r^n$.

3 Alexandre-Théophile Vandermonde (1735, Paris-1796, Paris).

4 This is the real Euclidean product of the m-th line of W_N, i.e. $(w_{m0}, w_{m1}, \dots, w_{m(N-1)})$ times the components $(z(0), z(1), \dots, z(n))$ of z.

W_N is the *change of basis matrix used to go from B to F*, and $W_N^{-1} = \frac{1}{N}\overline{W_N}$ is the *change of basis matrix used to go from F to B*.

OBSERVATIONS.– Using the definition of the DFT corresponding to equation [2.22], that is using the orthonormal Fourier basis, the associated matrix becomes $\tilde{W}_N = W_N/\sqrt{N}$. This is a unitary matrix, and thus its inverse matrix is $\overline{\tilde{W}_N}$.

Examples:

– $N = 2 : \omega_2 = e^{-2\pi i/2} = e^{-i\pi} = \cos(\pi) - i\sin(\pi) = -1$, thus:

$$W_2 = \begin{pmatrix} 1 & 1 \\ 1 & -1 \end{pmatrix}$$

hence:

$$W_2^{-1} = \frac{1}{2}\begin{pmatrix} 1 & 1 \\ 1 & -1 \end{pmatrix}$$

– $N = 4 : \omega_4 = e^{-2\pi i/4} = e^{-i\pi/2} = \cos(\pi/2) - i\sin(\pi/2) = -i$, thus:

$$W_4 = \begin{pmatrix} 1 & 1 & 1 & 1 \\ 1 & -i & (-i)^2 & (-i)^3 \\ 1 & (-i)^2 & (-i)^4 & (-i)^6 \\ 1 & (-i)^3 & (-i)^6 & (-i)^9 \end{pmatrix}$$

hence:

$$W_4 = \begin{pmatrix} 1 & 1 & 1 & 1 \\ 1 & -i & -1 & i \\ 1 & -1 & 1 & -1 \\ 1 & i & -1 & -i \end{pmatrix} \qquad [2.24]$$

The inverse matrix is:

$$W_4^{-1} = \frac{1}{4}\begin{pmatrix} 1 & 1 & 1 & 1 \\ 1 & i & -1 & -i \\ 1 & -1 & 1 & -1 \\ 1 & -i & -1 & i \end{pmatrix} \qquad [2.25]$$

Note that the columns of matrix W_4^{-1} consist of the *orthogonal* basis F of $\ell^2(\mathbb{Z}_4)$, as seen in formula [2.14]; this is coherent with the fact that this is the matrix used to change from the *orthogonal* basis F to the canonical basis of $\ell^2(\mathbb{Z}_4)$.

2.5.1. *The fast Fourier transform*

As we have seen, the action of the DFT on a signal $z \in \ell^2(\mathbb{Z}_N)$ can be represented as a matrix product.

We must therefore calculate N multiplications for each element $\hat{z}(m)$ in the sequence $\hat{z} \in \ell^2(\mathbb{Z}_N)$. Since $\hat{z}$ has N components, the complexity of the algorithm used to calculate the DFT is $\mathcal{O}(N^2)$.

This complexity means that the DFT is extremely time-consuming when working with signals of large dimension. In practice, the Fourier transform was almost never used outside of a theoretical context (that is, in real-world applications) before the 1960s.

A breakthrough came in 1965, when Cooley and Tukey used symmetries concealed within the DFT to construct a fast algorithm for calculating the DFT: this algorithm is known as the fast Fourier transform (FFT).

The complexity of the FFT is of the order of $\mathcal{O}(N \log N)$, and, using modern computers, it allows the Fourier transform of large dimension signals to be calculated in under a second.

The FFT is extremely efficient in cases where the signal dimension is a power of 2. This is the reason why a 512 or 1,024 format is typically used for digital images, enabling rapid and efficient processing using the FFT.

The development of the FFT is considered as *one of the greatest scientific breakthroughs of the 20th century, as it enables the use of Fourier transforms in a vast array of practical applications.*

2.6. The Fourier transform in signal processing

Fourier theory has applications in a wide range of domains, for example in solving ordinary and partial differential equations, classical and quantum physics, statistics and probabilities, and signal processing.

In this section, we shall highlight the crucial role of Fourier theory in signal processing in one dimension (1D).

2.6.1. *Synthesis formula for 1D signals: decomposition on the harmonic basis*

A discrete 1D signal of dimension N may be defined as the set of N samplings of a variable, which may be dependent on time, on a spatial dimension (x,y or z), or on another parameter with a single degree of freedom.

Two remarkable examples of discrete 1D signals, dependent on time or a single spatial dimension, are:

– the set of intensity values for a piece of music, sampled at N different moments in time;

– the set of grayscale values of a line or column in a simple image, corresponding to N different positions.

A discrete 1D signal can be processed using Fourier theory using the following basic identifications:

– the abstract mathematical representation of a discrete 1D signal is given by a sequence $z \in \ell^2(\mathbb{Z}_N)$;

– $n \in \mathbb{Z}_N = \{0, 1, \ldots, N-1\}$ represents the value of the parameter (time, spatial dimension, etc.) according to which the signal is sampled. The unit of measurement used for n is typically the second or meter;

– the energy of the signal z is associated with the square of the norm $\|z\|^2$.

The next step is to interpret the decomposition formula over the Fourier basis, the DFT and the IDFT, and Plancherel's theorem in the context of signal processing.

The interpretation of Plancherel's theorem in this case is simplest: the energy of the signal z is decomposed into the sum of the squared magnitudes of the Fourier coefficients.

The decomposition formula over the Fourier basis, equation [2.19], is known as the *synthesis formula* in the context of signal processing:

$$z(n) = \frac{1}{N} \sum_{m=0}^{N-1} \hat{z}(m) e^{2\pi i \frac{mn}{N}} \qquad \forall n \in \mathbb{Z}_N$$

Using this formula, the signal z can be reconstructed (or "synthesized") using the Fourier coefficients $\hat{z}(m)$ and the oscillating functions $e^{2\pi i \frac{mn}{N}}$.

The functions used in the signal synthesis operation are:

$$e^{2\pi i \frac{m}{N} n} = \cos\left(2\pi \frac{m}{N} n\right) + i \sin\left(2\pi \frac{m}{N} n\right). \qquad [2.26]$$

When $m = 0$, there is no oscillation; from $m = 1$ to $m = N - 1$ the functions $e^{2\pi i \frac{m}{N} n}$ oscillate at a certain frequency (m is therefore measured in *hertz* or rad/s). This will be discussed in detail in section 2.6.4. These functions are known as harmonics, a term derived from the field of music, as we see from Definition 2.8.

DEFINITION 2.8 (harmonics).– *The function* $n \mapsto e^{2\pi i \frac{1}{N} n}$ *is known as a fundamental (discrete) harmonic*[5] *and the functions* $n \mapsto e^{2\pi i \frac{m}{N} n}$ *for* $m = 2, \ldots, N-1$ *are (discrete) harmonics of (higher) order* m.

2.6.2. Signification of Fourier coefficients and spectrums of a 1D signal

The synthesis formula tells us that the signal z in the value n of its parameter can be reconstructed using a linear combination of harmonic waves of frequencies which are multiples of $1/N$ *via* the coefficient m: $\{0, 1/N, 2/N, \ldots, (N-1)/N\}$. The complex scalars of the linear combination are the Fourier coefficients $\hat{z}(m)$.

Each Fourier coefficient $\hat{z}(m) \in \mathbb{C}$ may be written as:

$$\hat{z}(m) = a(m) + ib(m) = |\hat{z}(m)| e^{i\text{Arg}(\hat{z}(m))}$$

where $|\hat{z}(m)| = \sqrt{a(m)^2 + b(m)^2}$ is the magnitude of the Fourier coefficient $\hat{z}(m)$ and $\text{Arg}(\hat{z}(m)) = \arctan\left(\frac{b(m)}{a(m)}\right)$ is its argument.

Evidently, the "weight" which measures the importance of each harmonic $e^{2\pi i \frac{mn}{N}}$ in reconstructing a signal z is the magnitude[6] of the Fourier coefficient $\hat{z}(m)$:

$|\hat{z}(m)|$: measures the importance of the harmonic $e^{2\pi i \frac{mn}{N}}$ in reconstructing z.

For this reason, in signal processing, the Fourier coefficient formula is known as the *analysis formula:*

$$\hat{z}(m) = \sum_{n=0}^{N-1} z(n) e^{-2\pi i \frac{mn}{N}} \qquad \forall m \in \mathbb{Z}_N$$

since $\hat{z}$ allows us to *analyze the frequency components of a signal.*

If the discrete signal z is dependent on the time t (or a spatial dimension x), then the transformation $z \to \hat{z}$ obtained using the DFT enables us to go from a *temporal (or spatial) representation of the signal* to a *frequential representation*, or the *Fourier space*.

The Fourier transform is often defined as *the equivalent of Newton's prism for mathematics*. Newton's prism breaks down light into "hidden" frequency components corresponding to the colors of the spectrum. The Fourier transform reveals the frequency components which are "hidden" in any signal.

This analogy explain the terms used in Definition 2.9.

5 It is important to specify that these harmonics are discrete; continuous harmonics are obtained using functions $t \mapsto e^{2\pi i m \nu t} = e^{i m \omega t}$, where ν is the frequency and $\omega = 2\pi\nu$ the pulse.

6 The magnitude must be used here due to the fact that complex numbers are not ordered.

DEFINITION 2.9.– *Given* $z \in \ell^2(\mathbb{Z}_N)$*:*

– $\{|\hat{z}(m)|,\ m \in \mathbb{Z}_N\}$ *is known as the* amplitude spectrum *of* z*, or simply the* spectrum *of* z*;*

– $\{|\hat{z}(m)|^2,\ m \in \mathbb{Z}_N\}$ *is the* power spectrum *of* z*;*

– $\{Arg(\hat{z}(m)),\ m \in \mathbb{Z}_N\}$ *is the* phase spectrum *of* z*.*

The signification of these spectra will be discussed in detail later.

Note the presence of one particularly special Fourier coefficient, $\hat{z}(0)$, which provides information concerning the average value of z:

$$\hat{z}(0) = \sum_{n=0}^{N-1} z(n)e^{2\pi i \frac{0n}{N}} = \sum_{n=0}^{N-1} z(n) = N\langle z\rangle \quad \Longrightarrow \quad \boxed{\hat{z}(0) = N\langle z\rangle}$$

where $\langle z\rangle = \frac{1}{N}\sum_{n=0}^{N-1} z(n)$ is the average value of the signal z.

Introducing this expression of $\hat{z}(0)$ into the synthesis formula and separating the first term from the rest of the sum, we obtain:

$$z(n) = \frac{1}{N}N\langle z\rangle + \frac{1}{N}\sum_{m=1}^{N-1} \hat{z}(m)e^{2\pi i \frac{mn}{N}}$$

$$\text{that is : } \boxed{z(n) = \langle z\rangle + \frac{1}{N}\sum_{m=1}^{N-1} \hat{z}(m)e^{2\pi i \frac{mn}{N}}}$$

The Fourier coefficient $\hat{z}(0)$ is known as the "DC" component of the synthesis formula, while the other terms constitute the "AC" component. This terminology is taken from the field of electronics, with DC standing for "direct current" (current of frequency zero) and AC standing for "alternating current".

One way of interpreting the formula set out above is to say that z *is decomposed into the sum of its mean value and the finer details reconstructed by higher order harmonics, weighted by the Fourier coefficients of* z.

2.6.3. *The synthesis formula and Fourier coefficients of the unit pulse*

It is helpful to compare the synthesis formula with formula [2.1], that is:

$$\sum_{n=0}^{N-1} e^{\pm 2\pi i n \frac{j-k}{N}} = N\delta_{j,k} = \begin{cases} N & j = k \\ 0 & j \neq k \end{cases}$$

Rewriting $j-k = m \in \mathbb{Z}_N$, switching m and n (an acceptable substitution, as both are arbitrary values of $\mathbb{Z}_N$) and normalizing by N, we obtain the following formula:

$$\frac{1}{N}\sum_{m=0}^{N-1} e^{\pm 2\pi i n \frac{m}{N}} = \begin{cases} 1 & n = 0 \\ 0 & n = 1, \dots, N-1 \end{cases} = e_0(n) \equiv \delta_0(n) \equiv \delta(n)$$

δ is known as the unit pulse. If we select the option "+" in the formula shown above, we obtain the synthesis formula for the unit pulse, in which all Fourier coefficients are unitary:

$$\widehat{\delta_0}(m) = 1 = \left|\widehat{\delta_0}(m)\right| \qquad \forall m \in \mathbb{Z}_N$$

This result is particularly informative: *the DFT transforms a signal which is completely "localized" at a value on its parameter into a signal which is fully "delocalized" across the spectrum: the harmonics for all frequencies have the same weight when reconstructing the signal.*

Let us now calculate the Fourier coefficients of the constant signal $z(n) = \frac{1}{N}$, $\forall n \in \mathbb{Z}_N$, we obtain:

$$\hat{z}(m) = \sum_{n=0}^{N-1} \frac{1}{N} e^{-2\pi i \frac{mn}{N}} = \frac{1}{N}\sum_{n=0}^{N-1} e^{-2\pi i \frac{mn}{N}} = \delta_0(m) = \begin{cases} 1 & m = 0 \\ 0 & m = 1, \dots, N-1 \end{cases}$$

We see that the DFT of a constant signal (which is completely delocalized in relation to its parameter) is therefore a unit pulse in the Fourier domain, meaning that it is completely localized in its frequencies.

The generalization of this behavior for spaces which are more complicated than $\ell^2(\mathbb{Z}_N)$ – notably $L^2(\Omega)$, $\Omega \subseteq \mathbb{R}^n$, which we will examine later – forms the basis for understanding the *Heisenberg uncertainty principle*, the conceptual core of quantum mechanics.

Thanks to the results that we have discussed above, we can give a physical interpretation of the formula [2.1] in Lemma 2.1: the superposition of harmonic functions with frequencies which are integer multiples of one another is subjected to a destructive interference everywhere, except at one value where the harmonics experience a constructive interference.

Moreover, according to the synthesis formula, harmonics must be weighted differently in order to reconstruct any signal which is not a pulse.

2.6.4. *High and low frequencies in the synthesis formula*

Let us take a closer look at the meaning of the frequency coefficients m in the set

$$\left\{e^{2\pi i \frac{mn}{N}} = \cos\left(2\pi \frac{mn}{N}\right) + i \sin\left(2\pi \frac{mn}{N}\right),\ n = 0, 1, \dots, N-1\right\},$$

which represents the value of the harmonics in each of the N parameters n.

For the sake of simplicity, we shall only consider the real part of the elements of the set above, that is

$$H_m = \left\{\cos\left(2\pi \frac{mn}{N}\right),\ n = 0, 1, \ldots, N-1\right\};$$

our remarks concerning the cosine are equally applicable to the sine.

Consider the behavior of $\cos\left(2\pi \frac{mn}{N}\right)$ when the value of m is between 0 and $N-1$, where N is *even* (the case where N is odd will be discussed later):

– $m = 0:$ As we have already seen, in this case, *there is no oscillation*, but simply a series of constant values, $\cos\left(2\pi \frac{0n}{N}\right) = 1$, so:

$$H_0 = \{1, 1, \ldots, 1\};$$

– $m = 1:$

$$H_1 = \left\{1, \cos\left(2\pi \frac{1}{N}\right), \cos\left(2\pi \frac{2}{N}\right), \ldots, \cos\left(2\pi \frac{N-1}{N}\right)\right\}$$

The values of H_1 represent N samples of a *cosine oscillation*. The cycle does not terminate as we do not consider the value $n = N$, which would give us $\cos\left(2\pi \frac{N}{N}\right) = \cos(2\pi) = 1$. Figure 2.1 shows the graph of H_m for $m = 1, N = 16$;

– $m = 2:$

$$H_2 = \left\{1, \cos\left(2\pi \frac{2}{N}\right), \cos\left(2\pi \frac{4}{N}\right), \ldots, \cos\left(2\pi \frac{2\frac{N}{2}}{N}\right)\right.$$

$$\left. = 1, \ldots, \cos\left(2\pi \frac{2(N-1)}{N}\right)\right\}$$

The values of H_2 represent N samples of *two cosine oscillations*. $n = N/2$ marks the end of a cosine cycle. Figure 2.2 shows the graph of H_m for $m = 2, N = 16$. We see that, for $n = 8 = 16/2$, the cosine value is 1.

Increasing m up to $N/2$, the oscillation frequency of the cosine increases. The maximum frequency is reached when $m = N/2$; in this case, $\cos\left(2\pi \frac{\frac{N}{2}n}{N}\right) = \cos(\pi n)$, thus:

$$H_{\frac{N}{2}} = \{(-1)^n,\ n = 0, 1, \ldots, N-1\}$$

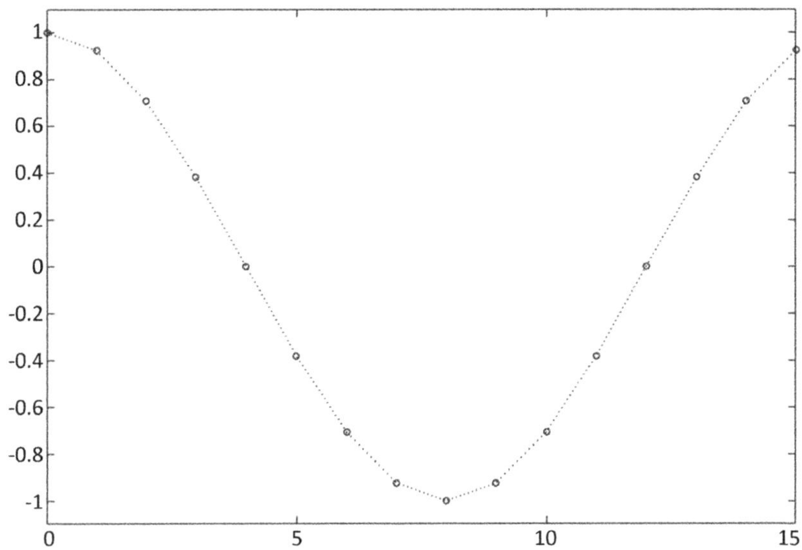

Figure 2.1. H_m *for m = 1, N = 16*

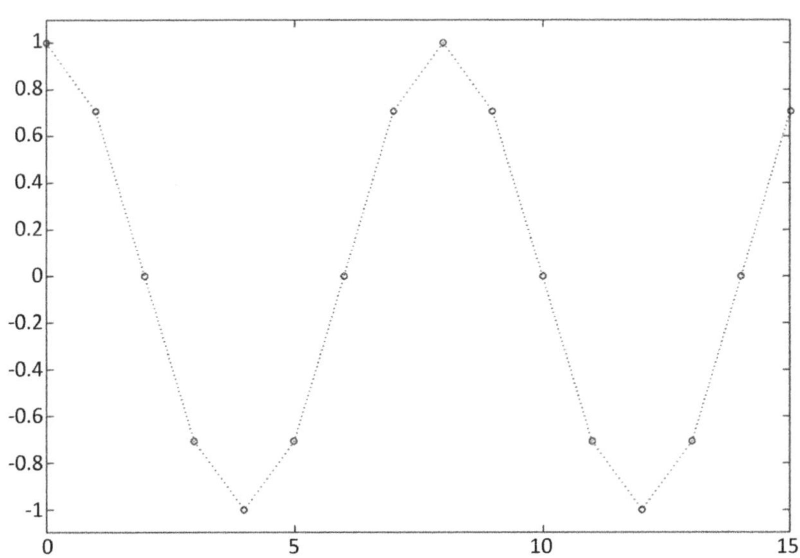

Figure 2.2. H_m *for m = 2, N = 16*

We might expect the cosine oscillation frequency to increase up to $N-1$, but this is not the case. In reality, *from* $m = N/2+1$, *the cosine oscillation frequency decreases*. To understand this behavior, consider $\cos\left(2\pi\frac{nm}{N}\right)$ when m belongs to the set $\left\{\frac{N}{2}+1, \frac{N}{2}+2, \ldots, N-1\right\}$, and apply the following change of variable:

$$k = N - m \Leftrightarrow m = N - k, \quad m \in \left\{\frac{N}{2}+1, \frac{N}{2}+2, \ldots, N-1\right\}$$
$$\Leftrightarrow k \in \left\{\frac{N}{2}-1, \ldots, 2, 1\right\},$$

then, when m increases from $\frac{N}{2}+1$ up to $N-1$, k decreases from $\frac{N}{2}-1$ down to 1. Applying this variable change to the cosine, we obtain:

$$\cos\left(2\pi\frac{n(N-k)}{N}\right) = \cos\left(2\pi n - 2\pi\frac{nk}{N}\right) = \cos\left(-2\pi\frac{nk}{N}\right) = \cos\left(2\pi\frac{nk}{N}\right),$$

having used the periodicity and parity of the cosine.

Consequently:

$$\boxed{\cos\left(2\pi\frac{nm}{N}\right),\ m \in \left\{\frac{N}{2}+1, \frac{N}{2}+2, \ldots, N-1\right\} \Longleftrightarrow \cos\left(2\pi\frac{nk}{N}\right),\ k \in \left\{\frac{N}{2}-1, \ldots, 1\right\}}$$

Thus, *the maximum number of harmonic oscillations is obtained when* $m = N/2$, *and is symmetrical about this value*. For example, the graph of H_m for $m = 9, N = 16$ is exactly equal to the graph of H_m with $m = 7, N = 16$. Similarly, the graph of $m = 15, N = 16$ is exactly equal to the graph in Figure 2.1, representing H_m with $m = 1, N = 16$;

– evidently, if N is *odd*, the considerations set out above are valid for $\left[\frac{N}{2}\right]$, the integer part of $\frac{N}{2}$, that is the integer closest to, but not greater than $\frac{N}{2}$.

The elements described above are the reasons for certain choices of terminology:

– *high frequencies*: values of m close to $\frac{N}{2}$;

– *low frequencies*: values of m close to 0 or $N-1$.

If the synthesis formula for a discrete signal $z \in \ell^2(\mathbb{Z}_N)$ includes Fourier coefficients $\hat{z}(m)$ with a high magnitude for values of m which are close to $N/2$, the signal will be characterized by relatively violent variations (as in the case of high sounds, such as those produced by cymbals). However, if the Fourier coefficients with the highest modulus correspond to values of m close to 0 and $N-1$, the signal will be characterized by "gentler" variations (as in the case of low sounds, such as those produced by bass drums).

The frequency $m = N/2$ is known as the *Nyquist frequency*[7]. This is the highest harmonic frequency which can appear in the synthesis formula for N samples of a signal.

2.6.5. *Signal filtering in frequency representation*

The DFT can be used to easily modify the frequency content of a signal, for example increasing the strength of the lowest or highest frequencies.

The standard approach is to obtain the Fourier space using the DFT then adjust the Fourier coefficients as required using a *filter* $f : \ell^2(\mathbb{Z}_N) \to \ell^2(\mathbb{Z}_N)$, which may be either a linear or a nonlinear transform. Finally, the IDFT is applied to the sequence of modified Fourier coefficients to reconstruct the original signal in its modified form.

The signal processing approach used in the frequency domain is shown in Figure 2.3.

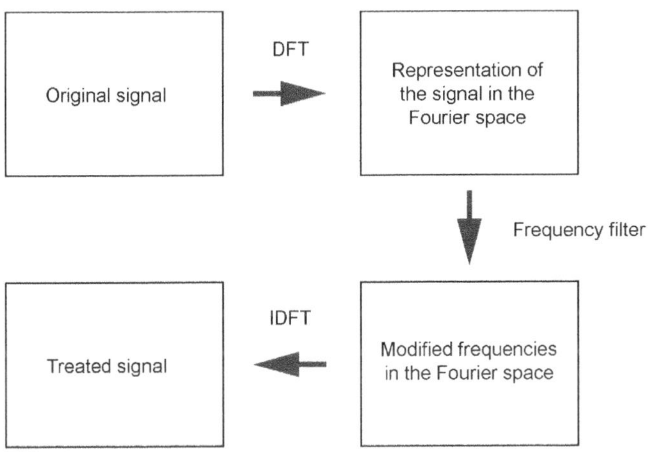

Figure 2.3. *Filtering approach in the Fourier domain*

Note that, in the IDFT $\circ f \circ$ DFT transform composition, only f has the capacity to change the energy of the signal: the composition of the Fourier transform with its inverse produces an identity, so the energy of the original signal is retained.

One particularly important example of a filter f, defined in section 2.6.6, can be used to define the concept of the Fourier multiplier, defined in section 2.6.7.

7 For the Swedish engineer Harry Nyquist (1889–1976).

2.6.6. *The multiplication operator and its diagonal matrix representation*

Let $w : \mathbb{Z}_N \to \mathbb{C}$ be a fixed sequence in $\ell^2(\mathbb{Z}_N)$.

DEFINITION 2.10.– *The linear application below is known as the multiplication operator by sequence w:*

$$\begin{aligned} M_w : \ell^2(\mathbb{Z}_N) &\longrightarrow \ell^2(\mathbb{Z}_N) \\ z &\longmapsto M_w(z) = w \cdot z \end{aligned}$$

where $M_w(z) = w \cdot z : \mathbb{Z}_N \to \mathbb{C}$ is the sequence defined by the point-wise (also called Hadamard) product of w and z:

$$\boxed{M_w z(n) = (w \cdot z)(n) = w(n) \cdot z(n)} \qquad \forall n \in \mathbb{Z}_N$$

Note that if z is represented as a column vector in the canonical basis of $\ell^2(\mathbb{Z}_N)$, then *the matrix associated with the operator M_w in relation to the canonical basis of $\ell^2(\mathbb{Z}_N)$ is a diagonal matrix D_w* with diagonal elements given by the components of sequence w:

$$D_w z = \begin{pmatrix} w(0) & & 0 \\ & \ddots & \\ 0 & & w(N-1) \end{pmatrix} \begin{pmatrix} z(0) \\ \vdots \\ z(N-1) \end{pmatrix} = \begin{pmatrix} w(0)z(0) \\ \vdots \\ w(N-1)z(N-1) \end{pmatrix}$$

EXAMPLE OF A MULTIPLICATION OPERATOR.–

Consider the sequence of $\ell^2(\mathbb{Z}_6)$ given by $z = (2, 3-i, 2i, 4+i, 0, 1)$ and the sequence $w(n) = i^n$, $n \in \mathbb{Z}_6$, then:

$$(w(0) = 1, w(1) = i, w(2) = -1, w(3) = -i, w(4) = 1, w(5) = i)$$

and thus:

$$(M_w z)(n) = (1{\cdot}2, i{\cdot}(3-i), -1{\cdot}2i, -i{\cdot}(4+i), 1{\cdot}0, i{\cdot}1) = (2, 3i+1, -2i, -4i+1, 0, i)$$

This provides the foundation for introducing the Fourier multiplier operator.

2.6.7. *The Fourier multiplier operator*

The Fourier multiplier operator – or multiplier – is one notable example of a frequency filter.

DEFINITION 2.11.– *Given a sequence* $w : \mathbb{Z}_N \to \mathbb{C}$, *the* Fourier multiplier *by sequence* w *is the following operator:*

$$\begin{aligned} T_{(w)} : \ell^2(\mathbb{Z}_N) &\longrightarrow \ell^2(\mathbb{Z}_N) \\ z &\longmapsto T_{(w)}(z) = \widecheck{w \cdot \hat{z}} \end{aligned}$$

that is, $T_{(w)}$ *is the operator given by the composition*

$$\boxed{T_{(w)} = \text{IDFT} \circ M_w \circ \text{DFT}},$$

that is,

$$\begin{array}{ccccccc} \ell^2(\mathbb{Z}_N) & \underset{DFT}{\longrightarrow} & \ell^2(\mathbb{Z}_N) & \underset{M_w}{\longrightarrow} & \ell^2(\mathbb{Z}_N) & \underset{IDFT}{\longrightarrow} & \ell^2(\mathbb{Z}_N) \\ z & \mapsto & DFT(z) = \hat{z} & \mapsto & M_w(DFT(z)) = w \cdot \hat{z} & \mapsto & IDFT(M_w(DFT(z))) = \widecheck{w \cdot \hat{z}} \end{array}$$

Applying the DFT to both sides of the definition of $T_{(w)}$, we see that the action of the Fourier multiplier is diagonal in the Fourier basis F:

$$\text{DFT}\, T_{(w)} z = [T_{(w)} z]_F = M_w \circ \text{DFT}\, z = M_w \hat{z}, \qquad \forall z \in \ell^2(\mathbb{Z}_N) \qquad [2.27]$$

Thus, $T_{(w)}$ multiplies the Fourier coefficients of z by the components of sequence w (making this operator a multiplier). This means that we can:

– *attenuate the low frequencies* of a signal z by selecting a sequence $w(m)$ with a low value of $|w(m)|$ when $m \simeq 0$ and $m \simeq N-1$;

– *attenuate the high frequencies* of a signal z by selecting a sequence $w(m)$ with a low value of $|w(m)|$ when $m \simeq N/2$;

– *amplify the low frequencies* of a signal z by selecting a sequence $w(m)$ with a high value of $|w(m)|$ when $m \simeq 0$ and $m \simeq N-1$;

– *amplify the high frequencies* of a signal z by selecting a sequence $w(m)$ with a high value of $|w(m)|$ when $m \simeq N/2$.

This information is used in graphic equalizers, used by musicians to adjust the level of high frequencies and bass notes in an audio signal.

2.7. Properties of the DFT

In this section, we shall demonstrate the most important properties of the DFT. We shall begin by recalling the translation property of a summation index:

$$\sum_{i=n_0}^{n} a_i = \sum_{i=n_0-k}^{n-k} a_{i+k} = \sum_{i=n_0+k}^{n+k} a_{i-k} \qquad [2.28]$$

This property will be used on several occasions, along with the following lemma.

LEMMA 2.2.– Let $f : \mathbb{Z} \to \mathbb{C}$ be an N-periodic function, with $N \in \mathbb{N}$:

$$f(n + aN) = f(n) \qquad \forall a, n \in \mathbb{Z}$$

Then, for all $m \in \mathbb{Z}$:

$$\boxed{\sum_{n=m}^{m+N-1} f(n) = \sum_{n=0}^{N-1} f(n)}$$

that is, *the sum of an N-periodic function across any interval of size N is constant.*

PROOF.– If $m = 0$, there is nothing to prove, so we may take $m \in \mathbb{Z}$, $m \neq 0$. Considering values of $m > 0$:

$$\sum_{n=m}^{m+N-1} f(n) = \sum_{n=0}^{m+N-1} f(n) - \sum_{n=0}^{m-1} f(n) = \sum_{n=0}^{N-1} f(n) + \sum_{n=N}^{m+N-1} f(n) - \sum_{n=0}^{m-1} f(n)$$

but, using [2.28]:

$$\sum_{n=N}^{m+N-1} f(n) = \sum_{n=0}^{m-1} f(n+N) = \sum_{n=0}^{m-1} f(n)$$

because of the N-periodicity of f, thus:

$$\sum_{n=m}^{m+N-1} f(n) = \sum_{n=0}^{N-1} f(n) + \sum_{n=0}^{m-1} f(n) - \sum_{n=0}^{m-1} f(n) = \sum_{n=0}^{N-1} f(n)$$

A similar demonstration may be used for cases where $m < 0$. □

2.7.1. *Periodicity of $\hat{z}$ and $\check{z}$*

In what follows, we shall examine the most important properties of the discrete Fourier theory, starting with the periodicity of $\hat{z}$ and $\check{z}$.

By direct calculation, if $a \in \mathbb{Z}$, then:

$$\hat{z}(m + aN) = \sum_{n=0}^{N-1} z(n)e^{-2\pi i\frac{(m+aN)n}{N}} = \sum_{n=0}^{N-1} z(n)e^{-2\pi i\frac{mn}{N}}e^{-2\pi i\frac{aNn}{N}}$$

$$= \sum_{n=0}^{N-1} z(n)e^{-2\pi i\frac{mn}{N}}e^{-2\pi ani} = \hat{z}(m)$$

since $e^{-2\pi ani} = \cos(2\pi an) - i\sin(2\pi an) = 1$. The same calculation is used to prove $\check{z}(n + aN) = \check{z}(n)\ \forall a \in \mathbb{Z}$.

Thanks to this property, the definitions of $\hat{z}$ and $\check{z}$ can be extended to $\mathbb{Z}$ by considering the two N*-periodic sequences*:

$$\begin{aligned} \hat{z} : \mathbb{Z} &\longrightarrow \mathbb{C} \\ m &\longmapsto \boxed{\hat{z}(m) = \hat{z}(m + aN)} \end{aligned}$$

and:

$$\begin{aligned} \check{z} : \mathbb{Z} &\longrightarrow \mathbb{C} \\ n &\longmapsto \boxed{\check{z}(n) = \check{z}(n + aN)} \end{aligned}$$

with $a \in \mathbb{Z}$ such that $m + aN \in \mathbb{Z}_N$, or $n + aN \in \mathbb{Z}_N$, respectively.

2.7.2. *DFT and shift*

We now wish to consider how the DFT of a signal $z \in \ell^2(\mathbb{Z}_N)$ varies in response to a shift in z by a quantity different to N. Another operator for $\ell^2(\mathbb{Z}_N)$ must be introduced in order to formalize this consideration.

DEFINITION 2.12.– *Take* $z \in \ell^2(\mathbb{Z}_N)$*. The following linear application is the* right shift operator *of the quantity* k*:*

$$\begin{aligned} R_k : \ell^2(\mathbb{Z}_N) &\longrightarrow \ell^2(\mathbb{Z}_N) \\ z &\longmapsto R_k(z) \end{aligned}$$

where $R_k(z) : \mathbb{Z}_N \to \mathbb{C}$ *is the sequence defined by the formula:*

$$\boxed{R_k z(n) = z(n - k)} \qquad \forall n \in \mathbb{Z}_N$$

EXAMPLE OF A SHIFT OPERATOR.–

$N = 6$, $k = 2$, $z = (2, 3 - i, 2i, 4 + i, 0, 1)$. Then:

$$\begin{cases} R_2 z(0) = z(0 - 2) = z(-2) = z(-2 + 6) = z(4) = 0 \\ R_2 z(1) = z(1 - 2) = z(-1) = z(-1 + 6) = z(5) = 1 \\ \quad \vdots \end{cases}$$

giving us: $R_2 z = (0, 1, 2, 3 - i, 2i, 4 + i)$. Evidently, the effect of R_2 on z is a simple displacement of each element in the sequence by two positions to the right (hence the notation R).

The final two elements "turn" into the first two positions, as though following a circle. For this reason, R_k is also known as a *circular shift operator* or *rotation operator*.

Now, consider the composition of this shift operator with the DFT and, inversely, that of the DFT with the shift operator. We shall begin with this latter composition: DFT$(z(n-k))$, that is:

$$\begin{array}{ccccc} \ell^2(\mathbb{Z}_N) & \underset{R_k}{\longrightarrow} & \ell^2(\mathbb{Z}_N) & \underset{\text{DFT}}{\longrightarrow} & \ell^2(\mathbb{Z}_N) \\ z & \longmapsto & R_k z & \longmapsto & (\text{DFT} \circ R_k)z = \text{DFT}(R_k z) = \widehat{R_k z} \end{array}$$

Theorem 2.4 shows that, due to the DFT, the action of the operator R_k is transformed into a multiplication by a complex exponential.

THEOREM 2.4.– Take $z \in \ell^2(\mathbb{Z}_N)$ and $k \in \mathbb{Z}$. Then:

$$\boxed{\widehat{R_k z}(m) = e^{-2\pi i \frac{mk}{N}}\, \hat{z}(m)} \qquad \forall m \in \mathbb{Z} \qquad [2.29]$$

that is, if we define the sequence $\omega_N^k \in \ell^2(\mathbb{Z}_N)$, $\omega_N^k(m) = \omega_N^{mk} = e^{-2\pi i \frac{mk}{N}}$ $\forall m \in \mathbb{Z}$, then:

$$\boxed{DFT \circ R_k = M_{\omega_N^k} \circ DFT} \qquad [2.30]$$

PROOF.–

$$\begin{aligned} \widehat{R_k z}(m) &= \textstyle\sum_{n=0}^{N-1} (R_k z)(n) e^{-2\pi i \frac{mn}{N}} \\ &= \sum_{n=0}^{N-1} z(n-k) e^{-2\pi i \frac{mn}{N}} \\ &= \sum_{n=-k}^{N-k-1} z(n-k+k) e^{-2\pi i \frac{m(n+k)}{N}} \\ &= \sum_{n=-k}^{N-k-1} z(n) e^{-2\pi i \frac{mn}{N}} e^{-2\pi i \frac{mk}{N}} \end{aligned}$$

Factor $e^{-2\pi i \frac{mk}{N}}$ is independent of the index n and can thus be left out of the summation:

$$\begin{aligned} \widehat{R_k z}(m) &= e^{-2\pi i \frac{mk}{N}} \sum_{n=-k}^{N-k-1} z(n) e^{-2\pi i \frac{mn}{N}} \\ &\underset{\text{(Lemma 2.2)}}{=} e^{-2\pi i \frac{mk}{N}} \sum_{n=0}^{N-1} z(n) e^{-2\pi i \frac{mn}{N}} \\ &= e^{-2\pi i \frac{mk}{N}}\, \hat{z}(m) \end{aligned}$$

Lemma 2.2 can be applied in this case as, by hypothesis, z is N-periodic and the exponential $e^{-2\pi i \frac{mn}{N}}$ is itself an N-periodic function. □

Note that, if we write $\hat{z}(m) = |\hat{z}(m)|e^{i\text{Arg}(\hat{z}(m))}$ then, since $\left|e^{-2\pi i \frac{mk}{N}}\right| = 1$, the product $e^{-2\pi i \frac{mk}{N}}\hat{z}(m)$ only changes the phase of $\hat{z}(m)$. This is the reason why we say that *the DFT transforms the shift into a phase shift*. The fact that the phase of the Fourier coefficients is modified by translations implies that *the phase spectrum contains information regarding the geometry of the original signal.*

2.7.2.1. *Shift invariance of the spectrum*

Theorem 2.4 highlights an important limitation of the Fourier transform. Since:

$$\left|e^{-2\pi i \frac{mk}{N}}\right| = 1 \quad \Longrightarrow \quad |\widehat{R_k z}(m)| = |\hat{z}(m)| \qquad \forall m, k \in \mathbb{Z},$$

the magnitudes of the Fourier coefficients of z and of all its shifts are equal. Consequently, *the magnitude of the Fourier coefficients $|\hat{z}(m)|$ informs us of the (global) importance of the harmonic of frequency m in the reconstruction of the signal z, but not of its (local) position within the signal.*

To gain a clearer understanding of this behavior, let us consider the unit pulse, to which an arbitrary shift is applied: $R_k\delta_0$. The spectrum of this signal is $|\widehat{R_k\delta_0}(m)| = |e^{-2\pi i \frac{mk}{N}}\hat{\delta}_0(m)|$, but, as we have seen, $\hat{\delta}_0(m) = 1$ for all $m \in \mathbb{Z}_N$, thus $|\widehat{R_k\delta_0}(m)| = |e^{-2\pi i \frac{mk}{N}}| = 1$. The difference between this case and that of the non-shifted unit pulse is that, in the latter case, the spectrum is real and thus $\left|\hat{\delta}_0(m)\right| = \hat{\delta}_0(m) = 1$ $\forall m \in \mathbb{Z}_N$.

The spectrum of the unit pulse is therefore exactly the same as that of any of its shifted forms. Knowledge of the spectrum alone is not sufficient to reconstruct the spatial location of a signal; to do this, we need information from the phase, which is not easy to interpret or handle.

One solution to this problem lies in using two transforms which "localize" the Fourier transform: the Gabor transform and the wavelet transform. These transforms lie outside the scope of this book, the interested reader can consult, for instance, Frazier (2001).

Now, let us analyze the composition of the shift operator and the DFT : $\hat{z}(m-k)$, that is:

$$\boxed{\begin{array}{ccccc} \ell^2(\mathbb{Z}_N) & \underset{\text{DFT}}{\longrightarrow} & \ell^2(\mathbb{Z}_N) & \underset{R_k}{\longrightarrow} & \ell^2(\mathbb{Z}_N) \\ z & \longmapsto & \text{DFT}(z) & \longmapsto & (R_k \circ \text{DFT})z = \hat{z}(m-k) \end{array}} \qquad [2.31]$$

THEOREM 2.5.– Using the hypotheses from Theorem 2.4, this is equivalent to the formula:

$$\boxed{(R_k\hat{z})(m) = \hat{z}(m-k) = \widehat{\left(e^{2\pi i \frac{nk}{N}} z\right)}(m)}, \qquad \forall m \in \mathbb{Z} \qquad [2.32]$$

that is:

$$\boxed{R_k \circ DFT = DFT \circ M_{\overline{\omega_N^k}}} \qquad [2.33]$$

PROOF.–

$$(R_k\hat{z})(m) = \hat{z}(m-k) = \sum_{n=0}^{N-1} z(n)e^{-2\pi i \frac{(m-k)n}{N}}$$

$$= \sum_{n=0}^{N-1} \left(e^{2\pi i \frac{kn}{N}} z(n)\right) e^{-2\pi i \frac{mn}{N}} = \widehat{\left(e^{2\pi i \frac{kn}{N}} z\right)}(m)$$

□

The properties analyzed above may be summarized in the form of Fourier pairs, shown in Table 2.2. This information shows that the shift operation in the original representation of z becomes a phase change in the Fourier space; conversely, the shift operation in the Fourier space corresponds to a phase change (with a conjugate phase) in the original representation of z.

The following situation illustrates a particularly *remarkable case*. If N is an even value and $k = N/2$, then:

$$e^{-\frac{2\pi i m \frac{N}{2}}{N}} = e^{-\pi i m} = (e^{-\pi i})^m = (-1)^m$$

and:

$$e^{\frac{2\pi i n \frac{N}{2}}{N}} = e^{\pi i n} = (e^{\pi i})^n = (-1)^n$$

so:

$$\text{DFT}\left(z\left(n - \frac{N}{2}\right)\right) = (-1)^m \hat{z}(m), \quad \hat{z}\left(m - \frac{N}{2}\right) = (\widehat{(-1)^n z})(m) \qquad [2.34]$$

Thus, multiplying sequence z by $(-1)^n$ corresponds to shifting the spectrum by $N/2$. This operation is often used to center a spectrum on $m = 0$.

Original representation	Fourier space
$z(n-k)$	$e^{-2\pi i \frac{km}{N}}\hat{z}(m)$
$e^{2\pi i \frac{kn}{N}}z(n)$	$\hat{z}(m-k)$

Table 2.2. *Fourier pairs and translation*

Finally, note the relation between formula [2.30] and the diagonal representation of the operator R_k. Composing the left and right members of formula [2.30] with the IDFT, we obtain:

$$\text{DFT} \circ R_k \circ \text{IDFT} = M_{\omega_N^k}$$

Using A_k and $D_{\omega_N^k}$ (diagonal, see section 2.6.6) to write the matrices associated with the operator R_k and with $M_{\omega_N^k}$ in relation to the canonical basis, the previous equation can be rewritten as:

$$W_N A_k W_N^{-1} = D_{\omega_N^k}.$$

This tells us that *the matrix A_k associated with the shift operator R_k is similar to the diagonal matrix $D_{\omega_N^k}$*.

The invertible matrix which produces the matrix conjugation of A_k and $D_{\omega_N^k}$ is the Sylvester matrix W_N, so we can say that the action of *the shift operator R_k is diagonal in the Fourier space.*

2.7.3. *DFT and conjugation*

Given a sequence $z \in \ell^2(Z_N)$, the conjugate sequence $\bar{z}$ is written as $\bar{z} = (\bar{z}(0), \bar{z}(1), \ldots, \bar{z}(N-1))$, that is, $\bar{z}(n) = \overline{z(n)}\ \forall n \in \mathbb{Z}_N$.

The relationship between the DFT and conjugation is shown in Theorem 2.6.

THEOREM 2.6.– For all $z \in \ell^2(Z_N)$:

$$\hat{\bar{z}}(m) = \overline{\hat{z}(-m)} = \overline{\hat{z}(N-m)} \qquad \forall m \in \mathbb{Z}_N$$

PROOF.–

$$\hat{\bar{z}}(m) = \sum_{n=0}^{N-1} \overline{z(n)} e^{-2\pi i \frac{mn}{N}} = \overline{\sum_{n=0}^{N-1} z(n) e^{2\pi i \frac{mn}{N}}} = \overline{\sum_{n=0}^{N-1} z(n) e^{-2\pi i \frac{(-m)n}{N}}} = \overline{\hat{z}(-m)}$$

$\overline{\hat{z}(N-m)} = \overline{\hat{z}(-m)}$, by periodicity. □

COROLLARY 2.1.– $z \in \ell^2(\mathbb{Z}_N)$ is real, that is, $z(n) \in \mathbb{R}\ \forall n \in \mathbb{Z}_N$, if and only if:

$$\hat{z}(m) = \overline{\hat{z}(-m)} = \overline{\hat{z}(N-m)}, \qquad \forall m \in \mathbb{Z}_N$$

$\hat{z} \in \ell^2(\mathbb{Z}_N)$ is real, that is, $\hat{z}(m) \in \mathbb{R}\ \forall m \in \mathbb{Z}_N$, if and only if:

$$z(n) = \overline{z(-n)} = \overline{z(N-n)}, \qquad \forall n \in \mathbb{Z}_N$$

PROOF.– As the DFT is an isomorphism of $\ell^2(\mathbb{Z}_N)$, z is real, that is, $z = \bar{z}$, if and only if $\hat{z} = \hat{\bar{z}}$, but, from Theorem 2.6, this also holds true when $\hat{z}(m) = \overline{\hat{z}(-m))} = \overline{\hat{z}(N-m)}$.

$\hat{z}$ is real if and only if $\hat{z} = \bar{\hat{z}}$, but the previous theorem states that $\overline{\hat{z}(-m)} = \hat{\bar{z}}(m)$ $\forall m \in \mathbb{Z}_N$, implying that $\overline{\hat{z}(m)} = \hat{\bar{z}}(-m)$, by simple substitution of the variable $m \leftrightarrow -m$. Hence:

$$\text{IDFT}(\overline{\hat{z}(m)}) = \text{IDFT}(\hat{\bar{z}}(-m)) = \text{IDFT}(\text{DFT}(\bar{z}(-m))) = \bar{z}(-n) = \overline{z(-n)}$$

Therefore $\hat{z}$ is real $\iff$ $\hat{z}(m) = \overline{\hat{z}(m)}$ $\iff$ $\text{IDFT}(\hat{z}(m)) = \text{IDFT}(\overline{\hat{z}(m)}) \iff z(n) = \overline{z(-n)} = \overline{z(N-n)}\ \forall n \in \mathbb{Z}_N$. □

Corollary 2.2 is an immediate consequence of the previous result.

COROLLARY 2.2.– $z, \hat{z} \in \ell^2(\mathbb{Z}_N)$ are simultaneously real if and only if they are symmetrical about 0, that is, $z(n) = z(-n)$ and $\hat{z}(m) = \hat{z}(-m)$, $\forall m, n \in \mathbb{Z}_N$.

2.7.4. *DFT and convolution*

One of the most important properties of the Fourier transform relates to the convolution operation.

To understand this operation, we first note the formula for polynomial products.

If $P(x) = a_0 + a_1 x + \ldots + a_n x^n = \sum_{i=0}^{n} a_i x^i$ and $Q(x) = b_0 + b_1 x + \ldots + b_m x^m = \sum_{j=0}^{m} b_j x^j$, then:

$$P(x)Q(x) = \sum_{\ell=0}^{n+m} c_\ell x^\ell, \quad \text{where} \quad c_\ell = \sum_{k=0}^{\ell} a_{\ell-k} b_k = \sum_{k=0}^{\ell} a_k b_{\ell-k} \qquad [2.35]$$

EXAMPLE.–

$P(x) = a_0 + a_1x + a_2x^2$, $Q(x) = b_0 + b_1x + b_2x^2$, so:

$$P(x)Q(x) = a_0b_0 + (a_0b_1 + a_1b_0)x + (a_0b_2 + a_1b_1 + a_2b_0)x^2$$

$$+(a_1b_2 + a_2b_1)x^3 + (a_2b_2)x^4$$

The coefficients of the powers of the variable x verify formula [2.35]. We see that the coefficients c_ℓ include a sum of the products of the coefficients a_i and b_j. Notably, the sum of the indices $i+j$ is always equal to ℓ; as the index of one variable increases, that of the other decreases.

These are the defining characteristics of the convolution operation (in its discrete form), which we shall introduce in the space $\ell^2(\mathbb{Z}_N)$.

DEFINITION 2.13.– *Take $z, w \in \ell^2(\mathbb{Z}_N)$. The convolution of z with w, written as $z*w$, is the sequence of $\ell^2(\mathbb{Z}_N)$ with components defined by:*

$$\boxed{(z * w)(n) = \sum_{k=0}^{N-1} z(n-k)w(k) = \sum_{k=0}^{N-1} w(n-k)z(k)}, \qquad \forall n \in \mathbb{Z}_N$$

Convolution is *symmetrical*, that is $z * w = w * z$, due to the commutative nature of the product in $\mathbb{C}$.

EXAMPLE.–

$z, w \in \ell^2(\mathbb{Z}_4)$, $z = (1, 1, 0, 2)$, $w = (i, 0, 1, 2)$, with canonical periodicity: $z(n + kN) = z(n)$ and $w(n + kN) = w(n)$ $\forall n \in \mathbb{Z}_N$ and $k \in \mathbb{Z}$. Then:

$$\begin{aligned}(z * w)(0) &= \sum_{k=0}^{4-1} z(0-k)w(k) = \sum_{k=0}^{3} z(-k)w(k)\\ &= z(0)w(0) + z(-1)w(1) + z(-2)w(2) + z(-3)w(3)\\ &= z(0)w(0) + z(4-1)w(1) + z(4-2)w(2) + z(4-3)w(3)\\ &= z(0)w(0) + z(3)w(1) + z(2)w(2) + z(1)w(3)\\ &= 1 \cdot i + 2 \cdot 0 + 0 \cdot 1 + 1 \cdot 2\\ &= i + 2\end{aligned}$$

We also have $(z * w)(1) = 2 + i$, $(z * w)(2) = 1 + 2i$, $(z * w)(3) = 1 + 3i$, hence $(z * w) = (i + 2, 2 + i, 1 + 2i, 1 + 3i)$.

The interaction between the DFT and convolution has a particularly elegant and useful property, described in Theorem 2.7.

THEOREM 2.7.– Take $z, w \in \ell^2(\mathbb{Z}_N)$. Then:

$$\boxed{DFT\,(z * w)(m) = \hat{z}(m) \cdot \hat{w}(m)} \iff \boxed{(z * w)(n) = IDFT\,(\hat{z} \cdot \hat{w})(n)} \quad \forall n, m \in \mathbb{Z} \tag{2.36}$$

$$\boxed{IDFT\,(\hat{z} * \hat{w})(n) = N z(n) \cdot w(n)} \iff \boxed{(\hat{z} * \hat{w})(m) = N\, DFT(z \cdot w)(m)} \quad \forall n, m \in \mathbb{Z} \tag{2.37}$$

In other words, the Fourier transform of the convolution of z and w is the pointwise product of the Fourier transforms and *vice versa*: the inverse Fourier transform of the convolution of $\hat{z}$ and $\hat{w}$ is N times the pointwise product of z and w. In other words, we obtain the Fourier pairs shown in Table 2.3.

Original representation	Fourier space
$z * w$	$\hat{z} \cdot \hat{w}$
$N z \cdot w$	$\hat{z} * \hat{w}$

Table 2.3. *Fourier pairs relative to convolution*

PROOF.– By definition :

$$\widehat{(z * w)}(m) = \sum_{n=0}^{N-1} (z*w)(n) e^{-2\pi i \frac{mn}{N}} = \sum_{n=0}^{N-1} \left(\sum_{k=0}^{N-1} z(n-k) w(k) \right) e^{-2\pi i \frac{mn}{N}}$$

The exponential is rewritten as:

$$e^{-2\pi i \frac{mn}{N}} = e^{-2\pi i \frac{m(n-k+k)}{N}} = e^{-2\pi i \frac{m(n-k)+mk}{N}} = e^{-2\pi i \frac{m(n-k)}{N}} e^{-2\pi i \frac{mk}{N}}$$

Then:

$$\begin{aligned}
\widehat{(z*w)}(m) &= \sum_{n=0}^{N-1}\sum_{k=0}^{N-1} z(n-k)w(k)e^{-2\pi i\frac{m(n-k)}{N}}e^{-2\pi i\frac{mk}{N}} \\
&= \sum_{k=0}^{N-1} w(k)e^{-2\pi i\frac{mk}{N}}\sum_{n=0}^{N-1} z(n-k)e^{-2\pi i\frac{m(n-k)}{N}} \\
&= \sum_{k=0}^{N-1} w(k)e^{-2\pi i\frac{mk}{N}}\sum_{n=-k}^{N-k-1} z(n-k+k)e^{-2\pi i\frac{m(n-k+k)}{N}} \\
&= \sum_{k=0}^{N-1} w(k)e^{-2\pi i\frac{mk}{N}}\sum_{n=-k}^{N-k-1} z(n)e^{-2\pi i\frac{mn}{N}} \\
&\underset{\text{(Lemma 2.2)}}{=} \sum_{k=0}^{N-1} w(k)e^{-2\pi i\frac{mk}{N}}\sum_{n=0}^{N-1} z(n)e^{-2\pi i\frac{mn}{N}} \\
&= \hat{w}(m)\hat{z}(m) = \hat{z}(m)\hat{w}(m)
\end{aligned}$$

Lemma 2.2 can be applied here as it is valid for any $k \in \mathbb{Z}$. Thus:

$$\widehat{(z*w)}(m) = \hat{z}(m)\hat{w}(m), \qquad \forall m \in \mathbb{Z}$$

The proof that the IDFT$(\hat{z} * \hat{w})(n) = z(n) \cdot w(n)$ is very similar, by definition :

$$\begin{aligned}
\text{IDFT}(\hat{z}*\hat{w})(n) &= \frac{1}{N}\sum_{m=0}^{N-1}(\hat{z}*\hat{w})(m)e^{2\pi i\frac{mn}{N}} \\
&= \frac{1}{N}\sum_{m=0}^{N-1}\left(\sum_{k=0}^{N-1}\hat{z}(m-k)\hat{w}(k)\right)e^{2\pi i\frac{mn}{N}}
\end{aligned}$$

The exponential is rewritten as:

$$e^{2\pi i\frac{mn}{N}} = e^{2\pi i\frac{n(m-k+k)}{N}} = e^{2\pi i\frac{n(m-k)+nk}{N}} = e^{2\pi i\frac{n(m-k)}{N}}e^{2\pi i\frac{nk}{N}}$$

Then:

$$\begin{aligned}
\text{IDFT}(\hat{z}*\hat{w})(n) &= \tfrac{1}{N}\textstyle\sum_{m=0}^{N-1}\sum_{k=0}^{N-1}\hat{z}(m-k)\hat{w}(k)e^{2\pi i\frac{n(m-k)}{N}}e^{2\pi i\frac{nk}{N}} \\
&= \frac{1}{N}\sum_{k=0}^{N-1}\hat{w}(k)e^{2\pi i\frac{nk}{N}}\sum_{m=0}^{N-1}\hat{z}(m-k)e^{2\pi i\frac{n(m-k)}{N}} \\
&= \frac{1}{N}\sum_{k=0}^{N-1}\hat{w}(k)e^{2\pi i\frac{nk}{N}}\sum_{m=-k}^{N-k-1}\hat{z}(m-k+k)e^{2\pi i\frac{n(m-k+k)}{N}} \\
&= \frac{1}{N}\sum_{k=0}^{N-1}\hat{w}(k)e^{2\pi i\frac{nk}{N}}\sum_{m=-k}^{N-k-1}\hat{z}(m)e^{2\pi i\frac{mn}{N}}
\end{aligned}$$

$$\underset{\text{(Lemma 2.2)}}{=} \frac{1}{N} \sum_{k=0}^{N-1} \hat{w}(k) e^{2\pi i \frac{nk}{N}} \sum_{m=0}^{N-1} \hat{z}(m) e^{2\pi i \frac{mn}{N}}$$

$$= N \left(\frac{1}{N} \sum_{k=0}^{N-1} \hat{w}(k) e^{2\pi i \frac{nk}{N}} \right) \cdot \left(\frac{1}{N} \sum_{m=0}^{N-1} \hat{z}(m) e^{2\pi i \frac{mn}{N}} \right)$$

$$= N \, \text{IDFT}\, \hat{w}(n) \cdot \text{IDFT}\, \hat{z}(n) = N\, w(n) \cdot z(n) = N\, z(n) \cdot w(n) \qquad \square$$

OBSERVATIONS.–

– In this proof, $\sum_{k=0}^{N-1} w(k) e^{-2\pi i \frac{mk}{N}}$ cannot be replaced with $\hat{w}(m)$ before the final step, as the index k is still present in the second sum. $\hat{w}(m)$ can only be substituted in once k has been eliminated.

– Formulas [2.36] demonstrate a sort of "distributive property" in connection with convolution and the pointwise product: *when the DFT is applied to a convolution product, it is distributed over the factors, and the convolution becomes a pointwise product*. Inversely, *when the IDFT is applied to a pointwise product, it is distributed over the factors, and the pointwise product becomes a convolution*. Thus:

$$\text{DFT}(\check{z} * \check{w}) = z \cdot w, \quad \text{IDFT}(z \cdot w) = \check{z} * \check{w} \qquad \forall z, w \in \ell^2(\mathbb{Z}_N) \qquad [2.38]$$

– Using the Fourier transform, *a complex operation such as convolution can be transformed into a simple product of Fourier transforms* (which can be calculated rapidly using the FFT). This result is particularly useful for signal processing applications. If we define the DFT using the normalization induced by the orthonormal Fourier basis, coefficients appear in the DFT formula of the convolution. These coefficients may be extremely large, particularly when dealing with DFTs in dimensions greater than 1 and/or large signals; this may result in calculation errors. The simplicity of formula [2.36] is the reason why many authors – and programmers – prefer the definition of Fourier coefficients used in this book to other definitions.

– Convolution is often carried out between a signal z and another signal w which is non-zero only on a support of size T. The value of T is important in choosing whether to apply the convolution operation directly or to use the FFT. The complexity of the direct convolution operation is $\mathcal{O}(NT)$; using the FFT, the complexity is $\mathcal{O}(N \log N)$. It is therefore helpful to transform the convolution into a pointwise product with FFT in cases where $T > \log(N)$. For example, taking $z \in \ell^2(\mathbb{Z}_N)$ with $N = 1{,}000$, then $\log(N) \simeq 7$ and it is thus preferable to carry out the convolution $z * w$ in the Fourier domain for all cases where the support of w is larger than 7.

If one of the vectors in the convolution is fixed, we can define an endomorphism of $\ell^2(\mathbb{Z}_N)$.

DEFINITION 2.14.– *Taking a fixed sequence* $w \in \ell^2(\mathbb{Z}_N)$, *the following linear transformation is the* convolution operator *with* w:

$$\begin{aligned} T_w : \ell^2(\mathbb{Z}_N) &\longrightarrow \ell^2(\mathbb{Z}_N) \\ z &\longmapsto T_w(z) = z * w \end{aligned}$$

As in the case of the shift operator, a diagonal representation of the convolution operator can be produced. To do this, we rewrite formula [2.36] without specifying the index m (as the representation is valid for any index), that is, $\text{DFT}(z * w) = \hat{z} \cdot \hat{w}$, but $\text{DFT}(z * w) = (\text{DFT} \circ T_w)z$ and $\hat{z} \cdot \hat{w} = \hat{w} \cdot \hat{z} = M_{\hat{w}}\, \hat{z} = (M_{\hat{w}} \circ \text{DFT})z$, that is, $(\text{DFT} \circ T_w)z = (M_{\hat{w}} \circ \text{DFT})z \; \forall z \in \ell^2(\mathbb{Z}_N)$, making it possible to write the operator relationship $\text{DFT} \circ T_w = M_{\hat{w}} \circ \text{DFT}$.

Applying a composition between the IDFT and the left and right sides of this expression, we obtain:

$$\text{DFT} \circ T_w \circ \text{IDFT} = M_{\hat{w}}$$

Let us consider this relationship in the context of the canonical basis B, just as we did in the case of the shift operator. The DFT and the IDFT become W_N and W_N^{-1}, and the multiplication operator $M_{\hat{w}}$ takes the form of the diagonal matrix $D_{\hat{w}} = \text{diag}(\hat{w}(0), \ldots, \hat{w}(N-1))$. If the matrix A_w is the representation of T_w in the basis B, that is, $A_w = [T_w]_B$, then:

$$W_N A_w W_N^{-1} = D_{\hat{w}}$$

which shows that the action of the convolution operator is diagonalized in the Fourier basis.

Shift and convolution operators are not unique in this regard: there is a whole specific category of operators which have a diagonal action in the Fourier basis. These operators are called *stationary* and they will be examined in greater detail in section 2.8.

2.8. The DFT and stationary operators

The relationship between the Fourier transform and the class of "stationary" operators is an important one. The DFT enables the diagonalization of these operators and they can be shown to be equivalent to convolutions and to Fourier multipliers. To prove these results, we shall also introduce the category of "circulant" matrices, which represent stationary operators in the canonical basis of $\ell^2(\mathbb{Z}_N)$.

Before giving the formal mathematical definition of stationary operator, let us introduce the idea behind such object by considering an audio signal z and a device T that acts linearly on it. If the signal z is transmitted to T with a delay Δt, and the only effect of this delay on T is that its output is delayed by the same quantity Δt, then the device T is said to be stationary.

Mathematically speaking, if R_k is the shift operator by the quantity $k \in \mathbb{Z}$, then the stationarity of T is translated as the following relationship:

$$T(R_k z) = R_k(Tz), \qquad \forall z \in \ell^2(\mathbb{Z}_N)$$

The left side represents the action of the operator T on the z shifted by a quantity k, while the right side represents the shift in the action of operator T on the original signal z. These notions are summarized in the commutative diagram below.

$$\begin{array}{ccc} \ell^2(\mathbb{Z}_N) & \xrightarrow{R_k} & \ell^2(\mathbb{Z}_N) \\ T\downarrow & & \downarrow T \\ \ell^2(\mathbb{Z}_N) & \xrightarrow[R_k]{} & \ell^2(\mathbb{Z}_N) \end{array}$$

These considerations justify Definition 2.15.

DEFINITION 2.15.– *An operator* $T : \ell^2(\mathbb{Z}_N) \to \ell^2(\mathbb{Z}_N)$ *is said to be* stationary *(or shift invariant) if:*

$$T(R_k z) = R_k(Tz), \qquad \forall z \in \ell^2(\mathbb{Z}_N),\ \forall k \in \mathbb{Z} \qquad [2.39]$$

that is, T *is stationary if it commutes with all shift operators* R_k*:*

$$\boxed{T \circ R_k = R_k \circ T}, \qquad \forall k \in \mathbb{Z} \qquad [2.40]$$

In section 2.8.5, we shall show that a linear operator $T \in \mathrm{End}(\ell^2(\mathbb{Z}_N))$ is stationary if and only if $(Tz)(n) = \sum_{k=0}^{N-1} a_k z(n-k) = \sum_{k=0}^{N-1} a_k R_k z(n)$, $n \in \{0, \ldots, N-1\}$, $a_k \in \mathbb{C}$.

The DFT provides an extremely important example of a non-stationary operator over $\ell^2(\mathbb{Z}_N)$. To prove that *the DFT is not a stationary operator,* we simply recall the way in which it interacts with shift operators R_k: using formulas [2.30] and [2.33] we obtain, respectively,

$$\mathrm{DFT} \circ R_k = M_{\omega_N^k} \circ \mathrm{DFT} \text{ and } R_k \circ \mathrm{DFT} = \mathrm{DFT} \circ M_{\overline{\omega_N^k}},$$

which shows that the DFT does not commute with the shift operators.

2.8.1. *The DFT and the diagonalization of stationary operators*

The most important properties of the DFT with regard to stationary operators can be summarized in a single theorem, but we prefer to highlight the fact that the Fourier transform diagonalizes stationary operators through a separate theorem.

THEOREM 2.8.– Let $T \in End(\ell^2(\mathbb{Z}_N))$ be a stationary operator. Then, T is diagonalizable, and each element of the orthogonal Fourier basis F_m of $\ell^2(\mathbb{Z}_N)$ is an Eigenvector of T.

PROOF.– For every *fixed* $m \in \{0, \dots, N-1\}$, let us consider the element m of the orthogonal Fourier basis: $F_m(n) = \frac{1}{N}e^{2\pi i \frac{mn}{N}}$.

As T is an endomorphism, $TF_m \in \ell^2(\mathbb{Z}_N)$, and thus TF_m *can be decomposed over the basis* $(F_0, \dots, F_{N-1})$ *itself*:

$$(TF_m)(n) = \sum_{k=0}^{N-1} a_k F_k(n) = \frac{1}{N} \sum_{k=0}^{N-1} a_k e^{2\pi i \frac{kn}{N}}, \qquad \forall n \in \mathbb{Z}_N \qquad [2.41]$$

Now, consider the action of the shift operator R_1 on F_m:

$$\begin{aligned} R_1 F_m(n) &= F_m(n-1) = \tfrac{1}{N} e^{2\pi i \frac{m(n-1)}{N}} = e^{-2\pi i \frac{m}{N}} \cdot \tfrac{1}{N} e^{2\pi i \frac{mn}{N}} \\ &= e^{-2\pi i \frac{m}{N}} \cdot F_m(n) \end{aligned}$$

Applying T to $R_1 F_m$, we obtain:

$$\begin{aligned} TR_1F_m(n) &= T\left(e^{-2\pi i \frac{m}{N}} \cdot F_m(n)\right) \\ &\underset{\text{Linearity of } T}{=} e^{-2\pi i \frac{m}{N}} \left(TF_m\right)(n) \\ &\underset{\text{equation [2.41]}}{=} e^{-2\pi i \frac{m}{N}} \sum_{k=0}^{N-1} a_k F_k(n) \\ &= \sum_{k=0}^{N-1} a_k e^{-2\pi i \frac{m}{N}} F_k(n) \end{aligned}$$

Now, we switch the order of composition of R_1 and T:

$$\begin{aligned} R_1TF_m(n) &= TF_m(n-1) \\ &\underset{\text{equation [2.41]}}{=} \frac{1}{N} \sum_{k=0}^{N-1} a_k e^{2\pi i \frac{k(n-1)}{N}} \\ &= \frac{1}{N} \sum_{k=0}^{N-1} a_k e^{-2\pi i \frac{k}{N}} \cdot e^{2\pi i \frac{kn}{N}} \\ &= \sum_{k=0}^{N-1} a_k e^{-2\pi i \frac{k}{N}} F_k(n) \end{aligned}$$

Since T is stationary, $TR_1F_m = R_1TF_m$, that is:

$$\sum_{k=0}^{N-1} a_k e^{-2\pi i \frac{m}{N}} F_k(n) = \sum_{k=0}^{N-1} a_k e^{-2\pi i \frac{k}{N}} F_k(n)$$

and due to the uniqueness of decomposition over a basis:

$$a_k e^{-2\pi i \frac{m}{N}} = a_k e^{-2\pi i \frac{k}{N}}, \qquad \forall k \in \mathbb{Z}_N, \ (m \text{ is fixed}) \qquad [2.42]$$

Let us analyze this equality. If $k = m$, then equation [2.42] is simply an identity and requires no further discussion. In the case where $k \neq m$, we begin by recalling that $m, k \in \{0, \ldots, N-1\}$, so the cosine and sine of the complex exponentials have their values in only one period, as the next period begins when $m, k = N$. Then:

$$k \neq m \quad \Longrightarrow \quad e^{-2\pi i \frac{m}{N}} \neq e^{-2\pi i \frac{k}{N}}$$

and equation [2.42] can be verified if and only if $a_k = 0 \ \forall k \neq m$.

Equation [2.41] thus becomes:

$$TF_m(n) = a_m F_m(n), \qquad \forall n \in \mathbb{Z}_N,$$

that is, F_m is an eigenvector of T with an eigenvalue a_m given by the m-th coefficient of the decomposition of TF_m on the orthogonal Fourier basis. Evidently, the coefficient a_m is dependent on T.

Given that we fixed an arbitrary index m, every element of the orthogonal Fourier basis is an eigenvector of T, and consequently $\ell^2(\mathbb{Z}_N)$ has a basis of eigenvectors of T. By definition, T is therefore diagonalizable. □

Theorem 2.9 shows how the eigenvalues a_m can be made explicit using the DFT.

The theorem shown above can be interpreted using matrices. We know that the action of the DFT is represented by the Sylvester matrix W_N defined in equation [2.23] and that W_N is the matrix used to pass from the canonical basis B of $\ell^2(\mathbb{Z}_N)$ to the Fourier basis F of $\ell^2(\mathbb{Z}_N)$; the inverse is $W_N^{-1} = \frac{1}{N}\overline{W_N}$, representing the matrix used to pass from basis F to basis B.

If A is the matrix associated with T with respect to the canonical basis of $\ell^2(\mathbb{Z}_N)$ and D is the diagonal matrix of the eigenvalues of A, then:

$$\boxed{D = W_N A W_N^{-1}}, \qquad \boxed{A = W_N^{-1} D W_N} \qquad [2.43]$$

If $[w]_F$ represents any vector $w \in \ell^2(\mathbb{Z}_N)$ with respect to the Fourier basis F, then:

$$W_N A z = [Az]_F \underset{(F \text{ diagonalizes } A)}{=} D[z]_F = DW_N z, \quad \forall z \in \ell^2(\mathbb{Z}_N)$$

so $W_N A = DW_N$, if and only if $W_N A W_N^{-1} = DW_N W_N^{-1} = D$.

2.8.2. *Circulant matrices*

Thanks to the introduction of the concept of circulant matrix, we will be able to prove the fundamental theorem concerning the link between the Fourier transform and stationary operators.

First, let us generalize the periodicity of sequences $\ell^2(\mathbb{Z}_N)$ to matrices: given a matrix $A = (a_{mn})_{m,n=0}^{N-1}$, we say that A is an N-*periodic matrix* if:

$$a_{m+kN,n} = a_{m,n} \text{ and } a_{m,n+kN} = a_{m,n}, \qquad \forall m, n, k \in \mathbb{Z}$$

EXAMPLE.

$$a_{0,2} = a_{N,2} = a_{N,N+2}$$

DEFINITION 2.16.– *Let $A = (a_{mn})_{m,n=0}^{N-1}$ be an $N \times N$ periodic matrix. A is said to be* circulant *if:*

$$\boxed{a_{m+1,n+1} = a_{m,n}}, \qquad \forall m, n \in \mathbb{Z}$$

Repeating the translation k times, the definition is rewritten as:

$$a_{m+k,n+k} = a_{m,n}, \qquad \forall m, n, k \in \mathbb{Z}$$

We see that, since $k \in \mathbb{Z}$, a circulant periodic matrix can also be defined with the property $a_{m-k,n-k} = a_{m,n}$, $k \in \mathbb{Z}$.

This definition is interpreted as follows. Line (column) $m + 1$ $(n + 1)$ is obtained from line (column) m (n) by shifting one position to the right (at the bottom), as follows:

$$\begin{pmatrix} a_0 & a_1 & a_2 & \cdots & a_{N-1} \\ a_{N-1} & a_0 & a_1 & \cdots & a_{N-2} \\ a_{N-2} & a_{N-1} & a_0 & \cdots & a_{N-3} \\ \vdots & \vdots & \vdots & \ddots & \vdots \\ a_1 & a_2 & a_3 & \cdots & a_0 \end{pmatrix}$$

EXAMPLE OF A CIRCULANT MATRIX.–

$$A = \begin{pmatrix} 3 & 2+i & -1 & 4i \\ 4i & 3 & 2+i & -1 \\ -1 & 4i & 3 & 2+i \\ 2+i & -1 & 4i & 3 \end{pmatrix}$$

EXAMPLE OF A NON-CIRCULANT MATRIX.–

$$B = \begin{pmatrix} 2 & i & 3 \\ 3 & 2 & i \\ i & 2 & 3 \end{pmatrix}$$

For this matrix to be circulant, the third line would have to be $(i, 3, 2)$.

2.8.3. *Exhaustive characterization of stationary operators*

Theorem 2.9 is the most important result of this chapter. It is used to produce the eigenvalues of a stationary operator T in a very simple manner; it can also be used to characterize T as a convolution operator, in the original representation of z, and as a multiplier, in the frequency representation.

THEOREM 2.9.– Let $T : \ell^2(\mathbb{Z}_N) \to \ell^2(\mathbb{Z}_N)$ be an endomorphism. The following properties are equivalent.

1) T is stationary.

2) The matrix A, which represents T in the canonical basis of $\ell^2(\mathbb{Z}_N)$, is circulant.

3) T is a convolution operator.

4) T is a Fourier multiplier.

5) The matrix D, which represents T in the orthogonal Fourier basis F, is diagonal.

Note that implication 1) $\Longrightarrow$ 5) has already been proved. The theorem will be proved using the following strategy:

$$\boxed{1) \Longrightarrow 2) \Longrightarrow 3) \Longrightarrow 1)} \text{ and } \boxed{3) \Longleftrightarrow 4)} \text{ and } \boxed{4) \Longleftrightarrow 5)}$$

The proof of this theorem is crucial, as it provides an explicit technique for finding the Eigenvalues of T and for constructing the convolution operator and Fourier multiplier which represent T.

PROOF.– $\boxed{1) \Longrightarrow 2)}$: let A be the associated matrix of T with respect to the canonical basis[8] $(e_n)_{n=0}^{N-1}$ of $\ell^2(\mathbb{Z}_N)$:

$$A = \begin{pmatrix} a_{0,0} & a_{0,1} & \cdots & a_{0,N-1} \\ a_{1,0} & a_{1,1} & \cdots & a_{1,N-1} \\ \vdots & \vdots & \ddots & \vdots \\ a_{N-1,0} & a_{N-1,1} & \cdots & a_{N-1,N-1} \end{pmatrix}$$

8 We recall that $e_n(m) = \delta_{n,m}$, $\forall n, m \in \mathbb{Z}_N$.

From the definition of the associated matrix, we have $a_{m,n} = (Te_n)(m)$, that is, the n-th column of A is the vector Te_n.

Using the fact that T is stationary, we wish to show that:

$$a_{m+1,n+1} = a_{m,n} \quad \Longleftrightarrow \quad (Te_{n+1})(m+1) = (Te_n)(m), \quad \forall m, n \in \mathbb{Z}_N$$

We see that:

$$(R_1 e_n)(m) = e_n(m-1) = \begin{cases} 1 & \text{if } n = m-1 \quad \Longleftrightarrow \quad m = n+1 \\ 0 & \text{if } n \neq m-1 \quad \Longleftrightarrow \quad m \neq n+1 \end{cases}$$

$$= e_{n+1}(m) \quad \forall m \in \mathbb{Z}_N$$

thus $e_{n+1} = R_1 e_n$ and, consequently:

$$a_{m+1,n+1} = (TR_1 e_n)(m+1) \underset{(T \text{ stationary})}{=} R_1(Te_n)(m+1) = (Te_n)(m+1-1)$$

$$= (Te_n)(m) = a_{m,n}$$

Since $a_{m+1,n+1} = a_{m,n}$ $\forall m, n \in \mathbb{Z}_N$, then A is circulant and the implication $\boxed{1) \implies 2)}$ is proved.

$\boxed{2) \implies 3)}$: let A be a periodic circulant matrix, that is, $a_{m,n} = a_{m-k,n-k}$ $\forall n, m, k \in \mathbb{Z}$. We wish to prove the existence of $h \in \ell^2(\mathbb{Z}_N)$ such that $Az = z * h = T_h(z)$ $\forall z \in \ell^2(\mathbb{Z}_N)$.

We shall prove that the sequence h which we are looking for is the first column in A, that is:

$$\boxed{h = Te_0} = \begin{pmatrix} a_{0,0} \\ \vdots \\ a_{N-1,0} \end{pmatrix}, \quad h(m) = a_{m,0}, \quad \forall m \in \mathbb{Z}_N$$

We see that $h(m-n) = a_{m-n,0} = a_{m-n,n-n} \underset{(A \text{ circulant})}{=} a_{m,n}$, and thus, from the definition of the matrix-vector product, we have:

$$(Az)(m) = \sum_{n=0}^{N-1} a_{m,n} z(n) = \sum_{n=0}^{N-1} h(m-n) z(n) = (h * z)(m)$$

and implication $\boxed{2) \implies 3)}$ is proved.

$\boxed{3) \implies 1)}$: we must prove that a convolution operator T_w is stationary, that is:

$$(T_w \circ R_k)(z) = (R_k \circ T_w)(z), \quad \forall z \in \ell^2(\mathbb{Z}_n), \ \forall k \in \mathbb{Z}$$

We begin by calculating the left side of the equation:

$$(T_w R_k z)(m) = (w * R_k z)(m) = \sum_{n=0}^{N-1} w(m-n) R_k z(n) = \sum_{n=0}^{N-1} w(m-n) z(n-k)$$

Making the index substitution $\ell = n - k \Leftrightarrow n = k + \ell$, the variability of ℓ is:

$$\begin{cases} n = 0 \Longrightarrow \ell = -k \\ \vdots \\ n = N-1 \Longrightarrow \ell = N-1-k \end{cases}$$

then:

$$\begin{aligned}(T_w R_k z)(m) &= \sum_{\ell=-k}^{N-1-k} w(m-k-\ell) z(\ell) \underset{\text{Lemma 2.2}}{=} \sum_{\ell=0}^{N-1} w((m-k)-\ell) z(\ell) \\ &= (z * w)(m-k) = R_k (z * w)(m) = (R_k T_w z)(m)\end{aligned}$$

and this proves the implication $\boxed{3) \implies 1)}$.

$\boxed{3) \iff 4)}$: we must prove that a linear operator $T : \ell^2(\mathbb{Z}_N) \to \ell^2(\mathbb{Z}_N)$ is a convolution operator if and only if T is a Fourier multiplier.

Taking an arbitrary fixed element $w \in \ell^2(\mathbb{Z}_N)$, we have:

$$\begin{aligned}T_w(z) &= z * w = \text{IDFT}(\text{DFT}(z * w)) \underset{\text{(Th. 2.36)}}{=} \text{IDFT}(\hat{z} \cdot \hat{w}) = \text{IDFT}(\hat{w} \cdot \hat{z}) \\ &= (\text{IDFT} \circ M_{\hat{w}} \circ \text{DFT})(z) = T_{(\hat{w})}(z), \quad \forall w, z \in \ell^2(\mathbb{Z}_N)\end{aligned}$$

where $M_{\hat{w}}$ is the multiplication operator by the sequence $\hat{w}$. Inversely:

$$\begin{aligned}T_{(w)}(z) &= (\text{IDFT} \circ M_w \circ \text{DFT})(z) = \text{IDFT}(w \cdot \hat{z}) \underset{\text{equation [2.38]}}{=} \check{w} * \check{\hat{z}} = \check{w} * z \\ &= T_{\check{w}}(z) \quad \forall w, z \in \ell^2(\mathbb{Z}_N)\end{aligned}$$

This shows us that *the convolution operator with w can be interpreted as the Fourier multiplier by $\hat{w}$* and *vice versa*, and that *the Fourier multiplier by w can be interpreted as the convolution operator with $\check{w}$*:

$$\boxed{T_w = T_{(\hat{w})}}, \quad \boxed{T_{(w)} = T_{\check{w}}} \qquad \forall w \in \ell^2(\mathbb{Z}_N)$$

The double implication $\boxed{3) \iff 4)}$ is thus proved.

Before continuing on to the final stage in our proof, let us summarize our findings. A stationary operator $T : \ell^2(\mathbb{Z}_N) \to \ell^2(\mathbb{Z}_N)$ is represented by a circulant matrix A with respect to the canonical basis $(e_0, \ldots, e_{N-1})$ of $\ell^2(\mathbb{Z}_N)$.

This matrix A can be represented by the convolution operator T_h with $h = Te_0$, the first column of A or, as we have just seen, by the Fourier multiplier $T_{(\hat{h})}$, where $\hat{h}$ is the sequence of Fourier coefficients of h.

$\boxed{4) \Longleftrightarrow 5)}$: we must prove that T is a Fourier multiplier $T_{(w)}$ if and only if the associated matrix of T with respect to the orthogonal Fourier basis F is diagonal.

The direct implication has already been proved in formula [2.27], so we simply need to prove the implication $5) \Longrightarrow 4)$. Stating that $D = \text{diag}(d_{n,n}), n = 0, \ldots, N-1$ is the diagonal matrix which represents an operator T in the Fourier basis F means that:

$$[T(z)]_F = D[z]_F \iff \text{DFT} \circ T(z) = M_w \circ \text{DFT}(z)$$

with M_w the multiplication operator by the sequence $w(n) = d_{n,n}$, $n = 0, \ldots, N-1$. Applying the IDFT to both sides:

$$T(z) = \text{IDFT} \circ M_w \circ \text{DFT}(z) \qquad \forall z \in \ell^2(\mathbb{Z}_N)$$

hence $T = T_{(w)}$ proving the implication $5) \Longrightarrow 4)$.

The proof of the theorem is now complete. □

The theorem demonstrated above provides a standard technique for studying stationary operators T over $\ell^2(\mathbb{Z}_N)$. We recall that the sequence:

$$\delta \in \ell^2(\mathbb{Z}_N), \quad \delta(n) = e_0(n) = \delta_{0,n} = \begin{cases} 1 & \text{if } n = 0 \\ 0 & \text{if } n \neq 0 \end{cases} \qquad \forall n \in \mathbb{Z}_N$$

is the *unit pulse*; thus, *operator T is completely determined by its action on δ*, $h = T\delta$, which is referred to as the *unit pulse response*. $\hat{h}$, the DFT of the unit pulse response, is called the *transfer function*.

The properties demonstrated in Theorems 2.8 and 2.9 can be used to summarize the analysis of stationary operators, as shown in Box 2.2.

– T is the stationary operator of $\ell^2(\mathbb{Z}_N)$.

– A is the *circulant* matrix associated with T with respect to the *canonical basis* of $\ell^2(\mathbb{Z}_N)$.

– h is the *unit pulse response* of T:

$$h = T\delta = \textit{first column of } A$$

– T_h is the *convolution operator with* h:

$$Tz = T_h z = h * z = z * h$$

– $T_{(\hat{h})}$ is the *Fourier multiplier by* $\hat{h}$, the *transfer function*:

$$Tz = T_{(\hat{h})} z = \text{IDFT}(\hat{h} \cdot \hat{z})$$

– Given $h = T\delta$, we obtain the Fourier pair in Table 2.4.

Original representation	Fourier space
$h * z$	$\hat{h} \cdot \hat{z}$

Table 2.4. *Fourier pair for the convolution between a signal z and the unit pulse response h of T*

– D is the diagonal matrix which represents T in the orthogonal Fourier basis F of $\ell^2(\mathbb{Z}_N)$:

$$D = \frac{1}{N} W_N A \overline{W_N} = \text{diag}(\hat{h}(0), \ldots, \hat{h}(N-1))$$

– The Eigenvalues of T (the spectrum, in the linear algebra sense) are the components of the transfer function, that is the Fourier coefficients of the unit pulse response, that is:

$$\textit{Eigenvalues of } T \ : \ \{\hat{h}(0), \ldots, \hat{h}(N-1)\}$$

Box 2.2. *Analysis of stationary operators over $\ell^2(\mathbb{Z}_N)$*

2.8.4. *High-pass, low-pass and band-pass filters*

The synthesis formula for any given signal $z \in \ell^2(\mathbb{Z}_N)$ transformed) *via* the action of a *stationary* operator $T \in \text{End}(\ell^2(\mathbb{Z}_N))$ is:

$$Tz(n) = \sum_{m=0}^{N-1} \widehat{Tz}(m) F_m(n) \qquad n \in \mathbb{Z}_N \tag{2.44}$$

where F_m is the vector with index m of the orthogonal Fourier basis of $\ell^2(\mathbb{Z}_N)$. Thus, $|\widehat{Tz}(m)|$ represents the importance of the harmonic of frequency m in the reconstruction of Tz, and $\{|\widehat{Tz}(m)|, \ m \in \mathbb{Z}_N\}$ represents the spectrum of the transformed signal Tz.

To understand how the spectrum of Tz is linked to that of the original signal z, let us apply the DFT to both sides of the formula $Tz = T_{(\hat{h})} z = \text{IDFT}(\hat{h} \cdot \hat{z})$, where $h = T\delta_0$:

$$\text{DFT}(Tz) = \text{DFT} \circ \text{IDFT}(\hat{h} \cdot \hat{z}) = \hat{h} \cdot \hat{z}$$

that is:

$$\boxed{\widehat{Tz}(m) = \hat{h}(m) \cdot \hat{z}(m)}, \qquad \forall m \in \mathbb{Z}_N$$

so the Fourier coefficients of Tz, the sequence transformed by the operator T, are given by the product of the Fourier coefficients of the original sequence z and the Fourier coefficients of the unit pulse response h.

Consequently, the *spectrum of the transformed sequence* Tz is:

$$\left\{ \left|\widehat{Tz}(m)\right| = |\hat{h}(m)| \cdot |\hat{z}(m)|,\ m \in \mathbb{Z}_N \right\} \qquad [2.45]$$

This allows us to *understand the action of stationary filters* T *on the frequency content of a signal* z:

– if $\hat{h}(0) = 0$, *the average of* Tz *is zero*, since $\left|\widehat{Tz}(0)\right| = 0 \cdot |\hat{z}(0)| = 0$ and we know that $\left|\widehat{Tz}(0)\right|$ is proportional to the average of Tz;

– if $|\hat{h}(0)| = 1$, then T *preserves the average* of z, that is $\langle Tz \rangle = \langle z \rangle$;

– if $|\hat{h}(m)| > 1$ for $m \simeq 0$ and $m \simeq N-1$, and $|\hat{h}(m)| \in [0,1[$ for $m \simeq N/2$, then T increases the low frequencies and reduces the high frequencies (*low-pass* filter);

– if $|\hat{h}(m)| > 1$ for $m \simeq N/2$ and $|\hat{h}(m)| \in [0,1[$ for $m \simeq 0$ and $m \simeq N-1$, then T increases the high frequencies and reduces the low frequencies (*high-pass* filter);

– if $|\hat{h}(m)| > 1$ for intermediate values of m, then T increases the mid-range frequencies (*band-pass* filter).

2.8.5. *Characterizing stationary operators using shift operators*

We now have all of the results we need to demonstrate the *characterization of a stationary operator as a linear combination of shift operators*, or, in an equivalent manner, as a *polynomial of the shift operator* R_1, since $R_k = R_1 \circ \cdots \circ R_1$ k times, that is, $R^k = R_1^k$, $\forall k \in \mathbb{Z}$.

THEOREM 2.10.– $T \in \mathrm{End}(\ell^2(\mathbb{Z}_N))$ is stationary if and only if the expression of T is:

$$(Tz)(n) = \sum_{k=0}^{N-1} a_k z(n-k) = \sum_{k=0}^{N-1} a_k R_k z(n) = \sum_{k=0}^{N-1} a_k (R_1)^k z(n) \qquad [2.46]$$
$$\forall n \in \{0, .., N-1\}$$

where $a_k \in \mathbb{C}$.

PROOF.–

$\boxed{\Longrightarrow}$: let T be stationary. We know that $T = T_h$, where T_h is the convolution operator with regard to the unit pulse response $h = T\delta$, that is:

$$(Tz)(n) = \sum_{k=0}^{N-1} h(k)z(n-k).$$

We must therefore simply identify the coefficients a_k of the formula $(Tz)(n) = \sum_{k=0}^{N-1} a_k z(n-k)$ with $h(k)$ to obtain our thesis.

$\boxed{\Longleftarrow}$: we can verify that T, written in the form used in formula [2.46], is stationary due to the linearity of T and R_k. We know that $\forall n \in \{0, \ldots, N-1\}$:

$$\begin{aligned}(TR_m z)(n) &= T(R_m z(n)) = T(z(n-m)) = \sum_{k=0}^{N-1} a_k z(n-k-m) \\ &= \sum_{k=0}^{N-1} a_k R_m z(n-k) \\ &\underset{(\text{linearity of } R_m)}{=} R_m\left(\sum_{k=0}^{N-1} a_k z(n-k)\right) = (R_m Tz)(n)\end{aligned}$$

hence: $T \circ R_m = R_m \circ T \; \forall m \in \mathbb{Z}$. □

Since $h(k) = T\delta(k)$, the proof of the theorem above also proves the validity of the formula:

$$\boxed{(Tz)(n) = \sum_{k=0}^{N-1} T\delta(k)z(n-k)} \qquad \forall T \text{ stationary}$$

2.8.6. *Frequency analysis of first and second derivation operators (discrete case)*

In this section, we shall analyze two stationary operators which represent the discrete version of the first and second derivatives. By comparing their eigenvalues, we see that the second derivation operator is more efficient for amplifying high frequencies in digital signals.

DEFINITION 2.17.– *Given a sequence $z \in \ell^2(\mathbb{Z}_N)$, we define:*

$$T_1 z(n) = z(n+1) - z(n) \qquad \textit{Discrete first derivative}$$

$$T_2 z(n) = z(n+1) - 2z(n) + z(n-1) \qquad \textit{Discrete second derivative}$$

The *discrete first derivative* is simply the *forward difference* of z, divided by the difference of the values of n, but since $(n+1)-n=1$ there is no need to write the denominator.

The *discrete second derivative* is the *backward difference of the discrete first derivative* of z, divided by the difference of the values of n, which – once again – is 1, so does not need to be written: $T_2z(n) = T_1z(n) - T_1z(n-1) = z(n+1)-z(n)-[z(n)-z(n-1)] = z(n+1)-2z(n)+z(n-1)$.

Let us begin by analyzing T_1. To calculate the pulse response, T_1 is applied to the unit pulse $\delta = e_0$:

$$h = T_1\delta = \begin{pmatrix} e_0(1) - \boxed{e_0(0)} & \longleftarrow n=0 \\ e_0(2) - e_0(1) & \longleftarrow n=1 \\ \vdots & \\ \boxed{e_0(N-1+1)} - e_0(N-1) & \longleftarrow n=N-1 \end{pmatrix} = \begin{pmatrix} -1 \\ 0 \\ 0 \\ \vdots \\ 0 \\ 1 \end{pmatrix}$$

using the fact that $e_0(0) = e_0(N) = 1$. The matrix which represents T_1 in the canonical basis of $\ell^2(\mathbb{Z}_N)$ is:

$$A_{T_1} = \begin{pmatrix} -1 & 1 & 0 & \dots & 0 \\ 0 & -1 & 1 & \dots & 0 \\ \vdots & & \ddots & \ddots & \vdots \\ 0 & 0 & \dots & -1 & 1 \\ 1 & 0 & \dots & 0 & -1 \end{pmatrix}$$

Now, let us calculate the DFT of h. For all $m \in \mathbb{Z}_N$, this is:

$$\begin{aligned} \hat{h}(m) &= \sum_{n=0}^{N-1} h(n)e^{-2\pi i \frac{mn}{N}} = -1 \cdot e^{-2\pi i \frac{m0}{N}} + 0 + \dots + 1 \cdot e^{-2\pi i \frac{m(N-1)}{N}} \\ &= -1 + e^{2\pi i \frac{m}{N}} e^{-2\pi i \frac{mN}{N}} = e^{2\pi i \frac{m}{N}} - 1 \end{aligned}$$

so the eigenvectors of T_1 are $\{\hat{h}(m) = e^{2\pi i \frac{m}{N}} - 1, m = 0,1,\dots,N-1\}$ and its diagonal representation is:

$$D = \operatorname{diag}\left(0, e^{2\pi i \frac{1}{N}} - 1, e^{2\pi i \frac{2}{N}} - 1 \dots, e^{2\pi i \frac{(N-1)}{N}} - 1\right)$$

The action of T_1 in terms of frequency can now be interpreted using formula [2.45]. We wish to calculate the magnitudes of the Eigenvalues $(\hat{h}(m))_{m\in\mathbb{Z}_N}$. We see that:

$$e^{2\pi i \frac{m}{N}} - 1 = e^{\pi i \frac{m}{N}}(e^{\pi i \frac{m}{N}} - e^{-\pi i \frac{m}{N}}) = e^{\pi i \frac{m}{N}} 2i \sin\left(\pi \frac{m}{N}\right)$$

Thus, $|\hat{h}(m)| = \left|e^{\pi i \frac{m}{N}}\right| \cdot \left|2i \sin\left(\pi \frac{m}{N}\right)\right| = 2\left|\sin\left(\pi \frac{m}{N}\right)\right|$, while $m \in \mathbb{Z}_N$, $\frac{m}{N} < 1$, so the sinus is always non-negative and the absolute value can be eliminated. To summarize:

$$\boxed{|\hat{h}(m)| = 2\sin\left(\pi \frac{m}{N}\right), \quad m \in \mathbb{Z}_N}$$

Specifically:

– $|\hat{h}(0)| = 0$: hence, the filtered signal $T_1 z$ averages to zero;

– $|\hat{h}(\frac{N}{2})| = 2$;

– $|\hat{h}(m)| < 2 \; \forall m \neq \frac{N}{2}$;

– $|\hat{h}(m)| \to 0$ if $m \to 0$ or $m \to N-1$;

– the action of the operator is symmetrical with regard to $\frac{N}{2}$.

Since $m = N/2$ represents the highest frequency of the signal and $m = 0$ and $m = N-1$ represent the lowest frequencies, we can deduce that T_1 reduces the low frequencies of z and increases the high frequencies by up to two times. Thus, *the discrete first derivative operator is a high-pass filter.*

Now, let us analyze T_2. Its pulse response is given by the vector:

$$h = T\delta = \begin{pmatrix} e_0(1) - 2\boxed{e_0(0)} + e_0(-1) \\ e_0(2) - 2e_0(1) + \boxed{e_0(0)} \\ \vdots \\ \boxed{e_0(N)} - 2e_0(N-1) + e_0(N-2) \end{pmatrix} = \begin{pmatrix} -2 \\ 1 \\ 0 \\ \vdots \\ 0 \\ 1 \end{pmatrix}$$

The matrix associated with T_2 in the canonical basis of $\ell^2(\mathbb{Z}_N)$ is:

$$A_{T_2} = \begin{pmatrix} -2 & 1 & 0 & \dots & 1 \\ 1 & -2 & 1 & \dots & 0 \\ \vdots & & \ddots & \ddots & \vdots \\ 0 & 0 & \dots & -2 & 1 \\ 1 & 0 & \dots & 1 & -2 \end{pmatrix}$$

Next, we calculate the DFT of h :

$$\hat{h}(m) = \sum_{n=0}^{N-1} h(n)e^{-2\pi i \frac{mn}{N}} = -2 \cdot e^{-2\pi i \frac{m0}{N}} + 1 \cdot e^{-2\pi i \frac{m}{N}} + 0 + \dots$$
$$+1 \cdot e^{-2\pi i \frac{m(N-1)}{N}}$$
$$= -2 + e^{-2\pi i \frac{m}{N}} + e^{-2\pi i m} e^{2\pi i \frac{m}{N}} = -2 + e^{2\pi i \frac{m}{N}} + e^{-2\pi i \frac{m}{N}}$$
$$= -2 + 2 \cdot \frac{e^{2\pi i \frac{m}{N}} + e^{-2\pi i \frac{m}{N}}}{2} = -2 + 2\cos\left(2\pi \frac{m}{N}\right)$$

These values of $\hat{h}(m)$ must now be compared with those of the first derivative operator. We do this by rewriting $\hat{h}(m) = -4\left[\frac{1}{2} - \frac{1}{2}\cos\left(2\pi\frac{m}{N}\right)\right]$ and using the trigonometric identity $\sin^2(\alpha) = \frac{1}{2} - \frac{1}{2}\cos(2\alpha)$ with $\alpha = \pi\frac{m}{N}$ to obtain $\hat{h}(m) = -4\sin^2\left(\pi\frac{m}{N}\right)$. The eigenvalues of T_2 are thus $\hat{h}(m) = -4\sin^2\left(\pi\frac{m}{N}\right)$, $m = 0, 1, \dots, N-1$, and its diagonal representation is:

$$D = \text{diag}\left(0, -4\sin^2\left(\frac{\pi}{N}\right), -4\sin^2\left(\frac{2\pi}{N}\right), \dots, -4\sin^2\left(\frac{(N-1)\pi}{N}\right)\right)$$

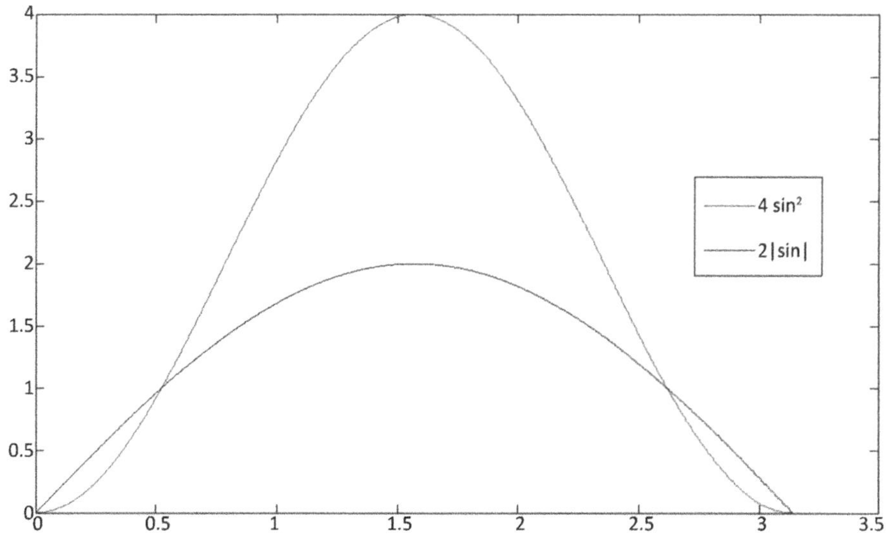

Figure 2.4. *Difference between the sine functions representing the spectrum values of the first and second derivative operators between 0 and π. For a color version of this figure, see www.iste.co.uk/provenzi/spaces.zip*

The effect of T_2 on the frequency is defined by the magnitudes of its Eigenvalues:

$$\boxed{|\hat{h}(m)| = 4\sin^2\left(\pi\frac{m}{N}\right), \quad m \in \mathbb{Z}_N}$$

We see that the magnitudes of the Eigenvalues of the second derivative operator are the squares of those of the first derivative operator. Hence:

– $|\hat{h}(0)| = 0$: thus, as in the case of the first derivative, the filtered signal $T_2 z$ averages to zero;

– $|\hat{h}(\frac{N}{2})| = 4$;

– $|\hat{h}(m)| < 4 \; \forall m \neq \frac{N}{2}$;

– $|\hat{h}(m)| \to 0$ if $m \to 0$ or $m \to N-1$ and the convergence to zero is faster than for the first derivative operator, as in this case, the sine is squared, as illustrated in Figure 2.4;

– The action of the operator is symmetrical about $\frac{N}{2}$.

Thus, *the discrete second derivative operator is also a high-pass filter*, amplifying high frequencies and reducing low frequencies in a way which is the square of the action of the discrete first derivative operator.

2.9. The two-dimensional discrete Fourier transform (2D DFT)

The Fourier transform considered up to now applies to signals $z(n)$ which depend on only one parameter n.

In practical contexts, signals are often very large and depend on multiple parameters. One classic example is that of *digital images*, which include two parameters: the *two spatial coordinates of a pixel*, as shown in Figure 2.5.

DFT theory can be generalized for signals which depend on any (finite) number of parameters. For simplicity's sake, we shall focus on the two-dimensional (2D) case, with parameters n_1, n_2.

The first step is to introduce the domain vector space: if $N_1, N_2 \in \mathbb{N}$, we define:

$$\ell^2(\mathbb{Z}_{N_1} \times \mathbb{Z}_{N_2}) = \{z : \mathbb{Z}_{N_1} \times \mathbb{Z}_{N_2} \implies \mathbb{C}\}$$

$z \in \ell^2(\mathbb{Z}_{N_1} \times \mathbb{Z}_{N_2})$ is a complex sequence which depends on two parameters:

$$\begin{cases} n_1 \in \{0, 1, \dots, N_1 - 1\} \\ n_2 \in \{0, 1, \dots, N_2 - 1\} \end{cases}$$

Figure 2.5. *The two coordinates of a pixel, n_1, n_2, in a digital image (image source: author). For a color version of this figure, see www.iste.co.uk/provenzi/spaces.zip*

$\ell^2(\mathbb{Z}_{N_1} \times \mathbb{Z}_{N_2})$ is a vector space of dimension $N_1 \cdot N_2$. The definitions used for summation and multiplication by a complex scalar are the same as those used for the 1D case and for inner products:

$$\langle z, w \rangle = \sum_{n_1=0}^{N_1-1} \sum_{n_2=0}^{N_2-1} z(n_1, n_2)\overline{w(n_1, n_2)}, \quad \forall z, w \in \ell^2(\mathbb{Z}_{N_1} \times \mathbb{Z}_{N_2})$$

To extend DFT theory from one to two dimensions, we use the procedure for generating bases in $\ell^2(\mathbb{Z}_{N_1} \times \mathbb{Z}_{N_2})$ from bases in $\ell^2(\mathbb{Z}_{N_1})$ and $\ell^2(\mathbb{Z}_{N_2})$.

THEOREM 2.11.– Let $\{B_0, B_1, \ldots, B_{N_1-1}\}$, $\{C_0, C_1, \ldots, C_{N_2-1}\}$ be orthonormal bases in $\ell^2(\mathbb{Z}_{N_1})$ and $\ell^2(\mathbb{Z}_{N_2})$, respectively.

For all $m_1 \in \{0, \ldots, N_1 - 1\}$ and $m_2 \in \{0, \ldots, N_2 - 1\}$, consider the sequences in $\ell^2(\mathbb{Z}_{N_1} \times \mathbb{Z}_{N_2})$ given by:

$$\boxed{D_{m_1,m_2}(n_1, n_2) = B_{m_1}(n_1) \cdot C_{m_2}(n_2)}$$

Then, D_{m_1,m_2} is an orthonormal basis of $\ell^2(\mathbb{Z}_{N_1} \times \mathbb{Z}_{N_2})$, known as the *tensor product basis* of the two original bases.

PROOF.– The sequences D_{m_1,m_2}, $m_1 \in \{0, \ldots, N_1 - 1\}$ and $m_2 \in \{0, \ldots, N_2 - 1\}$ are $N_1 \cdot N_2$ elements of $\ell^2(\mathbb{Z}_{N_1} \times \mathbb{Z}_{N_2})$, which is of dimension $N_1 \cdot N_2$. Proof that these constitute an orthonormal basis can be obtained by showing that:

$$\langle D_{m_1,m_2}, D_{k_1,k_2} \rangle = \delta_{(m_1,m_2),(k_1,k_2)} = \delta_{m_1,k_1}\delta_{m_2,k_2} = \begin{cases} 1 & \text{if } (m_1, m_2) = (k_1, k_2) \\ 0 & \text{if } (m_1, m_2) \neq (k_1, k_2) \end{cases}$$

$$
\begin{aligned}
\langle D_{m_1,m_2}, D_{k_1,k_2}\rangle &\underset{\text{def. of}\langle\,,\,\rangle}{=} \textstyle\sum_{n_1=0}^{N_1-1}\sum_{n_2=0}^{N_2-1} D_{m_1,m_2}(n_1,n_2)\overline{D_{k_1,k_2}(n_1,n_2)} \\
&\underset{\text{def. of } D}{=} \sum_{n_1=0}^{N_1-1}\sum_{n_2=0}^{N_2-1} B_{m_1}(n_1)C_{m_2}(n_2)\overline{B_{k_1}(n_1)C_{k_2}(n_2)} \\
&= \sum_{n_1=0}^{N_1-1} B_{m_1}(n_1)\overline{B_{k_1}(n_1)} \sum_{n_2=0}^{N_2-1} C_{m_2}(n_2)\overline{C_{k_2}(n_2)} \\
&= \underbrace{\langle B_{m_1}, B_{k_1}\rangle}_{\delta_{m_1,k_1}}\underbrace{\langle C_{m_2}, C_{k_2}\rangle}_{\delta_{m_2,k_2}} = \delta_{(m_1,m_2),(k_1,k_2)}.
\end{aligned}
$$

□

For $m_1 \in \{0,1,\dots,N_1-1\}$ and $m_2 \in \{0,1,\dots,N_2-1\}$, this theorem has the following corollaries:

– the canonical orthonormal basis of $\ell^2(\mathbb{Z}_{N_1}\times\mathbb{Z}_{N_2})$ is:

$$
\boxed{B = e_{m_1,m_2}(n_1,n_2) = \begin{cases} 1 & \text{if } (n_1,n_2) = (m_1,m_2) \\ 0 & \text{if } (n_1,n_2) \neq (m_1,m_2) \end{cases}}
$$

– the orthogonal Fourier basis of $\ell^2(\mathbb{Z}_{N_1}\times\mathbb{Z}_{N_2})$ is:

$$
\boxed{F_{m_1,m_2}(n_1,n_2) = \frac{1}{N_1N_2}e^{2\pi i\frac{m_1n_1}{N_1}}\cdot e^{2\pi i\frac{m_2n_2}{N_2}} = \frac{1}{N_1N_2}e^{2\pi i\left(\frac{m_1n_1}{N_1}+\frac{m_2n_2}{N_2}\right)}}
$$

– the orthonormal Fourier basis of $\ell^2(\mathbb{Z}_{N_1}\times\mathbb{Z}_{N_2})$ is:

$$
E_{m_1,m_2}(n_1,n_2) = \sqrt{N_1N_2}F_{m_1,m_2}(n_1,n_2)
$$

– the orthogonal basis of the complex exponentials in $\ell^2(\mathbb{Z}_{N_1}\times\mathbb{Z}_{N_2})$ is:

$$
\mathcal{E}_{m_1,m_2}(n_1,n_2) = N_1N_2F_{m_1,m_2}(n_1,n_2)
$$

Using the theory of complex inner product spaces, the definition of Fourier coefficients, the DFT and the IDFT can be generalized to $\ell^2(\mathbb{Z}_{N_1}\times\mathbb{Z}_{N_2})$. Taking $z \in \ell^2(\mathbb{Z}_{N_1}\times\mathbb{Z}_{N_2})$, we have:

$$
\begin{aligned}
\langle z, \mathcal{E}_{m_1,m_2}\rangle &= \sum_{n_1=0}^{N_1-1}\sum_{n_2=0}^{N_2-1} z(n_1,n_2)\overline{e^{2\pi i\frac{n_1m_1}{N_1}}e^{2\pi i\frac{n_2m_2}{N_2}}} \\
&= \sum_{n_1=0}^{N_1-1}\sum_{n_2=0}^{N_2-1} z(n_1,n_2)e^{-2\pi i\frac{m_1n_1}{N_1}}e^{-2\pi i\frac{m_2n_2}{N_2}} \\
&= \sum_{n_1=0}^{N_1-1}\sum_{n_2=0}^{N_2-1} z(n_1,n_2)e^{-2\pi i(\frac{m_1n_1}{N_1}+\frac{m_2n_2}{N_2})}
\end{aligned}
$$

thus the *Fourier coefficients* of $z \in \ell^2(\mathbb{Z}_{N_1} \times \mathbb{Z}_{N_2})$ are defined as follows:

$$\boxed{\hat{z}(m_1, m_2) = \sum_{n_1=0}^{N_1-1} \sum_{n_2=0}^{N_2-1} z(n_1, n_2) e^{-2\pi i \left(\frac{m_1 n_1}{N_1} + \frac{m_2 n_2}{N_2}\right)}}$$

As in the 1D case:

$$\boxed{\hat{z}(0,0) = N_1 N_2 \langle z \rangle}$$

where $\langle z \rangle$ is the average of z. Note that the quantity $N_1 N_2$ may be extremely large.

The synthesis formula can also be generalized to the 2D case, as follows:

$$\boxed{z(n_1, n_2) = \frac{1}{N_1 N_2} \sum_{m_1=0}^{N_1-1} \sum_{m_2=0}^{N_2-1} \hat{z}(m_1, m_2) e^{2\pi i \left(\frac{m_1 n_1}{N_1} + \frac{m_2 n_2}{N_2}\right)}}$$

The 2D DFT and 2D IDFT operators can therefore be written using the following formulas:

$$\begin{aligned} \hat{} : \ell^2(\mathbb{Z}_{N_1} \times \mathbb{Z}_{N_2}) &\longrightarrow \ell^2(\mathbb{Z}_{N_1} \times \mathbb{Z}_{N_2}) \\ z &\longmapsto \hat{z},\ \hat{z}(m_1, m_2) = \sum_{n_1=0}^{N_1-1} \sum_{n_2=0}^{N_2-1} z(n_1, n_2) e^{-2\pi i \left(\frac{m_1 n_1}{N_1} + \frac{m_2 n_2}{N_2}\right)} \end{aligned}$$

and:

$$\begin{aligned} \check{} : \ell^2(\mathbb{Z}_{N_1} \times \mathbb{Z}_{N_2}) &\longrightarrow \ell^2(\mathbb{Z}_{N_1} \times \mathbb{Z}_{N_2}) \\ z \longmapsto \check{z},\ \check{z}(n_1, n_2) &= \frac{1}{N_1 N_2} \sum_{m_1=0}^{N_1-1} \sum_{m_2=0}^{N_2-1} z(m_1, m_2) e^{2\pi i \left(\frac{m_1 n_1}{N_1} + \frac{m_2 n_2}{N_2}\right)} \end{aligned}$$

Clearly, if the dimension is increased from 2 to $2 < d < +\infty$, these formulas can be generalized in the following manner:

$$\hat{z}(m_1, \ldots, m_d) = \sum_{n_1=0}^{N_1-1} \cdots \sum_{n_d=0}^{N_d-1} z(n_1, \ldots, n_d) e^{-2\pi i \sum_{k=1}^{d} \frac{m_k n_k}{N_k}}$$

$$\check{z}(n_1, \cdots, n_d) = \left(\prod_{k=1}^{d} N_k\right)^{-1} \sum_{m_1=0}^{N_1-1} \cdots \sum_{m_d=0}^{N_d-1} z(m_1, \ldots, m_d) e^{2\pi i \sum_{k=1}^{d} \frac{m_k n_k}{N_k}}$$

2.9.1. *Matrix representation of the 2D DFT: Kronecker product versus iteration of two 1D DFTs*

The matrix representation of the 2D DFT in the canonical basis of $\ell^2(\mathbb{Z}_{N_1} \times \mathbb{Z}_{N_2})$ can be constructed using the Sylvester matrices W_{N_1} and W_{N_2} associated with the 1D DFT for $\ell^2(\mathbb{Z}_{N_1})$ and for $\ell^2(\mathbb{Z}_{N_2})$, respectively.

The operation used to obtain a matrix representation of the 2D DFT is the Kronecker product, which is defined below.

DEFINITION 2.18.– *Given two matrices, A of dimension $m \times n$ and B of dimension $p \times q$:*

$$A = \begin{pmatrix} a_{11} & \cdots & a_{1n} \\ \vdots & \ddots & \vdots \\ a_{m1} & \cdots & a_{mn} \end{pmatrix}, \quad B = \begin{pmatrix} b_{11} & \cdots & b_{1q} \\ \vdots & \ddots & \vdots \\ b_{p1} & \cdots & b_{pq} \end{pmatrix}$$

the Kronecker product *matrix $A \otimes B$ is the matrix of dimension $mp \times nq$ defined by:*

$$A \otimes B = \begin{pmatrix} a_{11}B & \cdots & a_{1n}B \\ \vdots & \ddots & \vdots \\ a_{m1}B & \cdots & a_{mn}B \end{pmatrix}$$

The matrix associated with the 2D DFT in the canonical basis of $\ell^2(\mathbb{Z}_{N_1} \times \mathbb{Z}_{N_2})$ can be shown, by direct calculation, to be the matrix of dimension $N_1N_2 \times N_1N_2$ given by:

$$\boxed{W_{N_1,N_2} = W_{N_1} \otimes W_{N_2}} \quad \Longrightarrow \quad \boxed{\hat{z}(m_1, m_2) = W_{N_1} \otimes W_{N_2} z(n_1, n_2)}$$

Unfortunately, the calculation needed to obtain the Kronecker product matrix becomes unfeasibly large for high values of N_1 and N_2. In practice, the 2D DFT is generally written as the iteration of two 1D DFTs.

To understand this approach, $z \in \ell^2(\mathbb{Z}_{N_1} \times \mathbb{Z}_{N_2})$ *must be interpreted as a matrix made up of N_2 column vectors with N_1 elements*:

$$z(n_1, n_2) = \underbrace{\left.\begin{pmatrix} \vdots & \vdots & \cdots & \vdots \\ z(\cdot, 0) & z(\cdot, 1) & \cdots & z(\cdot, N_2 - 1) \\ \vdots & \vdots & \cdots & \vdots \end{pmatrix}\right\}}_{N_2 \text{ column vectors}}$$

N_1 elements for each column vector

From the definition of the 2D DFT, we can write:

$$\begin{aligned} \hat{z}(m_1, m_2) &= \sum_{n_2=0}^{N_2-1} \underbrace{\left(\sum_{n_1=0}^{N_1-1} z(n_1, n_2) e^{-2\pi i \frac{n_1 m_1}{N_1}} \right) e^{-2\pi i \frac{n_2 m_2}{N_2}}} \\ &= \sum_{n_2=0}^{N_2-1} W_{N_1} z(n_1, n_2) e^{-2\pi i \frac{n_2 m_2}{N_2}} \\ \hat{z}(m_1, n_2) &= W_{N_1} z(n_1, n_2) \end{aligned} \quad [2.47]$$

In this formula, the sum with regard to index n_2 is the furthest out, so n_2 is *fixed* each time. Taking a fixed value for n_2, $z(n_1, n_2)$ is a column vector, so the highlighted parenthesis represents the 1D DFT of the column vector, which can be obtained by applying matrix W_{N_1} to $z(n_1, n_2)$, with a fixed value of n_2, as before.

The next problem is that n_1 is fixed, and the changing index is n_2, meaning that $W_{N_1} z(n_1, n_2)$ is a row vector. For this reason, the DFT cannot be obtained by applying W_{N_2}: as we saw in section 2.5, the 1D DFT is obtained from the product of the matrix W_N and a sequence represented using a *column vector.*

The solution to this problem consists of transposing the two sides of equation [2.47], transforming the row vector $\hat{z}(m_1, n_2)$ into a column vector:

$$\hat{z}(m_1, m_2)^t = \sum_{n_2=0}^{N_2-1} (W_{N_1} z(n_1, n_2))^t e^{-2\pi i \frac{n_2 m_2}{N_2}}$$

Now, $(W_{N_1} z(n_1, n_2))^t$ is a column vector, so the DFT can be calculated by applying W_{N_2}:

$$\hat{z}(m_1, m_2)^t = W_{N_2} (W_{N_1} z(n_1, n_2))^t \underset{(AB)^t = B^t A^t}{=} W_{N_2}\, z(n_1, n_2)^t\, (W_{N_1})^t$$
$$= W_{N_2} z(n_2, n_1) W_{N_1}$$

since $W_{N_1}^t = W_{N_1}$ (note that n_1 and n_2 have swapped places). Thus, $\hat{z}(m_1, m_2)^t = W_{N_2} z(n_2, n_1)\, W_{N_1}$, so to find $\hat{z}(m_1, m_2)$, we must simply transpose both sides again:

$$\hat{z}(m_1, m_2) = (\hat{z}(m_1, m_2)^t)^t = (W_{N_2}\, z(n_2, n_1)\, W_{N_1})^t = W_{N_1}\, z(n_1, n_2)\, W_{N_2}$$

The formula used to calculate the 2D DFT of a sequence $z \in \ell^2(\mathbb{Z}_{N_1} \times \mathbb{Z}_{N_2})$ is thus:

$$\boxed{\hat{z}(m_1, m_2) = W_{N_1} z(n_1, n_2) W_{N_2}} \qquad [2.48]$$

It is important to note that equation [2.48] is only meaningful if $\hat{z}(m_1, m_2)$ and $z(n_1, n_2)$ are interpreted as $N_1 \times N_2$ matrices in their entirety.

Formula [2.48] is not the same as $W_{N_1} W_{N_2} z(n_1, n_2)$ or $W_{N_2} W_{N_1} z(n_1, n_2)$, i.e. the formulas that one could have naively thought to use to implement 1D DFT over the columns and rows of z. The reason for this difference, as we have seen, is that the 1D matrix DFT requires the presence of a column vector, hence the transposition which results in formula [2.48].

2.9.2. *Properties of the 2D DFT*

The generalization of the properties of the 1D DFT, presented in section 2.7, to the 2D DFT is trivial.

The demonstrations of these properties in 1D and 2D are practically identical, notwithstanding certain differences in notation. For this reason, we shall not provide proofs for the 2D extensions presented below.

As in the 1D case, in order to examine the properties of the 2D DFT, we must first extend the definition of a sequence $z \in \ell^2(\mathbb{Z}_{N_1} \times \mathbb{Z}_{N_2})$ by periodicity to any interval of length N_1 with regard to the variable n_1 and of length N_2 with regard to the variable n_2.

This extension is possible if z is defined outside of $\mathbb{Z}_{N_1} \times \mathbb{Z}_{N_2}$ in the following manner:

$$\boxed{z(n_1 + j_1 N_1, n_2 + j_2 N_2) = z(n_1, n_2)}, \qquad \forall n_1, n_2, j_1, j_2 \in \mathbb{Z} \qquad [2.49]$$

The shift operator is also helpful in 2D cases.

DEFINITION 2.19.– *Take $z \in \ell^2(\mathbb{Z}_{N_1} \times \mathbb{Z}_{N_2})$, extended by periodicity as in formula [2.49], and $k_1, k_2 \in \mathbb{Z}$. The shift operator over $\ell^2(\mathbb{Z}_{N_1} \times \mathbb{Z}_{N_2})$ is defined by:*

$$\begin{aligned} R_{k_1,k_2} : \ell^2(\mathbb{Z}_{N_1} \times \mathbb{Z}_{N_2}) &\longrightarrow \ell^2(\mathbb{Z}_{N_1} \times \mathbb{Z}_{N_2}) \\ z &\longmapsto R_{k_1,k_2} z, \end{aligned}$$

$$(R_{k_1,k_2} z)(n_1, n_2) = z(n_1 - k_1, n_2 - k_2)$$

Taking $z \in \ell^2(\mathbb{Z}_{N_1} \times \mathbb{Z}_{N_2})$, extended by periodicity as in formula [2.49], then, for all $n_1, n_2, m_1, m_2 \in \mathbb{Z}$:

– *periodicity* of $\hat{z}$ and $\check{z}$:

$$\boxed{\hat{z}(m_1, m_2) = \hat{z}(m_1 + N_1, m_2) = \hat{z}(m_1, m_2 + N_2) = \hat{z}(m_1 + N_1, m_2 + N_2)}$$

and:

$$\boxed{\check{z}(n_1, n_2) = \check{z}(n_1 + N_1, n_2) = \check{z}(n_1, n_2 + N_2) = \check{z}(n_1 + N_1, n_2 + N_2)}$$

– 2D DFT and *shift*:

$$\boxed{\widehat{R_{k_1,k_2} z}(m_1, m_2) = e^{-2\pi i \left(\frac{m_1 k_1}{N_1} + \frac{m_2 k_2}{N_2}\right)} \hat{z}(m_1, m_2)} \qquad \forall k_1, k_2 \in \mathbb{Z}$$

that is, if we define the sequence $\omega_{N_1,N_2}^{k_1,k_2} \in \ell^2(\mathbb{Z}_{N_1} \times \mathbb{Z}_{N_2})$, $\omega_{N_1,N_2}^{k_1,k_2}(m_1, m_2) = e^{-2\pi i \left(\frac{m_1 k_1}{N_1} + \frac{m_2 k_2}{N_2}\right)}$ $\forall m_1, m_2 \in \mathbb{Z}$, then:

$$\boxed{\text{DFT 2D} \circ R_{k_1,k_2} = M_{\omega_{N_1,N_2}^{k_1,k_2}} \circ \text{DFT 2D}}$$

where $M_{\omega_{N_1,N_2}^{k_1,k_2}}$ is the multiplication operator by $\omega_{N_1,N_2}^{k_1,k_2}$ in $\ell^2(\mathbb{Z}_{N_1} \times \mathbb{Z}_{N_2})$. Permutating the direction of composition, we obtain:

$$\boxed{(R_{k_1,k_2} \hat{z})(m_1, m_2) = \hat{z}(m_1 - k_1, m_2 - k_2) = \text{DFT 2D}\left(e^{2\pi i \left(\frac{m_1 k_1}{N_1} + \frac{m_2 k_2}{N_2}\right)} z\right)(m_1, m_2)}$$

that is:

$$\boxed{R_{k_1,k_2} \circ \text{DFT 2D} = \text{DFT 2D} \circ M_{\left(\omega_{N_1,N_2}^{k_1,k_2}\right)^*}}, \qquad \forall k_1, k_2 \in \mathbb{Z}$$

The properties examined above are summarized by the Fourier pairs in Table 2.5.

Original representation	Fourier space
$z(n_1 - k_1, n_2 - k_2)$	$e^{-2\pi i\left(\frac{m_1 k_1}{N_1} + \frac{m_2 k_2}{N_2}\right)} \hat{z}(m_1, m_2)$
$e^{2\pi i\left(\frac{n_1 k_1}{N_1} + \frac{n_2 k_2}{N_2}\right)} z(n_1, n_2)$	$\hat{z}(m_1 - k_1, m_2 - k_2)$

Table 2.5. *Fourier pairs for 2D shifts*

As in the case of 1D DFT, considering $k_1 = \frac{N_1}{2}$ and $k_2 = \frac{N_2}{2}$, then $(-1)^{n_1+n_2} z(n_1, n_2)$ and $\hat{z}(m_1 - \frac{N_1}{2}, m_2 - \frac{N_2}{2})$. This transformation is used to obtain a centered visualization of the spectrum of z. Furthermore, as in the o1D case, the amplitude spectrum of a 2D signal $z(n_1, n_2)$ and of any shifted form $z(n_1 - k_1, n_2 - k_2)$ is strictly identical, as $\left| e^{-2\pi i\left(\frac{m_1 k_1}{N_1} + \frac{m_2 k_2}{N_2}\right)} \right| = 1$. Thus, the *amplitude spectrum* gives us the frequency content of the signal, but *does not tell us where these frequencies are located*;

– 2D DFT and *conjugation*:

$$\boxed{\widehat{\overline{z}}(m_1, m_2) = \overline{\hat{z}(-m_1, -m_2)} = \overline{\hat{z}(N_1 - m_1, N_2 - m_2)}}$$

– 2D DFT and *convolution*:

$$\boxed{\widehat{(z * w)}(m_1, m_2) = \hat{z}(m_1, m_2)\hat{w}(m_1, m_2)}$$

where 2D convolution is defined as:

$$\begin{aligned} T_z : \ell^2(\mathbb{Z}_{N_1} \times \mathbb{Z}_{N_2}) &\longrightarrow \ell^2(\mathbb{Z}_{N_1} \times \mathbb{Z}_{N_2}) \\ w &\longmapsto T_z w = z * w \end{aligned}$$

$$\boxed{\begin{aligned} (z * w)(n_1, n_2) &= \sum_{k_1=0}^{N_1-1} \sum_{k_2=0}^{N_2-1} z(n_1 - k_1, n_2 - k_2) w(k_1, k_2) \\ &= \sum_{k_1=0}^{N_1-1} \sum_{k_2=0}^{N_2-1} z(n_1, n_2) w(n_1 - k_1, n_2 - k_2) \end{aligned}}$$

Box 2.3. *Properties of 2D DFT*

2.9.3. *2D DFT and stationary operators*

The properties of 2D and 1D DFT with regard to stationary operators are the same.

Strictly speaking, an operator $T : \ell^2(\mathbb{Z}_{N_1} \times \mathbb{Z}_{N_2}) \to \ell^2(\mathbb{Z}_{N_1} \times \mathbb{Z}_{N_2})$ is stationary if:

$$T \circ R_{k_1,k_2} = R_{k_1,k_2} \circ T, \quad \forall k_1, k_2 \in \mathbb{Z}$$

In practice, if z is a digital image, *a stationary operator is a transformation whose action is independent of the position of a pixel in the spatial context of the image.*

As in the 1D case, stationary operators over $\ell^2(\mathbb{Z}_{N_1} \times \mathbb{Z}_{N_2})$ may be characterized as convolution operators or as Fourier multiplier operators.

The theorem formalizing this relation relies on definitions of the Fourier multiplier, the unit pulse and the pulse response in the 2D case.

DEFINITION 2.20.– *Taking a fixed $w \in \ell^2(\mathbb{Z}_{N_1} \times \mathbb{Z}_{N_2})$, the Fourier multiplier associated with w is defined as:*

$$\begin{array}{rccl} T_{(w)} : & \ell^2(\mathbb{Z}_{N_1} \times \mathbb{Z}_{N_2}) & \longrightarrow & \ell^2(\mathbb{Z}_{N_1} \times \mathbb{Z}_{N_2}) \\ & z & \longmapsto & T_{(w)}z = \widecheck{w \cdot \hat{z}} \end{array}$$

DEFINITION 2.21.– *The unit pulse δ in $\ell^2(\mathbb{Z}_{N_1} \times \mathbb{Z}_{N_2})$ is the first vector in the canonical basis: $\delta = e_{0,0}$.*

Given a linear operator T over $\ell^2(\mathbb{Z}_{N_1} \times \mathbb{Z}_{N_2})$, the pulse response is defined as the sequence $h = T\delta \in \ell^2(\mathbb{Z}_{N_1} \times \mathbb{Z}_{N_2})$.

THEOREM 2.12.– Let $T : \ell^2(\mathbb{Z}_{N_1} \times \mathbb{Z}_{N_2}) \longrightarrow \ell^2(\mathbb{Z}_{N_1} \times \mathbb{Z}_{N_2})$ be a linear operator. The following conditions are equivalent:

1) T is stationary;

2) T is the convolution operator with the pulse response $h = T\delta$:

$$Tz = T_h z = h * z = z * h \qquad \forall z \in \ell^2(\mathbb{Z}_{N_1} \times \mathbb{Z}_{N_2})$$

3) T is the Fourier multiplier associated with $\hat{h}$:

$$Tz = T_{(\hat{h})} z = \widecheck{\hat{h} \cdot \hat{z}} \qquad \in \ell^2(\mathbb{Z}_{N_1} \times \mathbb{Z}_{N_2})$$

4) T is diagonalizable, its eigenvectors are the orthogonal Fourier basis F_{m_1,m_2} of $\ell^2(\mathbb{Z}_{N_1} \times \mathbb{Z}_{N_2})$, and its eigenvalues are the components of $\hat{h}$.

OBSERVATIONS.– This result can be extended to circulant matrices, but their definition in the 2D case is more complex.

2.9.4. *Gradient and Laplace operators and their action on digital images*

Repeating the analysis of discrete derivative operators from section 2.8.6 for 2D cases, the first derivative gives us the *gradient*, that is $\vec{\nabla} = (\frac{\partial}{\partial x}, \frac{\partial}{\partial y})$, and the second derivative gives us the *Laplacian*, that is $\nabla^2 = \frac{\partial^2}{\partial x^2} + \frac{\partial^2}{\partial y^2}$.

The gradient is used to detect the edges of an image in a particular direction. For isotropic edge detection – that is detection which is uniform with regard to direction – the Laplacian is used; this approach is more efficient than using a gradient for intensifying fine details, as we saw in the 1D case.

Even in 2D cases, the differential operators above cancel out the average of an image, which is why the output is entirely black, except near the edges, as we see from Figure 2.6.

2.9.5. *Visualization of the amplitude spectrum in 2D*

Visualizations of the spectrum of a 2D signal can be produced on the condition that the signal is centered, for the same reasons presented in the 1D case. Centering may be carried out using the 2D equivalent of formula [2.34], considering $(-1)^{n_1+n_2} z(n_1, n_2)$ in the place of $z(n_1, n_2)$, as we saw in section 2.9.2.

Note that the 1D symmetry of the 1D DFT with regard to frequencies $m \in \{0, 1, \dots, N/2\}$ and $m \in \{N/2 + 1, N/2 + 2, \dots, N - 1\}$ is replaced by 2D *mirror symmetry* in the case of the 2D DFT.

Figure 2.7 shows three grayscale digital images with their amplitude spectrums. The brightest points correspond to high magnitude values of the Fourier coefficients, while the darkest points correspond to low values.

There are several notable characteristics here:

– the *symmetry* of the spectrum: frequency content is repeated in each quadrant by mirror symmetry;

– the brightest points are located toward the center of the spectrum: this is due to the fact that these spectrums are centered, so the coordinates of the central frequency are $(m_1, m_2) = (0, 0)$ and $|\hat{z}(0, 0)| = N_1 N_2 \langle z \rangle$, that is, $N_1 N_2$ times the average value of the image. *This is why a compressive function, such as a logarithm, must be used to visualize a spectrum*: the values of $|\hat{z}(0, 0)|$ are so much higher than the others that the variability range needs to be compressed;

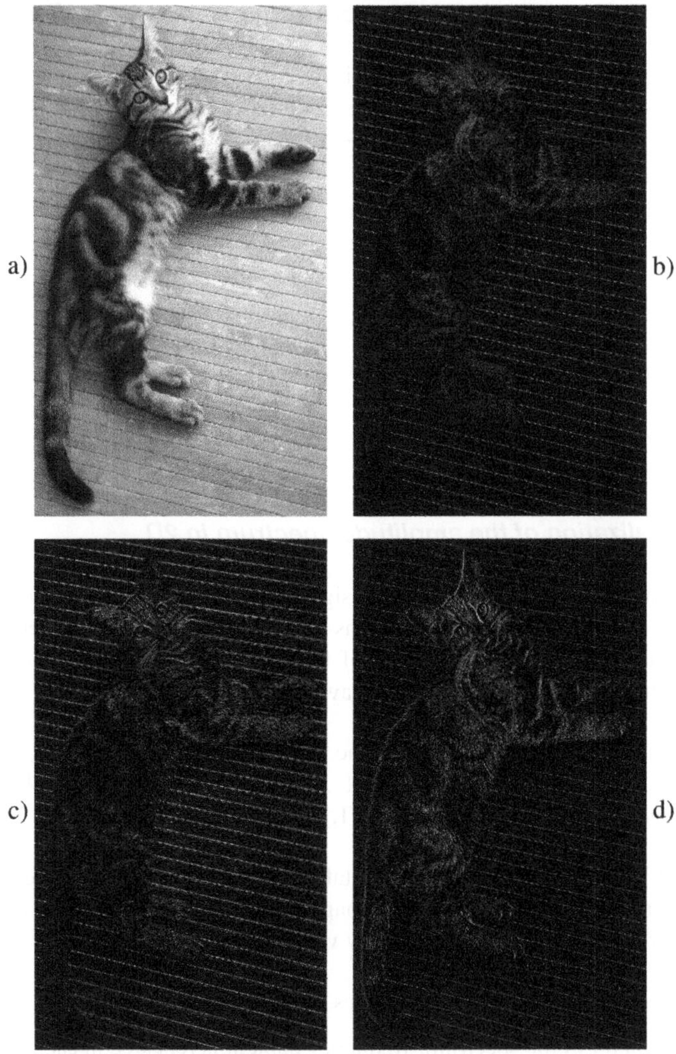

Figure 2.6. *a) Original image of Panko; b) image after Laplacian filter; c) image filtered using a gradient in the vertical direction; d) image filtered using a gradient in the horizontal direction (image source: author). For a color version of this figure, see www.iste.co.uk/provenzi/spaces.zip*

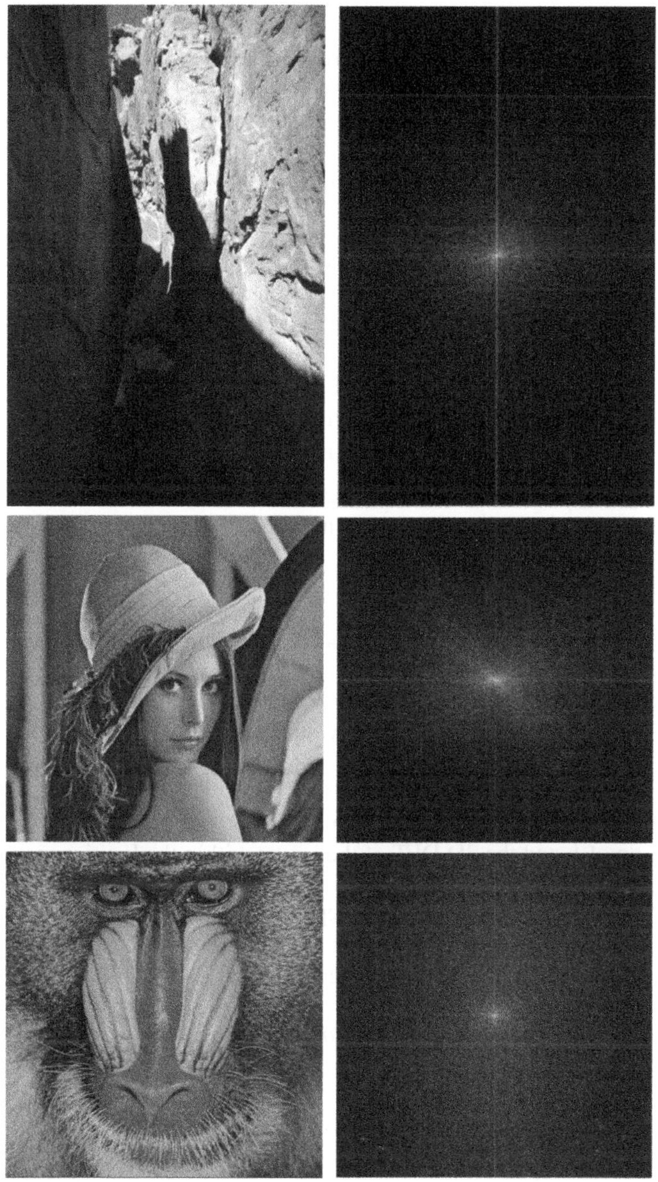

Figure 2.7. *Left column: original images. Right column: centered amplitude spectrums of the images in the left column, visualized using a logarithmic scale*

– moving out from the center, the spectrum shows the amplitude of the coefficients corresponding to the highest frequencies, up to the maximum frequencies $(N_1/2, N_2/2)$, if N_1, N_2 are even, or their integer parts $([N_1/2], [N_2/2])$ if N_1, N_2 are odd. The image with the highest frequency content is that of the mandrill: its spectrum is the widest of the three shown here. Note the particularly intense values near the edges, representing very high frequencies: these correspond to the fine details of the hairs near the animal's eyes;

– as m_1 and m_2 represent vertical and horizontal frequencies, the vertical and horizontal edges of the images produce Fourier coefficients which are localized on the corresponding axes. This is why the spectrum of the first image, which features strong vertical intensity gradients between the rocks and the sea, is heavily dominated by intense Fourier coefficients on the vertical axis. The second image ("Lena", a classic image used in image processing) features fine details in the hat area, at 45° and $-45°$. This results in evident diagonal structures in the spectrum;

– from this spectrum analysis, we see that the Fourier spectrum reveals the presence of geometric structures within an image, but does not tell us where in the image these structures are located.

2.9.6. *Filtering: an example of digital image filtering in a Fourier space*

Theorem 2.12 states that all stationary operators T acting on images (interpreted as finite 2D sequences) are "hidden" convolutions between the image and the pulse response $h = T\delta$.

Furthermore, these convolutions can be represented as Fourier multipliers (multiplication of $\hat{h}$ and the 2D DFT of the image within the Fourier space).

Different results will be obtained depending on the sequence h with which convolution is carried out. The effect of a convolution is often easier to interpret by examining the associated Fourier multiplier.

Let us consider the notion of *convolution with a discrete Gaussian*, noted $h(n_1, n_2)$.

As we shall see in Chapter 6, the Fourier transform of a Gaussian with a standard deviation σ is itself a Gaussian, but the standard deviation of the latter is inversely proportional to σ. Thus, we can further our understanding of the meaning of convoluting an image $z(n_1, n_2)$ with a Gaussian $h(n_1, n_2)$ by analyzing the multiplication $\hat{z}(m_1, m_2) \cdot \hat{h}(m_1, m_2)$ in the Fourier space.

Figure 2.8 features three images corresponding to 512×512 2D Gaussians. The intensity of the pixel in position (n_1, n_2) is $h(n_1, n_2) = \exp\left(-\frac{n_1^2+n_2^2}{2\sigma^2}\right)$ and the standard deviation is σ =1, 5 and 10, respectively.

Figure 2.8. *Two-dimensional Gaussian images with a standard deviation of (left - right) 1, 5 and 10*

As we stated above, the 2D DFTs of h are still Gaussians, but their standard deviations are proportional to 1, $\frac{1}{5}$, and $\frac{1}{10}$. Evidently, $h(0,0) = 1$ and the values of $\hat{h}(m_1, m_2)$ decrease as we move away from the center; thus, multiplication in the Fourier space $\hat{z}(m_1, m_2) \cdot \hat{h}(m_1, m_2)$ decreases the importance of the harmonics with $(m_1, m_2) \neq (0,0)$, which are associated with the finer details in the image. Applying the 2D IDFT to $\hat{z}(m_1, m_2) \cdot \hat{h}(m_1, m_2)$, we can reconstruct an image which is blurrier than the original.

In image processing, convolution with a Gaussian corresponds to a blurring operation, as we see in Figure 2.9.

Figure 2.9. *Blurred image of Lena obtained by multiplying DFTs and Gaussians with standard deviations of (left - right) 1, 5 and 10*

COMMENT CONCERNING FIGURE 2.9.– *Note that as the standard deviation of the DFT of a Gaussian is inversely proportional to the original standard deviation, the DFT of the Gaussian with a standard deviation of 10 has a small standard deviation in the latter case, and thus tends rapidly toward 0. So, when the DFT of the Gaussian with an SD of 10 is multiplied with the DFT of the image, much of the detail in the image is lost.*

Blurring has a number of uses; for example, in cases where the original image is noisy, blurring can make this noise less evident (although it also reduces edge sharpness).

Figure 2.10 shows a continuous version of the blurring frequency filter.

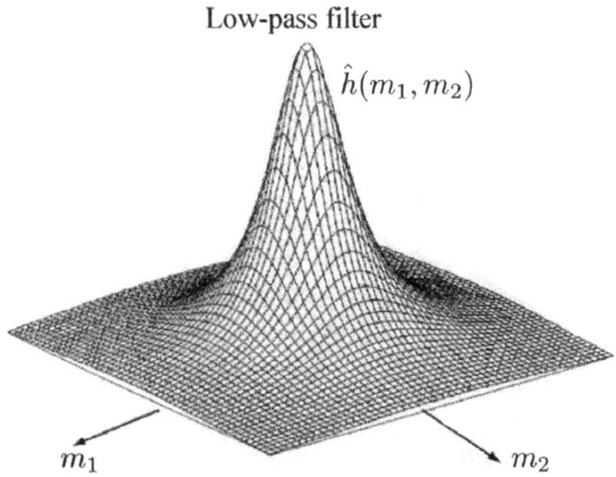

Figure 2.10. *Blurring filter/low-pass filter in the frequency domain*

NOTE.– Although convolution with a Gaussian results in a blurring effect, *it would be wrong to assume that convolution is always associated with a blurring action.* As we saw earlier, convolution, alongside the Fourier multiplier, constitutes a prototype for all stationary operators, which may blur a signal or enhance its contrast.

2.10. Summary

In this chapter, we considered the space $\ell^2(\mathbb{Z}_N)$ composed of N-periodic sequences with complex values, isomorphic to $\mathbb{C}^N$.

We introduced a special basis in this space, made up of the complex exponentials generated by the consecutive powers of the N-th complex roots of the unit. This basis is used to construct the Fourier basis of $\ell^2(\mathbb{Z}_N)$. We interpreted the elements of this basis as harmonic waves, oscillating at frequencies which are multiples of a fundamental one.

The Fourier coefficients of an element in $\ell^2(\mathbb{Z}_N)$ are its components with regard to the Fourier basis. As these coefficients are complex, their magnitude must be used to determine the importance of a harmonic in relation to a certain frequency when reconstructing (or synthesizing) the element itself. The set of magnitudes of the Fourier coefficients is known as the spectrum of an element in $\ell^2(\mathbb{Z}_N)$.

The DFT is the endomorphism of $\ell^2(\mathbb{Z}_N)$ which associates an element of $\ell^2(\mathbb{Z}_N)$ with the sequence of its Fourier coefficients. The DFT is actually an isomorphism, and its inverse is known as the IDFT.

The DFT may be associated with a matrix, known as a Sylvester matrix; this matrix is a Vandermonde matrix, that is, all of the lines and columns in the matrix can be obtained through geometric progressions.

We presented an interpretation of these concepts in the context of signal theory, notably highlighting the fact that the highest harmonic oscillation frequency in a discrete signal obtained from N samples is $N/2$ (or half of its integer part if N is odd); this is the Nyquist frequency.

The DFT transforms the shift operation into a multiplication by a phase factor, that is, a complex exponential with unit magnitude; this implies that the signal spectrum is shift-invariant.

Convolution is transformed by the DFT into a pointwise product, allowing the convolution operator to be expressed diagonally in the Fourier space.

Finally, we saw that the DFT can be used to diagonalize stationary operators, that is, operators which commutate with shift operators. Theorem 2.9 can be used to fully characterize a stationary operator as a convolution or as a Fourier multiplier and to determine the eigenvalues of this operator.

3

Lebesgue's Measure and Integration Theory

In this chapter, we shall present the most essential elements of measure theory and integration. Our aim here is simply to establish clear and unambiguous notation and a common vocabulary.

What follows is a deliberately brief summary. Readers who have not yet studied this important branch of mathematics may wish to look elsewhere for a more detailed introduction to measure theory and integration.

Two excellent reference works in this domain are Briane and Pagès (1998) and Bartle (1966).

3.1. Riemann versus Lebesgue

The main difference between the Riemann and Lebesgue approaches is shown in Figure 3.1.

The key to Riemann integration lies in approximating the area of the surface between the x axis and the curve of a function f using small rectangles $[a_{i-1}, a_i] \times [0, \Phi_i]$ with their base on the x axis, of a height Φ_i close to the average height of function f over $[a_{i-1}, a_i]$.

Lebesgue's integration theory differs in that the first stage involves breaking down the y axis into small intervals $[b_{j-1}, b_j]$; the surface below the curve f is then approximated using:

$$\int_a^b f \approx \sum_j \frac{b_{j-1} + b_j}{2} \cdot \text{length}\left(\{x \ : \ b_{j-1} \leqslant f(x) \leqslant b_j\}\right)$$

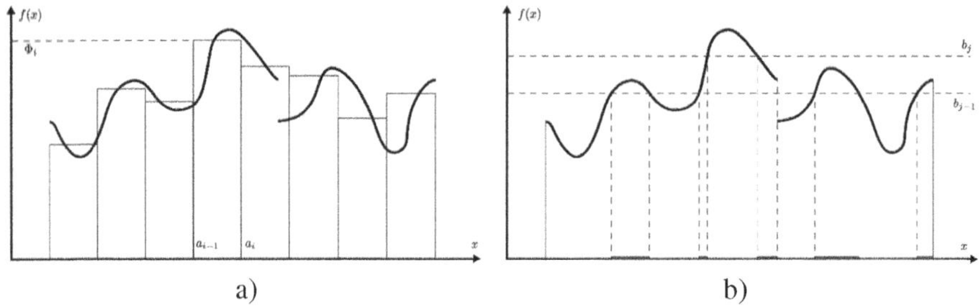

Figure 3.1. *Riemann and Lebesgue integration. For a color version of this figure, see www.iste.co.uk/provenzi/spaces.zip*

The main difficulty lies in the fact that the sets:

$$E_j = \{x \in [a,b] \ : \ b_{j-1} \leqslant f(x) \leqslant b_j\}$$

shown in red in Figure 3.1(b), are generally not intervals, and it can be complicated, if not (as in certain cases) impossible, to associate them with a length or measure.

The development of measure theory was motivated by the need to create a theory of integration using the strategy described above. This approach is far longer and more complicated than Riemann integration; however, Lebesgue integration presents a significant advantage in terms of generality, and the properties that can be proved are far more powerful.

3.2. σ-algebra, measurable space, measures and measured spaces

In order to define a Lebesgue integral, we must first define the sets and functions which can be measured. The definitions and results below, based on work carried out in the early 20th century, make up the necessary formalization.

Let X be a set. A *σ-algebra* on X is a collection $\mathcal{A}$ of subsets of X, that is $\mathcal{A} \subseteq \mathcal{P}(X)$, which verifies the following properties:

– $\varnothing, X \in \mathcal{A}$;

– $\mathcal{A}$ is closed under complementation: $E \in \mathcal{A} \implies E^c \in \mathcal{A}$;

– $\mathcal{A}$ is closed under countable unions: $(E_n)_{n\in\mathbb{N}} \in \mathcal{A} \Rightarrow \bigcup_{n\in\mathbb{N}} E \in \mathcal{A}$.

This definition implies that $\mathcal{A}$ is closed under countable intersection.

SIMPLE EXAMPLES.–

– $\mathcal{A} = \mathcal{P}(X)$: σ-algebra of the power set of X.
– $\mathcal{A} = \{\varnothing, X\}$: the minimal σ-algebra over X.

Mathematicians working on measure theory have proved that the defining properties of a σ-algebra are necessary and sufficient to "measure" the sets contained in the σ-algebra itself, in a sense which will be defined below. For this reason, the pair $(X, \mathcal{A})$ is called a *measurable space* and the elements of $\mathcal{A}$ are *measurable sets*.

One further concept must be introduced before we can examine a meaningful example of a measurable space: that of the *ordering relation between σ-algebras*. If every element in a σ-algebra $\mathcal{A}_1$ is contained in the σ-algebra $\mathcal{A}_2$, then $\mathcal{A}_1$ is said to be smaller than $\mathcal{A}_2$ and we write $\mathcal{A}_1 \subset \mathcal{A}_2$. This concept is used to define the smallest σ-algebra generated by a collection of power sets: taking $S \subset \mathcal{P}(X)$, the intersection of all σ-algebras which contain S is known as the *σ-algebra generated by S*.

The case of a topological set X is particularly interesting, and merits closer attention.

The existence of a topology means that we can define the concept of an open part of X. Taking $\tau \subseteq \mathcal{P}(X)$ to be the open sets of X, we clearly see that τ is not a σ-algebra, since the complement of an open set is a closed set. However, we can consider the σ-algebra generated by τ, called the *Borel σ-algebra*[1]and noted $\mathcal{B}(X)$. Each element in this algebra – which is a subset of X – is called a *Borel set*.

Once we have a measurable space $(X, \mathcal{A})$, the concept of a positive measure, or simply a *measure*, μ can be defined as a function $\mu : \mathcal{A} \to [0, +\infty]$ such that:

– $\mu(\varnothing) = 0$;

– μ is σ-additive (or countably additive): if $(E_n)_{n\in\mathbb{N}}$ is a countable family of two-by-two disjoint elements in $\mathcal{A}$, then:

$$\mu\left(\bigcup_{n\in\mathbb{N}} A_n\right) = \sum_{n\in\mathbb{N}} \mu(A_n)$$

The triple $(X, \mathcal{A}, \mu)$ is said to be a *measure space*. When the σ-algebra $\mathcal{A}$ and the measure μ are clearly specified, they are often omitted and one simply writes X.

One very simple, but meaningful, example of a measure is given by the Dirac measure in the measurable space $(\mathbb{R}, \mathcal{B}(\mathbb{R}))$, that is, $\mathbb{R}$ with the Borel σ-algebra. The *Dirac measure* centered on $x_0 \in \mathbb{R}$ is defined by: $\delta_{x_0} : \mathcal{B}(\tau) \to \{0, 1\}$:

$$\delta_{x_0}(E) = \begin{cases} 1 \text{ if } x_0 \in E \\ 0 \text{ if } x_0 \notin E \end{cases} \qquad \forall E \in \mathcal{B}(\mathbb{R})$$

1 This σ-algebra cannot be described explicitly.

Since $\mathbb{R}$ itself is an element in $\mathcal{B}(\mathbb{R})$, $\delta_{x_0}(\mathbb{R}) = 1$, and the Dirac measure of $\mathbb{R}$ is 1, independently of the starting point. It is therefore an example of a *finite measure*, that is, the measure of the entire space is finite.

Measures are generally σ-finite, rather than simply finite. Given a measure space $(X, \mathcal{A}, \mu)$, μ is said to be a *σ-finite measure* if X can be written as the countable union of measurable sets $(E_n)_{n\in\mathbb{N}} \subset X$ with a finite measure, that is:

$$X = \bigcup_{n\in\mathbb{N}} E_n, \quad \mu(E_n) < +\infty \quad \forall n \in \mathbb{N}$$

Several different techniques exist for constructing a measure, but these are not simple and cannot be described in short form. Readers may wish to consult the volume cited in the preface, or any other work on measure theory.

3.3. Measurable functions and almost-everywhere properties (a.e)

The next step is to introduce *the morphisms of measurable spaces*, that is, applications between measurable spaces which preserve measurability.

Let $(X_1, \mathcal{A}_1)$, $(X_2, \mathcal{A}_2)$ be two measurable spaces and $f : X_1 \to X_2$ an arbitrary function. f is a *measurable function* (with respect to the chosen σ-algebras $\mathcal{A}_1$ and $\mathcal{A}_2$) if the reciprocal image *via* f of any element of the σ-algebra $\mathcal{A}_2$ is included in $\mathcal{A}_1$, that is[2]:

$$E \in \mathcal{A}_2 \implies f^{-1}(E) \in \mathcal{A}_1.$$

This is equivalent, by definition, to stating that the reciprocal image *via* f of a measurable set of X_2 (with respect to $\mathcal{A}_2$) is a measurable set of X_1 (with respect to $\mathcal{A}_1$).

REMARKS.–

– Continuous functions between two topological spaces are clearly measurable with respect to their Borel σ-algebras.

– Without other specifications, whenever we consider real-valued functions, that is $f : X \to \mathbb{R}$, where $(X, \mathcal{A})$ is a measurable space, we fix the Borel σ-algebra on $\mathbb{R}$ and we test the measurability of f with respect to this choice.

– A complex-value function $f : X \to \mathbb{C}$ is measurable if both its real and imaginary parts are measurable.

2 Note the similarity between this definition and that of a continuous function, in the topological sense of the term.

Let us now recall the crucial concept of properties which are defined almost everywhere. A function f defined on a measure space $(X, \mathcal{A}, \mu)$ has a property which holds *almost everywhere* (written *a.e.*) if f possesses this property on $X \backslash E$, where $E \in \mathcal{A}$ has a measure of zero: $\mu(E) = 0$.

EXAMPLES.–

– f, g: measurable functions defined on $(X, \mathcal{A}, \mu)$, then $f = g$ a.e. if $f(x) = g(x)\ \forall x \in U \in \mathcal{A}$ and $\mu(X \backslash U) = 0$.

– f is the a.e. pointwise limit of the sequence $(f_n)_{n \in \mathbb{N}}$ if $\lim_{n \to +\infty} f_n(x) = f(x)$ $\forall x \in U \in \mathcal{A}$ and $\mu(X \backslash U) = 0$.

3.4. Integrable functions and Lebesgue integrals

Given a measure space $(X, \mathcal{A}, \mu)$, the integral of a measurable function defined by real or complex functions is relatively simple to obtain. We start by considering a special function, the *indicator (or characteristic) function* of a set $E \in \mathcal{A}$: $\chi_E : X \to \{0, 1\}$:

$$\chi_E(x) = \begin{cases} 1 & \text{if } x \in E \\ 0 & \text{if } x \notin E \end{cases}$$

An equivalent notation is $\mathbb{1}_E$.

Indicator functions are used to define *simple functions* or *step functions via* linear combination. More precisely, taking $(E_k)_{k=1}^n$ to be a finite and disjoint partition of X, that is, $E_k \cap E_{k'} = \varnothing\ \forall k \neq k'$ and

$$\bigcup_{k=1}^{n} E_k = X,$$

a simple function $s : X \to \mathbb{R}$ is defined as:

$$s = \sum_{k=1}^{n} c_k \chi_{E_k}$$

$s(x) = c_k\ \forall x \in E_k$; hence s can only take a finite number of values; if X is a subset of $\mathbb{R}$, then s is a piecewise constant function.

The natural definition of the Lebesgue integral of a simple function is:

$$\int_X s d\mu = \sum_{k=1}^{n} c_k \mu(E_k)$$

Note that, without the definition of the set measure E_k, the integral of s would not be correctly defined.

The importance of simple functions is expressed in Theorem 3.1.

THEOREM 3.1.– Let $(X, \mathcal{A}, \mu)$ be a measure space and $f : X \to [0, +\infty]$ a measurable and non-negative function. f can be approximated from below using a series of simple functions, that is, $\exists\ (s_n)_{n\in\mathbb{N}}$, with s_n a simple function; such that $(s_n)_{n\in\mathbb{N}} \nearrow f$, that is:

1) $0 \leqslant s_0(x) \leqslant s_1(x) \leqslant \ldots \leqslant s_n(x) \leqslant \ldots \leqslant f(x)\ \forall x \in X$;

2) $\lim\limits_{n\to+\infty} s_n(x) = f(x)\ \forall x \in X$ (pointwise limit). If f is bounded, then the convergence of the sequence $(s_n)_{n\in\mathbb{N}}$ toward f is uniform.

The proof of this theorem is both elegant and informative, showing that the sequence of simple functions is given by:

$$s_n(x) = \begin{cases} \frac{k}{2^n} \text{ if } \frac{k}{2^n} \leqslant f(x) \leqslant \frac{k+1}{2^n}, \text{ for } k = 0, 1, \ldots, 2^n - 1 \\ 2^n \text{ if } 2^n \leqslant f(x) \end{cases}$$

This fundamental theorem makes it possible to define the *integral of a measurable non-negative function* $f : X \to [0, +\infty]$ as:

$$\boxed{\int_X f d\mu = \sup_{0\leqslant s\leqslant f} \int_X s\, d\mu} \qquad s \text{ simple}$$

f is said to be (Lebesgue) *integrable* if $\int_X f d\mu < +\infty$.

If $f : X \to \overline{\mathbb{R}}$ is measurable, then its integral can be defined by considering its positive part:

$$f_+(x) = \begin{cases} f(x) & \text{if } f(x) \geqslant 0 \\ 0 & \text{if } f(x) < 0 \end{cases}$$

and its negative part:

$$f_-(x) = \begin{cases} -f(x) & \text{if } f(x) \leqslant 0 \\ 0 & \text{if } f(x) > 0, \end{cases}$$

note that both these functions are positive-valued. Since $f = f_+ - f_-$, if f_+ and f_- are integrable, then we can define the *integral of a measurable function with extended real values* as:

$$\boxed{\int_X f d\mu = \int_X f_+ d\mu - \int_X f_- d\mu}$$

The same strategy is used for measurable functions $f : X \to \mathbb{C}$, but using the positive and negative parts of the real part $\mathfrak{Re}(f)$ and the imaginary part $\mathfrak{Im}(f)$. The integral is thus defined as:

$$\int_X f d\mu = \int_X \mathfrak{Re}(f) d\mu + i \int_X \mathfrak{Im}(f) d\mu$$

Absolute integrability is a necessary and sufficient condition for integrability of a real or complex valued function:

$$\boxed{\int_X f d\mu < +\infty \iff \int_X |f| d\mu < +\infty}$$

3.5. Characterization of the Lebesgue measure on $\mathbb{R}$ and sets with a null Lebesgue measure

As we have seen, the construction of a measure is generally not trivial. However, given the importance of the Lebesgue measure on $\mathbb{R}$, it is helpful to provide a brief summary of the characteristics of this measure. Remarkably, a theorem exists which provides the characterization of the Lebesgue measure using certain properties. Before quoting the result, we recall some definitions.

– *Borel measure*: Let X be a topological space, and take the measurable space $(X, \mathcal{B}(X))$, where $\mathcal{B}(X)$ is the Borel σ-algebra. A measure μ defined on this space is said to be a *Borel measure* if it associates a finite number with each compact subset K of X;

– *Regular Borel measure*: A Borel measure is regular if, for any Borel set $E \in \mathcal{B}(X)$, we have:

1) $\mu(E) = \sup\{\mu(K),\ K \subset E,\ K \text{ compact}\}$;

2) $\mu(E) = \inf\{\mu(O),\ E \subset O,\ O \text{ open}\}$.

Consider now $(\mathbb{R}, \mathcal{B}(\mathbb{R}), \mu)$. μ is a *shift-invariant measure* if:

$$\mu(E + a) = \mu(E)$$

for any Borel set $E \in \mathcal{B}(\mathbb{R})$ and all $a \in \mathbb{R}$, where $E + a = \{x \in \mathbb{R} : x = e + a, \text{ where } e \in E\}$.

We can now quote the theorem that provides the characterization of the Lebesgue measure on $\mathbb{R}$, noted m.

THEOREM 3.2.– If a measure on $(\mathbb{R}, \mathcal{B}(\mathbb{R}), \mu)$ has the following properties:

1) μ is a regular Borel measure;
2) μ is shift-invariant;
3) μ is normalized, that is $\mu[0, 1] = 1$;

then μ is the Lebesgue measure m.

Thus, we can say that the *Lebesgue measure* on $(\mathbb{R}, \mathcal{B}(\mathbb{R}))$ *is a regular, shift-invariant, normalized Borel measure*; this also implies that $m[a, b] = b - a$.

A further consequence of this theorem is that *the Lebesgue measure is σ-finite*: $\mathbb{R}$ can be covered by a partition of compact intervals $[-n, n]$ with $n \in \mathbb{N}$, all of which possess finite measures ($\mu[-n, n] = 2n$).

Generalization of the Lebesgue measure on $\mathbb{R}$ to $\mathbb{R}^n$ is straightforward, and we can prove that, if a function is Riemann-integrable on $\mathbb{R}^n$, it is also Lebesgue-integrable and the two integrals coincide.

Important examples of sets with null Lebesgue measure are given by hypersurfaces of dimension $n - 1$ in $\mathbb{R}^n$, such as two-dimensional (2D) surfaces in $\mathbb{R}^3$ and curves in $\mathbb{R}^2$. Regarding $\mathbb{R}$, since $\mathbb{R}$ has the cardinality of continuous, the subsets of $\mathbb{R}$ with lower cardinality, that is, countable or finite subsets, have null Lebesgue measure, in particular:

$$m(\mathbb{N}) = m(\mathbb{Z}) = m(\mathbb{Q}) = 0$$

This means that even if we eliminated from a measurable set in $\mathbb{R}$, for example an interval $[a, b]$, a countably infinite number of points, its Lebesgue measure would not change.

This property means that the class of Lebesgue-integrable functions is much broader than that of Riemann-integrable functions. Take the case of a piecewise continuous function on a set with a finite or countable number of jump discontinuities: this function has no Riemann integral. It does, however, have a Lebesgue integral, which is the algebraic sum of the Riemann integrals of each section for which the function is continuous. As the number of discontinuities is finite or countable, we can simply ignore them, since they constitute a set of null Lebesgue measure and therefore have no effect on the final result of integration.

It is important to remember that Lebesgue integration theory does not provide more advanced tools for the *explicit calculation* of integrals, except in certain very specific cases; however, as just discussed, it allows us to give a meaningful sense to integrals of functions which are much less regular than is required for Riemann integration.

This result, along with the crucial theorems presented in section 3.6, gives Lebesgue integration theory a significant advantage over that of Riemann.

3.6. Three theorems for limit operations in integration theory

In this section, we shall summarize the three most important theorems concerning the limit operation in integration theory. These will be used in Chapter 4.

In these theorems, we shall take $(X, \mathcal{A}, \mu)$ to be an arbitrary fixed measure space.

THEOREM 3.3 (Monotone convergence theorem – Beppo Levi).– Let $(f_n)_{n\in\mathbb{N}}$, with $f_n : X \to \mathbb{R}$, be a *monotonically increasing* sequence of integrable functions. If the sequence of integrals is *bounded*, that is:

$$\forall n \in \mathbb{N},\ \exists K \in \mathbb{R} \text{ such that } \int_X f_n \, d\mu < K$$

then $\exists \lim\limits_{n\to+\infty} f_n(x) < +\infty$ a.e. Furthermore, if we define the limit function $f : X \to \mathbb{R}$ as:

$$f(x) = \begin{cases} \lim\limits_{n\to+\infty} f_n(x) & \textit{if the limit is finite} \\ 0 & \textit{otherwise} \end{cases}$$

then f is integrable, and the limit and integral commute:

$$\int_X f(x) \, d\mu = \lim_{n\to+\infty} \int_X f_n(x) \, d\mu.$$

Let us now pass to Fatou's lemma by first recalling that, given an arbitrary sequence $(x_n)_{n\in\mathbb{N}}$ of real numbers, $\liminf$ is the *limit inferior* of the sequence, that is:

$$\liminf_{n\in\mathbb{N}}(x_n) = \inf\{x \in \mathbb{R} \ : \ x \text{ limit point for } (x_n)_{n\in\mathbb{N}}\}$$

THEOREM 3.4 (Fatou's lemma).– Let $(f_n)_{n\in\mathbb{N}}$, with $f_n : X \to \mathbb{R}$, be a sequence of positive integrable functions and let $\liminf\limits_{n\to+\infty} \int_X f_n \, d\mu < +\infty$. The function f defined by:

$$f(x) = \begin{cases} \liminf\limits_{n\to+\infty} f_n(x) & \textit{if the limit inferior is finite} \\ 0 & \textit{otherwise} \end{cases}$$

is integrable, moreover, the following inequality holds:

$$\int_X f \, d\mu \leqslant \liminf_{n\to+\infty} \int_X f_n \, d\mu.$$

THEOREM 3.5 (Dominated convergence theorem – Henri Lebesgue).– Let $(f_n)_{n\in\mathbb{N}}$, where $f_n : X \to \mathbb{R}$, be sequence of measurable functions, and let $\Phi : X \to \mathbb{R}$ be a positive and integrable function such that:

$$|f_n| \leqslant \Phi \quad e.a \ \forall n \in \mathbb{N}$$

If the real sequence $(f_n(x))_{n\in\mathbb{N}}$ is convergent $\forall x \in X$ and if $f(x) = \lim_{n\to+\infty} f_n(x)$, then f_n and f are integrable and the limit and the integral commute, that is:

$$\int_X f(x)\, d\mu = \lim_{n\to+\infty} \int_X f_n(x)\, d\mu$$

3.7. Summary

In this chapter, we provided a brief overview of key elements of measure theory and Lebesgue integration, touching on subjects such as σ-algebra, measurable sets, measures, measure spaces and measurable functions.

Particular attention was paid to the Borel σ-algebra in a topological space: this σ-algebra is generated by the open subsets of the space in question.

Almost-everywhere (a.e.) properties play an important role in measure theory: a property is verified a.e. if it is valid on a measurable subset such that the measure of its complementary set is null.

We saw that the Lebesgue measure m on $\mathbb{R}$ can be characterized with respect to the Borel σ-algebra as a regular, normalized and shift-invariant measure. Remarkable examples of null Lebesgue measures in $\mathbb{R}$ include countable sets, specifically $m(\mathbb{N}) = m(\mathbb{Z}) = m(\mathbb{Q}) = 0$.

Given a measure, the definition of the integral of a measurable function is straightforward and follows a standard approach. We begin by considering simple (or step) functions, which are linear combinations of characteristic functions of measurable sets. Simple functions approach any non-negative measurable function from below. This result is essential, allowing us to define the integral of non-negative measurable functions as the *sup* of the integral of simple functions which are not greater than the function itself. This definition is extended to arbitrary real-valued functions by using their positive and negative parts, and to complex-valued functions by using their real and imaginary parts.

Finally, we outlined the three fundamental theorems concerning the relation between limits and integrals in Lebesgue theory: the monotonic and dominated convergence theorems (developed by Levi and Lebesgue, respectively) and Fatou's lemma.

4

Banach Spaces and Hilbert Spaces

In this chapter, we shall consider normed or inner product spaces of infinite dimensions. Particular attention will be paid to "complete" spaces, for which several crucial theorems – which do not hold for non-complete spaces of infinite dimensions – can be formulated.

Before we can begin our analysis, it is important to note that all of the properties described previously for inner product spaces of finite dimension which rely solely on the algebraic nature of the inner product remain valid for infinite-dimensional vector spaces. For example:

– a family of orthogonal vectors is free;

– if $\langle x, z\rangle = \langle y, z\rangle \; \forall z$, then vectors x and y necessarily coincide;

– the null vector is the only vector which is orthogonal to all other vectors;

– the Gram-Schmidt orthonormalization procedure can be iterated, guaranteeing that an infinite system of mutually orthogonal vectors with a unitary norm will be obtained from any given infinite set of vectors.

The proofs for the first three properties are identical to those used for finite-dimensional vector spaces. The proof of the final property relies on the *Zorn lemma*.

Results for finite sums are harder to generalize; in this case, we need to take account of topological arguments in addition to algebraic considerations.

As we shall see, the definition and analysis of Banach and Hilbert spaces rely primarily on the analysis of the compatibility between the linear and topological structures of a normed or inner product vector space. For this reason, we start by recalling the concept of topology in such spaces.

4.1. Metric topology of inner product spaces

As we have seen, all inner product spaces V can be assigned a norm, which is canonically induced from the scalar product. Using this norm, it will always, canonically, be possible to define a *distance* or *metric* on V:

$$d(x,y) = \|x-y\| = \sqrt{\langle x-y, x-y\rangle}, \qquad \forall x,y \in V$$

Function d possesses the following properties, $\forall u,x,y \in V$:

1) $d(x,y) \geqslant 0$ and $d(x,y) = 0 \iff x = y$;

2) $d(x,y) = d(y,x)$ (symmetry);

3) $d(x,y) \leqslant d(x,z) + d(z,y)$ (triangular inequality).

DEFINITION 4.1 (Metric vector space).– *A metric vector space is a pair (V,d) given by a vector space V and a function, the distance $d : V \times V \to \mathbb{R}_0^+ = [0,+\infty)$, which satisfies the three properties given above.*

An inner product space is thus automatically a normed vector space and possesses a distance, independently of whether the scalar product is real or complex. As we shall see in this chapter, the converse is true if and only if the norm satisfies the parallelogram formula.

The existence of a metric means that it is possible to establish relationships between points and subsets in a space which go further than simple notion of a point belonging to a set. As we know, these relationships form the basis for constructing a topology. Reminders of a number of common definitions are given below, establishing clear notation and naming conventions for the rest of this chapter.

DEFINITION 4.2.– *Let $(V, \langle , \rangle)$ be an inner product space and $\| \ \|$ the associated norm. Then:*

– *a* neighborhood (open) of $x \in V$ of radius ε *is the subset of V defined by:*

$$U_\varepsilon(x) = \{y \in V \ : \ \|x-y\| < \varepsilon\}$$

if $\| \ \|$ is the Euclidean norm, then $U_\varepsilon(x)$ is a sphere (open) centered in x and of radius ε. By extension, $U_\varepsilon(x)$ is often called a ball or sphere (open), *and we write $B(x,\varepsilon)$, for any norm $\| \ \|$;*

– *a subset* $O \subseteq V$ *is said to be* open *if:*

$$\forall x \in O\ \exists \varepsilon > 0 \text{ such that } y \in U_\varepsilon(x) \implies y \in O$$

– *a subset* $F \subseteq V$ *is said to be* closed *if its complement* $F^c = V \backslash F$ *is open. Remember that this is the same as saying that any convergent sequence of elements in* F *will reach its limit within* F*;*

– *the* closure *of* $E \subset V$ *is* $\overline{E} = \bigcap_{\alpha \in I} E_\alpha$*, where* E_α *is a closed subset of* V *containing* E*.* $\overline{E}$ *is the smallest closed subset of* V *which contains* E*;*

– *the* border *(or* spherical surface*) of* $U_\varepsilon(x)$ *is the subset of* V *defined by:*

$$\partial U_\varepsilon(x) = \{y \in V \ :\ \|x - y\| = \varepsilon\}$$

using the symbol $B(x, \varepsilon)$*, the border is noted* $\partial B(x, \varepsilon)$*;*

– *the closed* neighborhood (or ball, or sphere) of radius ε of $x \in V$ *is the subset of* V *defined by:*

$$\overline{U_\varepsilon(x)} = \{y \in V \ :\ \|x - y\| \leqslant \varepsilon\}$$

using the symbol $B(x, \varepsilon)$*, we can write* $\overline{B(x, \varepsilon)}$*.*

We also recall that a topology on V is a set of parts of V containing V itself and $\varnothing$, which is stable with respect to arbitrary unions and finite intersections. The topology generated by the opens in V is the smallest topology which contains the open sets in V. Using this topology, with respect to the opens defined above, V is a topological space.

The topology of V is *metric*, that is the open sets are defined using a distance function. We recall that this guarantees that the topology will be *separated*, that is, for all pairs $x, y \in V$, $x \neq y$, there exist two neighborhoods $U(x)$ and $V(y)$, of different or equal radius, such that $U(x) \cap V(y) = \varnothing$, and we say that the points are separated by these neighborhoods.

A standard result in topology guarantees the *uniqueness of the limit* of sequences in a separated topology; hence, if sequences of vectors in V converge, they have a single limit.

We now recall the definition of convergence for a sequence in the topology of V.

DEFINITION 4.3.– *Let* $(V, \|\ \|)$ *be a normed vector space. A sequence of vectors* $(x_n)_{n \in \mathbb{N}} \subset V$ *is* convergent, *or* convergent in norm $\|\ \|$, *toward the limit* x *if:*

$$\forall \varepsilon > 0\ \exists N_\varepsilon > 0 \ :\ n \geqslant N_\varepsilon \implies \|x_n - x\| < \varepsilon$$

that is if, from $n = N_\varepsilon$, $x_n \in U_\varepsilon(x)$. *This can be represented using the more compact notation:*

$$\lim_{n\to+\infty} x_n = x, \quad \|x_n - x\| \underset{n\to+\infty}{\longrightarrow} 0$$

REMARK.– Requiring the inequality $\|x_n - x\| < \varepsilon$ to be valid $\forall\varepsilon > 0$ enables us to reformulate the definition of convergent, adding a *strictly positive, finite multiplication constant* to ε, that is, $x_n \underset{n\to+\infty}{\longrightarrow} x$ if:

$$\forall\varepsilon > 0\, \exists N_\varepsilon > 0 \text{ and } \exists m \in (0, +\infty) \,:\, n \geqslant N_\varepsilon \implies \|x_n - x\| < m\varepsilon$$

If the property is valid for all positive and arbitrarily small ε, then we can consider that $\tilde{\varepsilon} = \frac{\varepsilon}{m}$ and redefine the convergence with respect to $\tilde{\varepsilon}$:

$$\forall\tilde{\varepsilon} > 0\, \exists N_{\tilde{\varepsilon}} > 0 \,:\, n \geqslant N_{\tilde{\varepsilon}} \implies \|x_n - x\| < \tilde{\varepsilon}$$

This is possible because using the symbol ε or $\tilde{\varepsilon}$ is insignificant; the two quantities can be as small as we wish, so the two definitions are equivalent.

In a metric topology, the uniqueness of the limit follows simply from the triangular inequality. If $x_n \underset{n\to+\infty}{\longrightarrow} x$ *and* $x_n \underset{n\to+\infty}{\longrightarrow} y$*, then:*

$$0 \leqslant d(x, y) \leqslant d(x, x_n) + d(x_n, y) \underset{n\to+\infty}{\longrightarrow} 0 + 0 = 0$$

In what follows, this consideration will be referred to using the standard expression "*due to the arbitrarity of* ε...".

It is also helpful to recall the concept of density of a subset in a normed vector space. Proof of the equivalence of the properties expressed in Definition 4.4 can be found in most works on the subject of topology.

DEFINITION 4.4 (density).– *Let* $(V, \|\ \|)$ *be a normed vector space. A subset* $E \subset V$ *is dense in* V *if one of the following propositions is verified:*

1) $\forall x \in V, \exists (x_n)_{n\in\mathbb{N}} \subset E \,:\, x_n \underset{n\to+\infty}{\longrightarrow} x$ *(i.e.* $\|x_n - x\| \underset{n\to+\infty}{\longrightarrow} 0$*), that is: any subset in* V *can be indefinitely approached by a sequence of elements in E, and is the limit of this sequence;*

2) $\forall x \in V$ *and* $\forall\varepsilon > 0$, $\exists y \in E : \|x - y\| < \varepsilon$*, that is, for every element* x *in* X *there exists an element* y *in* E *with an arbitrarily small distance from* x*;*

3) V *is the closure of E:* $\boxed{\overline{E} = V}$.

We end the recap of classical notions with the concept of continuity of a function between metric spaces, along with a classic result which says that we can characterize continuity of a function *via* its action on sequences.

DEFINITION 4.5 (limits and continuity of functions between metric spaces).– *Let X and Y be two arbitrary metric spaces, $\bar{x} \in X$ and $\ell \in Y$, then:*

$$\boxed{\lim_{x \to \bar{x}} f(x) = \ell} \iff \forall \varepsilon > 0 \; \exists \delta_\varepsilon > 0 \; : \; x \in U_{\delta_\varepsilon}(\bar{x}) \cap X \quad \Longrightarrow \quad f(x) \in U_\varepsilon(\ell) \cap Y$$

that is, the limit of f in $\bar{x}$ is ℓ if f transforms the points of X which are arbitrarily close to $\bar{x}$ into points of Y which are arbitrarily close to ℓ.

If $\ell = f(\bar{x})$, then the function $f : X \to Y$ is said to be continuous in $\bar{x} \in V$. In explicit terms:

$$\forall \varepsilon > 0 \; \exists \delta_\varepsilon > 0 \; : \; x \in U_{\delta_\varepsilon}(\bar{x}) \cap X \quad \Longrightarrow \quad f(x) \in U_\varepsilon(f(\bar{x})) \cap Y$$

f is continuous on X if it is continuous at every point in X.

THEOREM 4.1 (Sequential continuity).– The function $f : X \to Y$, with (X, d_X) and (Y, d_Y) arbitrary metric spaces, is continuous in $\bar{x} \in X$ if and only if:

$$\forall (x_n)_{n \in \mathbb{N}} \subseteq X \;\; \textit{such that}$$

$$\lim_{n \to +\infty} x_n = \bar{x} \implies \lim_{n \to +\infty} f(x_n) = f\left(\lim_{n \to +\infty} x_n\right) = f(\bar{x})$$

that is:

$$\forall (x_n)_{n \in \mathbb{N}} \subseteq X \; : \; d_X(x_n, \bar{x}) \underset{n \to +\infty}{\longrightarrow} 0 \quad \Longrightarrow \quad d_Y(f(x_n), f(\bar{x})) \underset{n \to +\infty}{\longrightarrow} 0$$

We see that the limit operation on the sequence $(x_n)_{n \in \mathbb{N}}$ is carried out in the metric space (X, d_X), while the operation on the sequence $(f(x_n))_{n \in \mathbb{N}}$ is carried out in the metric space (Y, d_Y).

The possibility of switching the order of the limit and the (continuous) function in the expression:

$$\boxed{\lim_{n \to +\infty} f(x_n) = f\left(\lim_{n \to +\infty} x_n\right)}$$

is essential for proving many of the results presented later.

PROOF.–

$\boxed{\Longrightarrow}$: let f be continuous in $\bar{x}$ and let $(x_n)_{n \in \mathbb{N}} \subseteq X$ be an arbitrary sequence of elements of X such that $\lim\limits_{n \to +\infty} x_n = \bar{x}$. Then, by definition of the limit of a sequence, for sufficiently large values of n, x_n belongs to a neighborhood of $\bar{x}$ of arbitrarily small radius $\delta > 0$: in other words, there exists $N \in \mathbb{N}$ such that $n \geqslant N \implies x_n \in U_\delta(\bar{x})$. On the other side, due to the continuity of f, the elements x_n belonging to the neighborhood $U_\delta(\bar{x})$ are transformed by f into points belonging to a neighborhood of $f(\bar{x})$ of arbitrarily small radius $\varepsilon > 0$, i.e. $n \geqslant N \implies f(x_n) \in U_\varepsilon(f(\bar{x})) \cap Y$, that is $\lim_{n\to+\infty} f(x_n) = f(\bar{x})$.

$\boxed{\Longleftarrow}$: we shall assume that, for all sequences $(x_n)_{n\in\mathbb{N}} \subseteq X$ such that $\lim_{n\to+\infty} x_n = \bar{x} \in X$, it holds that $\lim_{n\to+\infty} f(x_n) = f(\bar{x})$; we need to prove that this implies the continuity of f in $\bar{x} \in X$.

Using reductio ad absurdum, suppose that f is not continuous in $\bar{x}$: as we shall see, this results in a contradiction. Negation[1] of the continuity of f in $\bar{x}$ is equivalent to saying that $\forall\delta > 0$, $\exists\varepsilon_\delta > 0$ such that $x \in U_\delta(\bar{x}) \subset X$ implies $f(x) \notin U_{\varepsilon_\delta}(f(\bar{x}))$.

Since the values of δ are arbitrary, we may consider the sequence $(\delta_n)_{n\geqslant 1}$ defined by $\delta_n = \frac{1}{n}$ $\forall n \geqslant 1$, which implies the existence of a sequence $(x_n)_{n\geqslant 1} \subset X$ such that $x_n \in U_{\delta_n}(\bar{x})$ and $f(x_n) \notin U_{\varepsilon_{\delta_n}}(f(\bar{x}))$.

This leaves us with a contradiction: on the one hand, when $n \to +\infty$, $\delta_n \to 0$ and thus $x_n \to \bar{x}$, while on the other hand, $f(x_n) \underset{n\to+\infty}{\not\to} f(\bar{x})$, that is, the hypothesis that f is not continuous results in a sequence of elements in X which converges to $\bar{x}$ without $f(x)$ being convergent to $f(\bar{x})$. This contradicts our initial hypothesis, and thus the possibility that f is not continuous must be rejected. □

If V, W are two normed vector spaces, then they automatically constitute two metric spaces with respect to the distances canonically induced by the norm and definitions; the results presented above therefore remain valid.

4.2. Continuity of fundamental operations in inner product spaces

Given an inner product space with both a linear structure and metric topology, the question about the *compatibility* of these two structures is evidently important; in other words, we wish to know whether the linear operations of the vector space V, together with inner product and norm, are continuous in the topology of V generated by its inner product. The response to this question is affirmative, as Theorem 4.2 states.

1 Note that the negation of a mathematical proposition is performed by exchanging the universal and existential quantifiers and by considering the complementary affirmation of the initial proposition: thus, the negation of $(\forall A\ \exists B \implies C)$ is $(\forall B\ \exists A \implies \bar{C})$, where $\bar{C}$ is the negation of the affirmation C.

THEOREM 4.2.– Let $(V, \langle\ ,\ \rangle)$ be an inner product space on $\mathbb{K}$. We shall consider the topology induced by the inner product on V, the usual Euclidean topology on $\mathbb{K}$ and the product topology on $V \times V$ and $\mathbb{K} \times V$. Then:

– inner product:

$$\begin{aligned} \langle\ ,\ \rangle : V \times V &\longrightarrow \mathbb{K} \\ (x, y) &\longmapsto \langle x, y\rangle \end{aligned}$$

– norm:

$$\begin{aligned} \|\ \| : V &\longrightarrow \mathbb{R}_0^+ \\ x &\longmapsto \|x\| \end{aligned}$$

– sum:

$$\begin{aligned} + : V \times V &\longrightarrow V \\ (x, y) &\longmapsto x + y \end{aligned}$$

– and scalar multiplication:

$$\begin{aligned} \cdot_{\mathbb{K}} : \mathbb{K} \times V &\longrightarrow V \\ (k, x) &\longmapsto kx \end{aligned}$$

are *continuous functions.*

PROOF.– All of the proofs shown below involve majorizing a selected norm using an expression which contains the norm of the difference between a sequence and its bound, which evidently converges to 0.

– *Continuity of inner product*: we must prove that if $(x_n)_{n\in\mathbb{N}}$ and $(y_n)_{n\in\mathbb{N}}$ are any two sequences of elements in V which converge to x and y, respectively, then the sequence of scalars $(\langle x_n, y_n\rangle)_{n\in\mathbb{N}}$ converges to $\langle x, y\rangle$. To do this, we first write a simple algebraic manipulation which holds for all $n \in \mathbb{N}$:

$$\begin{aligned} &\langle x_n, y_n\rangle - \langle x, y\rangle = \langle x_n - x + x, y_n - y + y\rangle - \langle x, y\rangle \\ &= \langle x_n - x, y_n - y\rangle + \langle x_n - x, y\rangle + \langle x, y_n - y\rangle + \cancel{\langle x, y\rangle} - \cancel{\langle x, y\rangle} \\ &= \langle x_n - x, y_n - y\rangle + \langle x_n - x, y\rangle + \langle x, y_n - y\rangle \end{aligned}$$

We can write the following majorization:

$$|\langle x_n, y_n\rangle - \langle x, y\rangle| \leqslant |\langle x_n - x, y_n - y\rangle| + |\langle x_n - x, y\rangle| + |\langle x, y_n - y\rangle| \quad \forall n \in \mathbb{N}$$

and, from the Cauchy-Schwarz inequality:

$$|\langle x_n, y_n\rangle - \langle x, y\rangle| \leqslant \|x_n - x\|\|y_n - y\| + \|x_n - x\|\,\|y\| + \|x\|\,\|y_n - y\| \quad \forall n \in \mathbb{N}$$

As the equality holds for all $n \in \mathbb{N}$, the limit $n \to +\infty$ may be considered on both sides: by hypothesis, $\|x_n - x\| \underset{n\to+\infty}{\longrightarrow} 0$ and $\|y_n - y\| \underset{n\to+\infty}{\longrightarrow} 0$, so the right-hand side tends to 0, hence:

$$|\langle x_n, y_n\rangle - \langle x, y\rangle| \underset{n\to+\infty}{\to} 0$$

which proves the continuity of the inner product.

– *Continuity of the norm*: we must prove that if $(x_n)_{n\in\mathbb{N}}$ is an arbitrary sequence of elements in V which converges to x, then the sequence of positive real numbers $(\|x_n\|)_{n\in\mathbb{N}}$ converges to $\|x\|$. This can be done using the majorization of the norm provided by formula [1.3]:

$$|\|x_n\| - \|x\|| \leqslant \|x_n - x\|$$

but $\|x_n - x\| \underset{n\to+\infty}{\to} 0$, hence $\|x_n\| \underset{n\to+\infty}{\to} \|x\|$.

– *Continuity of the sum*: we must show that if $(x_n)_{n\in\mathbb{N}}$ and $(y_n)_{n\in\mathbb{N}}$ are any two sequences of elements in V which converge to x and y, respectively, then the sequence $(x_n + y_n)_{n\in\mathbb{N}}$ converges to $x + y$. To do this, we write:

$$\|(x_n+y_n)-(x+y)\| = \|(x_n-x)+(y_n-y)\| \leqslant \|x_n-x\|+\|y_n-y\| \underset{n\to+\infty}{\to} 0$$

– *Continuity of scalar multiplication*: we must show that if $(x_n)_{n\in\mathbb{N}}$ and $(k_n)_{n\in\mathbb{N}}$ are any two sequences of elements in V and $\mathbb{K}$, respectively, which converge to x and k, respectively, then the product sequence $(k_n x_n)_{n\in\mathbb{N}}$ converges to kx. Once again, an algebraic manipulation is involved:

$$\| \; k_n x_n - kx\| = \|k_n(x_n - x + x) - kx\| = \|k_n(x_n - x) + k_n x - kx\|$$

$$= \|k_n(x_n - x) + x(k_n - k)\| \leqslant |k_n|\|x_n - x\| + \|x\||k_n - k| \underset{n\to+\infty}{\to} 0 \quad \square$$

Let us consider the immediate consequences of this theorem. First, the continuity of the sum and scalar multiplication implies that the difference is also continuous, since $x - y = x + (-1)y$.

If $(x_n)_{n\in\mathbb{N}}$ and $(y_n)_{n\in\mathbb{N}}$ are two sequences in $(V, \langle\;,\;\rangle)$ which converge to elements of x and y, respectively, then the continuity of the inner product and the norm, taken alongside Theorem 4.1, give us the following formulas that will be used later:

$$\boxed{\lim_{n\to+\infty}\langle x_n, y_n\rangle = \langle \lim_{n\to+\infty} x_n, \lim_{n\to+\infty} y_n\rangle = \langle x, y\rangle} \qquad [4.1]$$

$$\boxed{\lim_{n\to+\infty}\|x_n\| = \left\|\lim_{n\to+\infty} x_n\right\| = \|x\|}. \qquad [4.2]$$

The case of series needs to be considered separately. First of all, let us recall the definitions of a series and of a convergent series.

DEFINITION 4.6.– *Given a sequence of vectors $(x_n)_{n\in\mathbb{N}} \subset V$, the* series *of general term x_n is the sequence of partial sums $(S_n)_{n\in\mathbb{N}}$, where $S_n = \sum_{k=0}^{n} x_k$, and we write:*

$$\sum_{n\in\mathbb{N}} x_n = \sum_{n=0}^{\infty} x_n = (S_n)_{n\in\mathbb{N}}$$

The series *$\sum_{n\in\mathbb{N}} x_n$ is said to be* convergent, *or convergent in norm $\| \ \|$, to the sum x if the sequence of partial sums $(S_n)_{n\in\mathbb{N}}$ is convergent to x, that is:*

$$x = \lim_{n\to+\infty} \sum_{k=0}^{n} x_k = \sum_{n\in\mathbb{N}} x_n \iff \lim_{n\to+\infty} \|S_n - x\| = 0 \iff \|S_n - x\| \underset{n\to+\infty}{\to} 0$$

The series $\sum_{n\in\mathbb{N}} x_n$ is said to be absolutely convergent[2] *if the sequence of the partial sums of the norms, that is $\left(\sum_{k=0}^{n} \|x_k\|\right)_{n\in\mathbb{N}}$, is convergent. In this case, we write: $\sum_{n\in\mathbb{N}} \|x_n\| < +\infty$.*

We observe that, since $S_n - x = \sum_{k=0}^{n} x_k - \sum_{k=0}^{\infty} x_n = -\sum_{k=n+1}^{\infty} x_k$, then:

$$\|S_n - x\| = \left\| \sum_{k=n+1}^{\infty} x_k \right\| \qquad [4.3]$$

hence the explicit definition of a convergent series in a normed vector space is:

$$\forall \varepsilon > 0\, \exists N_\varepsilon > 0 \,:\, n \geqslant N_\varepsilon \implies \left\| \sum_{k=n+1}^{\infty} x_k \right\| < \varepsilon$$

Given convergent series $\sum_{n\in\mathbb{N}} x_n$, $\sum_{m\in\mathbb{N}} y_m$, the fact that a series is the sequence of its partial sums means that we can write:

$$\boxed{\langle \sum_{n\in\mathbb{N}} x_n, \sum_{m\in\mathbb{N}} y_m \rangle = \langle \lim_{N\to+\infty} \sum_{n=0}^{N} x_n, \lim_{K\to+\infty} \sum_{m=0}^{K} y_m \rangle = \lim_{N,K\to+\infty} \sum_{n=0}^{N} \sum_{m=0}^{K} \langle x_n, y_m \rangle} \qquad [4.4]$$

2 The absolute convergence defined here becomes the normal convergence for the modulus of S_n when $V = \mathbb{R}$ or $V = \mathbb{C}$.

and:

$$\boxed{\left\|\sum_{n\in\mathbb{N}} x_n\right\| = \left\|\lim_{N\to+\infty}\sum_{n=0}^{N} x_n\right\| = \lim_{N\to+\infty}\left\|\sum_{n=0}^{N} x_n\right\|} \qquad [4.5]$$

Squaring the members of equation [4.5], we obtain:

$$\left\|\sum_{n\in\mathbb{N}} x_n\right\|^2 = \left\|\lim_{N\to+\infty}\sum_{n=0}^{N} x_n\right\|^2 = \left(\lim_{N\to+\infty}\left\|\sum_{n=0}^{N} x_n\right\|\right)^2 = \lim_{N\to+\infty}\left\|\sum_{n=0}^{N} x_n\right\|^2$$

having used the fact that $\left\|\sum_{n=0}^{N} x_n\right\| \in \mathbb{R}$ and the continuity of the square operation in $\mathbb{R}$ to exchange the limit with the square.

If we consider an *orthogonal family of vectors* $(u_n)_{n\in\mathbb{N}}$ in place of an arbitrary sequence $(x_n)_{n\in\mathbb{N}}$, the generalized Pythagorean theorem (Theorem 1.8) can be used to write $\lim_{N\to+\infty}\left\|\sum_{n=0}^{N} u_n\right\|^2 = \lim_{N\to+\infty}\sum_{n=0}^{N}\|u_n\|^2 = \sum_{n\in\mathbb{N}}\|u_n\|^2$, giving the following very helpful formula:

$$\boxed{\left\|\sum_{n\in\mathbb{N}} u_n\right\|^2 = \sum_{n\in\mathbb{N}}\|u_n\|^2} \qquad (u_n)_{n\in\mathbb{N}}\text{: orthogonal family of vectors} \qquad [4.6]$$

Formula [4.6] will be used extensively in Chapter 5. It is important to note that this formula does not generally hold if $(u_n)_{n\in\mathbb{N}}$ is not an orthogonal system of vectors and if we consider the norm rather than its square.

The *possibility to exchange the limit and inner product and norm operations* is crucial for proving many of the theorems that we shall see later. This consideration emphasized the *importance of the compatibility of the linear and topological structures* in an inner product space.

The result below is a first example of the usefulness of the continuity of the norm. In Chapter 1, we saw that the parallelogram law can be used to characterize the norms generated by an inner product, that is, Hilbertian norms. We now have all of the tools we need to formalize this affirmation, which Yosida (1995) refers to as the Fréchet-von Neumann-Jordan theorem.

THEOREM 4.3 (Fréchet-von Neumann-Jordan theorem).– Let V be a vector space on $\mathbb{K}$ (of finite or infinite dimension) and let $\|\ \|$ be a norm on V. $\|\ \|$ is a Hilbertian norm if and only if it satisfies the parallelogram law.

If the norm satisfies the parallelogram law, then the inner product from which it is induced is necessarily determined by the polarization formulas for real and complex cases, respectively:

$$\langle v, w\rangle = \frac{1}{4}(\|v + w\|^2 - \|v - w\|^2)$$

$$\langle v, w\rangle = \frac{1}{4}\left[\|v + w\|^2 - \|v - w\|^2 + i\left(\|v + iw\|^2 - \|v - iw\|^2\right)\right]$$

PROOF.– The direct implication is obvious, so we only need to prove the reverse implication, that is, if a norm $\| \ \|$ satisfies the parallelogram law, then it is induced by an inner product in the canonical manner: $\| \cdot \| = \sqrt{\langle \cdot, \cdot \rangle}$.

Let us begin by considering the real case. If an inner product exists which induces the norm, then it must take the following form:

$$p(v, w) = \frac{1}{4}(\|v + w\|^2 - \|v - w\|^2), \qquad \forall v, w \in V$$

Note that p is a composition of algebraic functions (sum and difference), the norm and its squared power, all of which are continuous functions; p itself is thus a continuous function of its arguments.

The next step is to verify that this definition satisfies the defining properties of a real inner product.

First, we note that the symmetry $p(v, w) = p(w, v)$ is obvious, as is definite positiveness, given that $p(v, v) = \frac{1}{4}\|2v\|^2 = \|v\|^2 \geqslant 0$ and $p(v, v) = 0$ if and only if $v = 0_V$.

Second, we must verify bilinearity. Given that the symmetry condition is satisfied, any property of p which is demonstrated with respect to the first argument also holds for the second argument, meaning that we can focus on the first entry of p.

Using the parallelogram law, we can write $\forall v, w, z \in V$:

$$\|(v+z)+w\|^2+\|(v-z)+w\|^2 = \|(v+w)+z\|^2+\|(v+w)-z\|^2 = 2\|v+w\|^2+2\|z\|^2$$

and:

$$\|(v+z)-w\|^2+\|(v-z)-w\|^2 = \|(v-w)+z\|^2+\|(v-w)-z\|^2 = 2\|v-w\|^?+2\|z\|^?$$

thus:

$$\begin{aligned} p(v+z,w)+p(v-z,w) &= \tfrac{1}{4}(\|v+z+w\|^2 - \|v+z-w\|^2 \\ &\quad + \|v-z+w\|^2 - \|v-z-w\|^2) \\ &= \frac{1}{4}[2(\|v+w\|^2 + \|z\|^2) - 2(\|v-w\|^2 + \|z\|^2)] \qquad [4.7] \\ &= 2\frac{1}{4}(\|v+w\|^2 - \|v-w\|^2) \\ &= 2p(v,w) \end{aligned}$$

Taking $v = z$, we obtain $p(v+v,w) + \overbrace{p(v-v,w)}^{0} = p(2v,w) \underset{[4.7]}{=} 2p(v,w)$ $\forall v, w \in V$, that is:

$$2p(v,w) = p(2v,w), \qquad \forall v, w \in V \qquad [4.8]$$

Now, take $v_1, v_2 \in V$ such that $v = \frac{1}{2}(v_1+v_2)$ and $z = \frac{1}{2}(v_1-v_2)$, thus $v+z = v_1$ and $v - z = v_2$, then:

$$\begin{aligned} p(v_1,w)+p(v_2,w) &= p(v+z,w)+p(v-z,w) \underset{[4.7]}{=} 2p(v,w) \underset{[4.8]}{=} p(2v,w) \\ &\underset{(\text{def. } v)}{=} p(v_1+v_2,w) \end{aligned}$$

Since v, w, z are arbitrary vectors, v_1, v_2 are also arbitrary, therefore, the demonstration that $p(v_1+v_2,w) = p(v_1,w)+p(v_2,w)$ proves the additivity of p.

Now, let us prove the property of homogeneity. We start by observing that if the reasoning which gave us $p(2v,w) = 2p(v,w)$ is iterated $n \in \mathbb{N}$ times, we obtain $p(nv,w) = np(v,w)$.

Furthermore, for all $m \in \mathbb{N}$, $m \neq 0$, it not only holds that $p(v,w) = p(m\frac{v}{m}, w)$, but also $p(m(\frac{v}{m}), w) = mp(\frac{v}{m}, w)$; combined with the formula $p(nv,w) = np(v,w)$, this gives us $p(\frac{n}{m}v, w) = \frac{n}{m}p(v,w)$ $\forall n, m \in \mathbb{N}$, $m \neq 0$, that is, p is homogeneous with respect to any number $r \in \mathbb{Q}, r \geqslant 0 : p(rv,w) = rp(v,w)$.

In order to extend this homogeneity to all rational numbers, we use the argument that if $r < 0$, then, by rewriting $rv = -|r|v = |r|(-v)$, we obtain:

$$\begin{aligned} & rp(v,w) - p(rv,w) = rp(v,w) - p(|r|(-v),w) = rp(v,w) - |r|p(-v,w) \\ & = rp(v,w) + rp(-v,w) \\ & = r(p(v,w)+p(-v,w)) \underset{(\text{additivity})}{=} rp(v-v,w) = rp(0_V,w) = 0 \end{aligned}$$

Hence, the property of homogeneity also holds for negative rational numbers, and thus for all rational numbers. Now, using the fact that $\mathbb{Q}$ is dense in $\mathbb{R}$, we know that for all $\alpha \in \mathbb{R}$ there exists a sequence of rational numbers $(r_n)_{n\in\mathbb{N}} \subset \mathbb{Q}$ such that $r_n \underset{n\to+\infty}{\longrightarrow} \alpha$. By the continuity of p, we have:

$$\alpha p(v,w) = \lim_{n\to+\infty} r_n p(v,w) = p(\lim_{n\to+\infty} r_n v, w) = p(\alpha v, w), \quad \forall \alpha \in \mathbb{R},\ v,w \in V$$

In summary, p is an inner product on V which is compatible with its norm if $\mathbb{K} = \mathbb{R}$.

Now, let us consider the complex case: $\mathbb{K} = \mathbb{C}$. As we saw in the real case, if there is an inner product which induces the norm, it must take the following form:

$$\begin{aligned}\tilde{p}(v,w) &= \tfrac{1}{4}\left[\|v+w\|^2 - \|v-w\|^2 + i\left(\|v+iw\|^2 - \|v-iw\|^2\right)\right] \\ &= p(v,w) + ip(v,iw)\end{aligned}$$

$\forall v,w \in V$.

From the observations presented in section 1.1, to prove that $\tilde{p}(v,w)$ is a complex inner product, we must simply verify the Hermitian property, that is, $\tilde{p}(v,w) = \overline{\tilde{p}(w,v)}$, since the linearity of the first variable and the definite positiveness of p imply that these properties also hold for $\tilde{p}$.

$\tilde{p}$ is an Hermitian form if and only if $\tilde{p}(w,v) = \overline{\tilde{p}(v,w)} = p(v,w) - ip(v,iw)$, since $p(v,w)$ and $p(v,iw) \in \mathbb{R}$. Furthermore, $\tilde{p}(w,v) = p(w,v) + ip(iw,v) = p(v,w) + ip(iw,v)$, given that $p(v,w) = p(w,v)$, thus $\tilde{p}(w,v) = p(v,w) + ip(iw,v)$. Comparing the formulas:

$$\tilde{p}(w,v) = \overline{\tilde{p}(v,w)} = p(v,w) - ip(v,iw) \quad \text{and} \quad \tilde{p}(w,v) = p(v,w) + ip(iw,v)$$

we see that $\tilde{p}$ is an Hermitian form if and only if $p(v,iw) = -p(iw,v)\ \forall v,w \in V$.

Now, we calculate:

$$\begin{aligned}\tilde{p}(v,iw) &= \frac{1}{4}(\|v+iw\|^2 - \|v-iw\|^2) = \frac{1}{4}(|i|\,\|v+iw\|^2 - |i|\,\|v-iw\|^2) \\ &= \frac{1}{4}(\|iv-w\|^2 - \|iv+w\|^2) = -\frac{1}{4}(\|w+iv\|^2 - \|w-iv\|^2) \\ &= -p(w,iv)\end{aligned}$$

using the fact that $\|w - iv\| = \|iv - w\|$. In short, $\tilde{p}$ is the inner product associated with our norm in the complex case. □

The mathematical object below is crucial in mathematics.

DEFINITION 4.7 (Topological vector space).– *A topological vector space (T.V.S.) is a vector space V with a topology which is compatible with the linear structure of V, that is such that the linear operations of the sum and of scalar multiplication are continuous functions.*

The continuity of fundamental operations in an inner product space implies that these spaces are always T.V.S. The same can be said of normed vector spaces; the continuity of linear operations is proved in exactly the same way. In terms of topological arguments, there is no difference between an inner product space and a normed vector space, as the norm is the mathematical object used to prove continuity in both cases.

The major difference between an inner product space and a normed vector space is related to the underlying geometric structure of the space itself, which is much richer in the former case.

4.2.1. *Equivalence of separated topologies in finite-dimension vector spaces*

The dimension of the vector space played no part in the proofs of Theorem 4.2, so the considerations presented in the previous section hold true for any vector space, whether of finite or infinite dimensions.

In finite dimension, however, the topology (separable) of a T.V.S. can be guaranteed to be essentially unique.

THEOREM 4.4 (Tychonoff).– Let V be a *separated* T.V.S. of finite dimension n on the field $\mathbb{K}$. Given an arbitrary fixed basis $B = (b_1, \ldots, b_n)$ in V, the linear isomorphism defined by:

$$\begin{aligned} I: \quad & V \longrightarrow \mathbb{K}^n \\ & x = [x]_B = \sum_{i=1}^{n} x_i b_i \longmapsto \begin{pmatrix} x_1 \\ \vdots \\ x_n \end{pmatrix} \end{aligned}$$

is a homeomorphism (or topological isomorphism), that is a bicontinuous application (continuous, inversible, and of which the inverse is continuous) considering the usual Euclidean topology on $\mathbb{K}^n$.

As we have seen, all inner product spaces, whether normed or metric, are separated T.V.S, so one immediate consequence of Tychonoff's theorem is that *all inner products, norms and distances which can be defined on a finite-dimensional vector space are topologically equivalent*, that is, they generate the same topology, which, up to an isomorphism, is the Euclidean topology. This does not hold for infinite dimensions, as shown by a number of counter-examples.

The simplest example of topological independence with respect to the choice of a norm in finite-dimension concerns vector spaces of dimension 1, as we see from the result below.

THEOREM 4.5.– If V is a normed, one-dimensional vector space on the field $\mathbb{K}$, any two norms defined over V are multiples of each other by a real, strictly positive scalar.

PROOF.– Let $\| \ \|_1$, $\| \ \|_2$ be two norms on V. By definition, $\|0_V\|_1 = \|0_V\|_2 = 0$; so we just concentrate on an arbitrary $v \in V$ different from the null vector. Let $\|v\|_1 = k_1$ and $\|v\|_2 = k_2$, then we can write $\frac{\|v\|_1}{\|v\|_2} = \frac{k_1}{k_2} = k \in \mathbb{R}^+$, and thus $\|v\|_1 = k\|v\|_2$.

Since V is of dimension 1, for any other vector $w \in V$ there exists $\lambda \in \mathbb{K}$ such that $w = \lambda v$. Thus, by the homogeneity of the norm, we can write:

$$\|w\|_1 = \|\lambda v\|_1 = |\lambda|\|v\|_1 = |\lambda|k\|v\|_2 = k\|\lambda v\|_2 = k\|w\|_2$$

that is, for all $w \in V$ and for any pair of norms $\| \ \|_1$, $\| \ \|_2$ on V, there exists a constant $k \in \mathbb{R}^+$ such that $\|w\|_1 = k\|w\|_2$. □

4.3. Cauchy sequences and completeness: Banach and Hilbert

Mathematicians working in the late 19th and early 20th centuries showed that the infinite-dimensional metric, normed and inner product vector spaces, which were most "similar" to finite-dimensional Euclidean spaces, can be characterized using a relatively simple property: converging sequences can be identified with Cauchy sequences.

DEFINITION 4.8.– *Given a generic metric space* (X, d)*, a sequence* $(x_n)_{n\in\mathbb{N}}$ *is a* Cauchy *sequence if:*

$$\forall \varepsilon > 0 \, \exists N_\varepsilon > 0 \, : \, n, m \geqslant N_\varepsilon \implies d(x_n, x_m) < \varepsilon$$

that is, the elements in the sequence become arbitrarily close to each other as the indices of the elements increase, that is, as the sequence progresses.

(X, d) *is said to be a* complete *metric space if all Cauchy sequences converge to limits contained within* X.

We shall see many examples of complete metric spaces in this chapter. Simple examples of non-complete metric spaces can be built by using the following basic result concerning Cauchy sequences.

THEOREM 4.6.– Any convergent sequence in a metric space is necessarily a Cauchy sequence.

PROOF.– If $x_n \underset{n \longrightarrow +\infty}{\longrightarrow} \bar{x}$, then, by the arbitrary nature of ε and the triangular inequality:

$$\forall \varepsilon > 0 \, \exists N_\varepsilon > 0 \, : \, n, m \geqslant N_\varepsilon \implies d(x_n, x_m) \leqslant d(x_n, \bar{x}) + d(\bar{x}, x_m) < \frac{\varepsilon}{2} + \frac{\varepsilon}{2} = \varepsilon$$

□

Using this result, we can prove that the metric spaces[3] $(\mathbb{Q}, |\ \ |)$ and $((0,1), |\ \ |)$ are not complete. To verify that $(\mathbb{Q}, |\ \ |)$ is not complete, consider the sequence $((1 + \frac{1}{n})^n)_{n \in \mathbb{N}}$: this sequence is rational, since $\mathbb{Q}$ is stable with respect to sum, division and power operators and to their composition. Furthermore, the sequence is known to converge to e, the basis of natural logarithms, so, by Theorem 4.6, it is a Cauchy sequence in $\mathbb{Q}$, interpreted as a subset of $\mathbb{R}$.

However, e is an irrational number, that is $e \in \mathbb{R} \backslash \mathbb{Q}$, implying the existence of at least one Cauchy sequence in $\mathbb{Q}$ which converges outside of $\mathbb{Q}$ itself.

Similarly, in $((0,1), |\ \ |)$, consider the sequence $(\frac{1}{n})_{n \geqslant 1}$; this is evidently contained within $(0,1)$ and converges to 0, making it a Cauchy sequence on $(0,1) \subset \mathbb{R}$, but $0 \notin (0,1)$.

Now, let us consider the relationship between complete and closed metric spaces.

THEOREM 4.7.– If (X, d) is a complete metric space and (E, d), $E \subseteq X$ a closed metric subspace in X, then (E, d) is complete.

PROOF.– Let $(x_n)_{n \in \mathbb{N}} \subseteq E$ be a Cauchy sequence, since $E \subset X$ we have that $(x_n)_{n \in \mathbb{N}} \subseteq X$, and thus, since X is complete, $(x_n)_{n \in \mathbb{N}}$ converges to a limit $x \in X$. However, the limits of sequences in E belong to $\overline{E}$, and, since E is closed, $\overline{E} = E$, hence $x \in E$, that is all Cauchy sequences of elements of E converge in E itself. □

THEOREM 4.8.– If (X, d) is an arbitrary metric space and (E, d), $E \subseteq X$ is a complete metric subspace of X, then (E, d) is closed.

PROOF.– Taking $x \in \overline{E}$, there exists a sequence $(x_n)_{n \in \mathbb{N}} \subseteq E$ which converges to x. Given that the sequence converges, it is a Cauchy sequence in E. As E is complete,

3 Remember that $\mathbb{Q}$ is not a real or complex vector space, as it is not stable with regard to its product by a real or complex scalar; thus, Tychonoff's theorem cannot be applied for $\mathbb{Q}$.

$(x_n)_{n \in \mathbb{N}}$ must converge to an element $y \in E$. By uniqueness of the limit, $x = y$ and thus $x \in E$, that is $\overline{E} = E$. □

An inner product vector space, or a normed vector space, is also a metric vector space; consequently, the definition of a Cauchy sequence can be rewritten as:

$$\forall \varepsilon > 0 \; \exists N_\varepsilon > 0 \; : \; n, m \geqslant N_\varepsilon \implies \|x_n - x_m\| < \varepsilon$$

Some authors use an even shorter form:

$$\lim_{n,m \to +\infty} \|x_n - x_m\| = 0$$

A standard result of Calculus guarantees that $(\mathbb{R}^n, |\;|)$ *and* $(\mathbb{C}^n, |\;|)$ *are complete metric spaces for all finite* $n \in \mathbb{N}$. Using Tychonoff's theorem (Theorem 4.4), we known that real or complex separated topological vector spaces of finite dimension n are topologically equivalent to the Euclidean spaces $\mathbb{R}^n$ or $\mathbb{C}^n$, respectively; it follows that *completeness is never a problem for pre-Hilbert vector spaces (or normed spaces) of finite dimension*: *converging sequences in these spaces are all, and only, Cauchy sequences.*

If the dimension of the vector space is not finite, then while it remains true that convergent sequences are necessarily Cauchy sequences, the inverse is not always true. For this reason, we shall introduce a definition to characterize spaces in which the Cauchy condition is necessary and sufficient for convergence[4].

DEFINITION 4.9 (Hilbert and Banach spaces).– *Let V be a vector space of finite or infinite dimension.*

– *If $(V, \|\;\|)$ is complete, then it is called a Banach space.*

– *If $(V, \langle\;,\;\rangle)$ is complete, then it is called a Hilbert space.*

One consequence of Tychonoff's theorem is that *real or complex normed vector spaces of finite dimension* are *all Banach spaces*, while *real or complex inner product spaces of finite dimension* are *all Hilbert spaces*. Finite or infinite-dimension *Hilbert spaces are also Banach spaces*, due to the fact that they are normed and complete vector spaces; the inverse is not generally true, as the existence of an inner product in a Banach space is guaranteed if and only if the parallelogram law holds.

Two results related to Cauchy sequences are presented below. These will be extremely useful in what follows. Before proving them, we recall that a sequence in a metric space is said to be bounded if all elements of the sequence fall within a finite neighborhood of one element of the space, as described in Definition 4.9.

4 One can also Fréchet spaces: locally convex topological vector spaces which are complete with respect to a shift-invariant topology.

DEFINITION 4.10.– *A sequence* $(x_n)_{n \in \mathbb{N}}$ *in a metric space* (X, d) *is said to be* bounded *if there exists* $x^* \in X$ *and* $M \geqslant 0$ *such that* $d(x_n, x^*) \leqslant M \; \forall n \in \mathbb{N}$.

THEOREM 4.9.– All Cauchy sequences are bounded.

PROOF.– By definition, if $(x_n)_{n \in \mathbb{N}}$ is a Cauchy sequence, there exists $N_\varepsilon > 0$ such that the distance between x_{N_ε} and all elements x_n of the sequence with $n \geqslant N_\varepsilon$ is less than ε, that is $d(x_n, x_{N_\varepsilon}) < \varepsilon \; \forall n \geqslant N_\varepsilon$. x_{N_ε} is thus a good candidate to take the place of x^* in the definition of a bounded sequence.

To prove this, we note that the elements of the sequence corresponding to an index value n lower than N_ε belong to X, thus their distance from x_{N_ε} is finite, and we can define the following value:

$$r = \max\{d(x_{N_\varepsilon}, x_0), d(x_{N_\varepsilon}, x_1), \ldots, d(x_{N_\varepsilon}, x_{N_\varepsilon - 1})\}$$

Now, defining $M = \max\{\varepsilon, r\}$, we obtain $d(x_n, x_{N_\varepsilon}) \leqslant M \; \forall n \in \mathbb{N}$. □

The second result relates to subsequences.

DEFINITION 4.11.– *Let* $(x_n)_{n \in \mathbb{N}}$ *be a sequence in a metric space* (X, d) *and let* $\varphi : \mathbb{N} \to \mathbb{N}$ *be a strictly increasing function, that is* $\varphi(n+1) > \varphi(n)$ *for all* $n \in \mathbb{N}$. *The sequence defined by* $(x_{\varphi(n)})_{n \in \mathbb{N}}$ *is a* subsequence *of the initial sequence* $(x_n)_{n \in \mathbb{N}}$.

As a very simple exercise, readers are invited to prove that, if a sequence $(x_n)_{n \in \mathbb{N}}$ in a metric space (X, d) is convergent, then all of its subsequences also converge, and converge to the same limit.

The following important result shows that, for Cauchy sequences, the order of this implication can be reversed.

THEOREM 4.10.– Any Cauchy sequence in a metric space (X, d) which possesses at least one convergent subsequence is itself convergent to the same limit.

PROOF.– Let $(x_n)_{n \in \mathbb{N}}$ be a Cauchy sequence in (X, d) which admits a convergent subsequence $(x_{\varphi(n)})_{n \in \mathbb{N}}$, where $\varphi : \mathbb{N} \to \mathbb{N}$ is the strictly increasing application which defines this subsequence. Let a be the limit of the subsequence, that is $a = \lim\limits_{n \to +\infty} x_{\varphi(n)}$.

For all $n \in \mathbb{N}$, by the triangular inequality, we have $d(x_n, a) \leqslant d(x_n, x_{\varphi(n)}) + d(x_{\varphi(n)}, a)$; if we can majorized both terms on the right by an arbitrarily small quantity ε, then the thesis of the theorem will be proven.

To show that this is possible, we shall use the definition of a Cauchy sequence for $(x_n)_{n\in\mathbb{N}}$ to write:

$$\forall \varepsilon > 0 \; \exists N_\varepsilon \in \mathbb{N} \text{ such that } m, n \geqslant N_\varepsilon \implies d(x_m, x_n) < \frac{\varepsilon}{2}$$

but, as φ is strictly increasing, $\varphi(n) \geqslant N_\varepsilon$, hence $d(x_n, x_{\varphi(n)}) < \frac{\varepsilon}{2}$.

Since the subsequence $(x_{\varphi(n)})_{n\in\mathbb{N}}$ is presumed to converge to a, this implies that:

$$\forall \varepsilon > 0 \; \exists K_\varepsilon \in \mathbb{N} \text{ such that: } n \geqslant K_\varepsilon \implies d(x_{\varphi(n)}, a) < \frac{\varepsilon}{2}$$

and, by considering $n \geqslant \max\{N_\varepsilon, K_\varepsilon\}$, we obtain $d(x_n, a) \leqslant d(x_n, x_{\varphi(n)}) + d(x_{\varphi(n)}, a) < \frac{\varepsilon}{2} + \frac{\varepsilon}{2} = \varepsilon \; \forall \varepsilon > 0$, that is $x_n \underset{n\to+\infty}{\longrightarrow} a$. □

This theorem has notable applications in pure and applied mathematics. We shall see a theoretical use in the next section; here we mention its usefulness in *optimization*, where one seeks to identify the optimal solution to a problem by minimizing an appropriate function. In many cases, the function is too complicated for an analytical description of its minima to be possible, so the solution must be approximated using an *iterative algorithm*: in this way, a minimum point is attained after passing through a *sequence of points*. Theorem 4.10 is often used to demonstrate that the iterative algorithm converges, proving that the sequence of points defined by the algorithm is a Cauchy sequence and proving that it admits a (wisely chosen) converging subsequence.

4.3.1. *Completeness of vector spaces*

Metric vector spaces can always be completed in an essentially unique way to complete spaces, as Theorem 4.11 establishes.

THEOREM 4.11 (Completion of a non-complete metric vector space).– If (V, d) is a non-complete metric vector space, then there exists a complete metric vector space $(\hat{V}, \hat{d})$ and an isometric injective function $\iota : V \longrightarrow \hat{V}$, that is:

$$\begin{cases} x_1, x_2 \in V, \; x_1 \neq x_2 \implies \iota(x_1) \neq \iota(x_2) \\ \forall x_1, x_2 \in V, \; d(x_1, x_2) = \hat{d}(\iota(x_1), \iota(x_2)) \end{cases}$$

such that $\overline{\iota(V)} = \hat{V}$, that is the image of V *via* ι is dense in $\hat{V}$.

COROLLARY 4.1.– Any pre-Hilbert space V can be completed to a Hilbert space $\mathcal{H}$.

PROOF.– This proof will focus on the case of pre-Hilbert spaces, which is most relevant for our purposes. The general proof follows a similar approach, except for the fact that the norm of the difference between two vectors is replaced by their distance.

The completion of a pre-Hilbert space V is, by definition, the space $\mathcal{H}_1$ of all the Cauchy sequences $(x_n)_{n\in\mathbb{N}}$ modulo the equivalency relationship $\sim$, defined as follows: two Cauchy sequences $(x_n)_{n\in\mathbb{N}}$ and $(y_n)_{n\in\mathbb{N}}$ of elements in V are equivalent if $\lim_{n\to+\infty} \|x_n - y_n\| = 0$.

The completion of V is written as $\mathcal{H} = \mathcal{H}_1/\sim$, and its elements are noted $[x]$. We define a norm on $\mathcal{H}$ as follows:

$$\forall [x] \in \mathcal{H}, \quad \|[x]\| = \lim_{n\to+\infty} \|x_n\|$$

where $(x_n)_{n\in\mathbb{N}}$ is any Cauchy sequence in the equivalence class $[x]$. This definition does not depend on the choice of the Cauchy sequence used to represent the equivalence class, since, given that $|\,\|x_n\| - \|y_n\|\,| \leqslant \|x_n - y_n\|$, if $(y_n)_{n\in\mathbb{N}} \in [x]$, then at the limit we have:

$$\lim_{n\to+\infty} |\,\|x_n\| - \|y_n\|\,| \leqslant \lim_{n\to+\infty} \|x_n - y_n\| = 0$$

that is: $\lim_{n\to+\infty} \|x_n\| = \lim_{n\to+\infty} \|y_n\|$.

Now, let us define an inner product on $\mathcal{H}$ which is compatible with this norm:

$$\langle [x], [y] \rangle = \lim_{n\to+\infty} \langle x_n, y_n \rangle$$

where $(x_n)_{n\in\mathbb{N}}$ and $(y_n)_{n\in\mathbb{N}}$ are any two Cauchy sequences in the equivalence classes $[x]$ and $[y]$, respectively.

To verify that this inner product is well defined, we must verify the existence of the limit used to define it, and show that it does not depend on the chosen representative elements.

The first step is to prove the existence of the limit. To do this, we must simply show that $\langle x_n, y_n \rangle$ (a sequence in $\mathbb{K}$) is Cauchy; given that $\mathbb{K}$ is complete, the limit must exist. Note that $\forall n, m \in \mathbb{N}$, by the triangular inequality and the Cauchy-Schwarz inequality, we can write:

$$\begin{aligned}
&|\langle x_n, y_n \rangle - \langle x_m, y_m \rangle| = |\langle x_n, y_n \rangle - \langle x_n, y_m \rangle + \langle x_n, y_m \rangle - \langle x_m, y_m \rangle| \\
&= |\langle x_n, y_n - y_m \rangle + \langle x_n - x_m, y_m \rangle| \leqslant |\langle x_n, y_n - y_m \rangle| + |\langle x_n - x_m, y_m \rangle| \\
&\leqslant \|x_n\|\,\|y_n - y_m\| + \|x_n - x_m\|\,\|y_m\| \underset{n,m\to+\infty}{\longrightarrow} 0
\end{aligned}$$

since $\|x_n\|$ and $\|y_n\|$ are bounded, given that $(x_n)_{n\in\mathbb{N}}$ and $(y_n)_{n\in\mathbb{N}}$ are Cauchy sequences.

Now, we must verify that the limit is independent of the choice of representative elements: let $(\xi_n)_{n\in\mathbb{N}}$ and $(\eta_n)_{n\in\mathbb{N}}$ be two other representatives of the equivalence classes $[x]$ and $[y]$, respectively.

Using direct algebraic manipulations, we can write:

$$\langle x_n, y_n\rangle = \langle x_n - \xi_n + \xi_n, y_n - \eta_n + \eta_n\rangle = \langle x_n - \xi_n, y_n\rangle + \langle \xi_n, y_n - \eta_n\rangle + \langle \xi_n, \eta_n\rangle$$

and:

$$|\langle x_n-\xi_n, y_n\rangle+\langle \xi_n, y_n-\eta_n\rangle| \leqslant \|x_n - \xi_n\| \, \|y_n\|+\|\xi_n\| \, \|y_n - \eta_n\| \underset{n,m\to+\infty}{\longrightarrow} 0$$

since $(x_n)_{n\in\mathbb{N}}, (\xi_n)_{n\in\mathbb{N}} \in [x]$ and $(y_n)_{n\in\mathbb{N}}, (\eta_n)_{n\in\mathbb{N}} \in [y]$, hence:

$$\langle [x], [y]\rangle = \lim_{n\to+\infty} \langle x_n, y_n\rangle = \lim_{n\to+\infty} \langle \xi_n, \eta_n\rangle$$

Due to the continuity of the inner product on V, all of these properties are transferred onto $\mathcal{H}$ by the limit operation.

The final step is to verify the isometry:

$$\|[x]\| = \lim_{n\to+\infty} \|x_n\| = \lim_{n\to+\infty} \sqrt{\langle x_n, x_n\rangle} = \sqrt{\langle [x], [x]\rangle}, \qquad \forall [x] \in \mathcal{H} \quad \square$$

An alternative proof may be found in El Hage Hassan (2011).

4.3.2. *Characterizing the completeness of normed vector spaces using series*

In this section, we shall consider a completeness criterion for normed vector spaces which draws on series and is highly useful in practice.

The explicit definition of the Cauchy condition for the sequence of partial sums of a series $\sum_{n\in\mathbb{N}}$ is:

$$\forall \varepsilon > 0 \; \exists N_\varepsilon > 0 \; : \; n, m \geqslant N_\varepsilon \implies \left\| \sum_{k=0}^{n} x_k - \sum_{k=0}^{m} x_k \right\| < \varepsilon$$

The two indices n and m vary *independently* of one another, and we can suppose, without loss of generality, that one is always greater than the other. For instance, supposing that $n > m$: $\left\| \sum\limits_{k=0}^{n} x_k - \sum\limits_{k=0}^{m} x_k \right\| = \left\| \sum\limits_{k=m+1}^{n} x_k \right\|$, implying that the *Cauchy condition for series* can be rewritten as:

$$\forall \varepsilon > 0 \, \exists N_\varepsilon > 0 \, : \, n > m \geqslant N_\varepsilon \implies \left\| \sum_{k=m+1}^{n} x_k \right\| < \varepsilon \qquad [4.9]$$

Instead, the Cauchy condition for the series of norms of x_k, that is, the sequence $\left(\sum\limits_{k=0}^{n} \|x_k\| \right)_{n \in \mathbb{N}}$, is:

$$\forall \varepsilon > 0 \, \exists N_\varepsilon > 0 \, : \, n > m \geqslant N_\varepsilon \implies \sum_{k=m+1}^{n} \|x_k\| < \varepsilon \qquad [4.10]$$

This observation will be used in proving the following result.

THEOREM 4.12 (Characterizing the completeness of normed spaces using series).– A normed vector space $(V, \| \, \|)$ is complete if and only if all absolutely convergent series of elements in V are also (simply) convergent in V.

PROOF.– The proof of the direct implication is extremely simple, while that of the inverse is much more complicated and it involves techniques that are very commonly used in functional analysis.

$\boxed{\implies}$: Let us suppose $(V, \| \, \|)$ to be complete, and let us demonstrate that if $\sum\limits_{n \in \mathbb{N}} \|x_n\|$ is convergent, then $\sum\limits_{n \in \mathbb{N}} x_n$ is also convergent in V.

By completeness, the convergence of $\sum\limits_{n \in \mathbb{N}} \|x_n\|$ is equivalent to the Cauchy condition [4.10], that is:

$$\forall \varepsilon > 0 \, \exists N_\varepsilon > 0 \, : \, n > m \geqslant N_\varepsilon \implies \sum_{k=m+1}^{n} \|x_k\| < \varepsilon \qquad [4.11]$$

and since $\left\| \sum\limits_{k=m+1}^{n} x_k \right\| \leqslant \sum\limits_{k=m+1}^{n} \|x_k\|$, the sequence of partial sums $\left(S_n = \sum\limits_{k=0}^{n} x_k \right)_{n \in \mathbb{N}}$ is also Cauchy, that is $\sum\limits_{n \in \mathbb{N}} x_n$ is convergent.

$\boxed{\Leftarrow}$: Now, let us suppose that all absolutely convergent series of elements in V are also simply convergent in V. We must prove that this implies that V is complete, that is any Cauchy sequence $(x_n)_{n \in \mathbb{N}} \subset V$, that is:

$$\forall \varepsilon > 0 \, \exists N_\varepsilon > 0 \, : \, n, m \geqslant N_\varepsilon \implies \|x_n - x_m\| < \varepsilon$$

converges in V, that is, there exists $\bar{x} \in V$ such that $(x_n)_{n \in \mathbb{N}} \underset{n \to +\infty}{\longrightarrow} \bar{x}$.

The Cauchy condition must be valid for all values of $\varepsilon > 0$, and consequently for $\varepsilon_k = \frac{1}{2^k}$, $k \in \mathbb{N}$; thus, any Cauchy sequence in V must verify:

$$\forall k \geqslant 0 \; \exists \tilde{N}_k > 0 \; : \; n, m \geqslant \tilde{N}_k \implies \|x_n - x_m\| < \frac{1}{2^k}, \qquad [4.12]$$

note that all the objects contained in the expression above are discrete.

Since $\frac{1}{2^{k+1}} < \frac{1}{2^k}$, it follows that $\tilde{N}_{k+1} \geqslant \tilde{N}_k$; this simple consideration allows us to define a strictly increasing sequence of natural numbers $(N_k)_{k \geqslant 0}$ simply by defining $N_k := \inf_{\ell \in \mathbb{N}} \{\tilde{N}_{k+\ell} > \tilde{N}_k\}$ for all $k \in \mathbb{N}$. Using this result, we can define the subsequence $(x_{N_k})_{k \in \mathbb{N}} \subset V$ of $(x_n)_{n \in \mathbb{N}}$ which, by its own definition, satisfies [4.12], that is:

$$\forall k \geqslant 0, \; \left\| x_{N_k} - x_{N_{k+1}} \right\| < \frac{1}{2^k} \qquad [4.13]$$

The interest of using this subsequence is that, if it converges in V, that is, if there exists $\bar{x} \in V$ such that $\lim_{k \to +\infty} x_{N_k} = \bar{x}$, then by Theorem 4.10 the initial Cauchy sequence $(x_n)_{n \in \mathbb{N}}$ also converges to $\bar{x} \in V$.

To complete the proof, we must therefore demonstrate that the subsequence $(x_{N_k})_{k \in \mathbb{N}}$ is convergent in V. In the absence of information concerning the convergence of the original sequence $(x_n)_{n \in \mathbb{N}}$, the convergence of $(x_{N_k})_{k \in \mathbb{N}}$ cannot be proved directly; instead, we must use the hypothesis that absolutely convergent series in V imply the simple convergence of series in V.

The *link to series* is obtained using a startlingly simple technique: *rewriting the subsequence $(x_{N_k})_{k \in \mathbb{N}}$ as a sequence of telescopic partial sums*. To do this, we use $(x_{N_k})_{k \in \mathbb{N}}$ to define a new sequence $(y_k)_{k \in \mathbb{N}} \subset V$ as follows:

$$\begin{cases} y_0 = x_{N_0} \\ y_k = x_{N_k} - x_{N_{k-1}}, \; \forall k \geqslant 1, \end{cases} \qquad \text{hence: } (y_k)_{k \in \mathbb{N}} = (x_{N_0}, x_{N_1} - x_{N_0}, x_{N_2} - x_{N_1}, \ldots)$$

then:

$$\sum_{j=0}^{k} y_j = \underbrace{\cancel{x_{N_0}}}_{y_0} + \underbrace{\cancel{x_{N_1}} - \cancel{x_{N_0}}}_{y_1} + \underbrace{\cancel{x_{N_2}} - \cancel{x_{N_1}}}_{y_2} + \ldots + \underbrace{x_{N_k} - \cancel{x_{N_{k-1}}}}_{y_k} = x_{N_k}$$

and this holds $\forall k \in \mathbb{N}$, thus $\left(\sum_{j=0}^{k} y_j \right)_{k \in \mathbb{N}} = (x_{N_k})_{k \in \mathbb{N}}$.

To resume, the completeness of V, that is, the convergence of an arbitrary Cauchy sequence $(x_n)_{n \in \mathbb{N}}$ in V, is implied by the convergence of the sequence $(x_{N_k})_{k \in \mathbb{N}}$ in

V; this is equivalent to the convergence of $\left(\sum_{j=0}^{k} y_j\right)_{k\in\mathbb{N}}$ in V, that is, the simple convergence of the series $\sum_{k=0}^{\infty} y_k$. By the starting hypothesis, if we can prove that $\sum_{k\in\mathbb{N}} \|y_k\|$ is convergent, this would be enough to prove the whole theorem. We begin by setting out the terms of the series $\sum_{k\in\mathbb{N}} \|y_k\|$:

$$\sum_{k\in\mathbb{N}} \|y_k\| = \|y_0\| + \sum_{k=1}^{\infty} \|y_k\| = \|y_0\| + \sum_{k=1}^{\infty} \|x_{N_k} - x_{N_{k-1}}\|$$
$$= \|y_0\| + \sum_{k=0}^{\infty} \|x_{N_{k+1}} - x_{N_k}\|$$

From inequality [4.13], it holds that $\|x_{N_{k+1}} - x_{N_k}\| < \frac{1}{2^k} = \left(\frac{1}{2}\right)^k \forall k \geqslant 0$, thus:

$$\sum_{k\in\mathbb{N}} \|y_k\| = \|y_0\| + \sum_{k=0}^{\infty} \|x_{N_{k+1}} - x_{N_k}\| < \|y_0\| + \sum_{k=0}^{\infty} \left(\frac{1}{2}\right)^k$$
$$= \|y_0\| + \frac{1}{1-\frac{1}{2}} = \|y_0\| + 2 < +\infty$$

using the geometric series formula.

Hence, $\sum_{k\in\mathbb{N}} \|y_k\|$ is a bounded series of positive real terms, and, from a classic result in series theory, we know that it converges. □

If the space $(V, \| \ \|)$ in the previous theorem is complete, then the Cauchy sequence $(x_n)_{n\in\mathbb{N}} \subset V$ seen at the start of the proof is convergent; consequently, we know that the subsequence $(x_{N_k})_{k\in\mathbb{N}}$ also converges to the same limit. This remark is formalized in Corollary 4.2.

COROLLARY 4.2.– *Taking $(V, \| \ \|)$ to be a complete normed vector space and $(x_n)_{n\in\mathbb{N}} \subset V$ a sequence which converges to $x_0 \in V$, there exists a subsequence $(x_{n_k})_{k\in\mathbb{N}}$ which converges to x_0.*

4.3.2.1. *The matrix exponential*

In this section, we shall examine a particularly important application of the previous theorem: the definition of the matrix exponential.

DEFINITION 4.12 (Matrix exponential).– *Let $A \in M(n, \mathbb{K})$ be a square matrix*[5] *with coefficients in the field $\mathbb{K} = \mathbb{R}$ or $\mathbb{C}$. The exponential of A is the matrix defined by:*

$$\boxed{e^A = \sum_{k=0}^{\infty} \frac{A^k}{k!}}$$

The proof that e^A is well defined is trivial using the theorem proved above. For instance, let us consider the *Frobenius norm* of A:

$$\|A\| = \left(\sum_{i=1}^{n} \sum_{j=1}^{n} |a_{ij}|^2 \right)^{1/2}$$

This is the Euclidean norm of a vector in $\mathbb{K}^{n^2}$ obtained by A using lexicographical order, that is, by sequencing the lines (or rows) of A one after another. We shall prove that the series $I_n + A + \frac{A}{2} + \frac{A^2}{3!} + \frac{A^k}{k!} + \ldots$ converges in the topology of $\mathrm{M}(n, \mathbb{K})$ generated by this norm, implying, by Tychonoff's theorem, its convergence with respect to any other norm.

$\mathrm{M}(n, \mathbb{K})$ is homeomorphic to the Euclidean $\mathbb{K}^{n^2}$, which we know to be a complete normed space. To show that e^A is well defined, we must show that the series defining e^A is absolutely convergent; simple convergence is implied by Theorem 4.12.

The proof of absolute convergence is extremely simple: consider the inequality $\|A^k\| \leqslant \|A\|^k$, verified by the Frobenius norm for all $k \in \mathbb{N}$ and for any matrix $A \in \mathrm{M}(n, \mathbb{K})$, then:

$$\sum_{k=0}^{\infty} \frac{\|A^k\|}{k!} \leqslant \sum_{k=0}^{\infty} \frac{\|A\|^k}{k!} = e^{\|A\|}$$

using the fact that $\|A\|$ is a real number $\geqslant 0$ and that the convergence radius of the exponential series in $\mathbb{R}$ is infinite.

4.3.3. *Banach fixed-point theorem*

The result presented in this section is highly significant for many different fields of mathematics, such as analysis, topology, solving differential equations, etc.

5 A must be square as we will be working with powers of A; for dimensional reasons, these are not defined if A is not square.

We begin by recalling Definition 4.12.

DEFINITION 4.13 (Contraction mapping).– *Let* (X_1, d_1) *and* (X_2, d_2) *be any two metric subspaces and let* $k \in (0,1)$ *be a real constant. The application* $f : X_1 \to X_2$ *is a contraction with coefficient* k *if, for all* $x, y \in X_1$*:*

$$d_2(f(x), f(y)) \leqslant k d_1(x, y) \qquad [4.14]$$

The smallest value of k for which [4.14] holds is called the Lipschitz constant *of* f.

Verification that *a contraction mapping is always a continuous function* is immediate: for any value $\varepsilon > 0$, let us take an arbitrary fixed element $\bar{x} \in X_1$ and consider the elements $y \in X_1$ such that $d_1(\bar{x}, y) < \varepsilon$. Then, by the definition of a contraction mapping, $d_2(f(\bar{x}), f(y)) \leqslant k d_1(\bar{x}, y) < k\varepsilon < \varepsilon$, since $k \in (0,1)$, so function f is continuous in $\bar{x}$. As $\bar{x}$ is an arbitrary element in X_1, f is continuous on all X_1.

REMARK.– It is evident from the definition that the distance (in the codomain) between the images of a pair of elements *via* a contraction mapping is smaller than the initial distance. However, contraction mapping cannot be redefined using this property alone; the definition given above is not the same as stating that, for all $x, y \in X_1$, $x \neq y$, $d_2(f(x), f(y)) < d_1(x, y)$. If f satisfies this condition, it is said to be a *weak contraction mapping*, or an *application which reduces the distance between points*.

To understand the subtle difference between these two definitions, we begin by noting that if f is a weak contraction mapping, then for any pair $x, y \in X_1$, $x \neq y$, there exists $k_{x,y} \in (0,1)$ such that $d_2(f(x), f(y)) \leqslant k_{x,y}\, d_1(x, y)$, that is, $k_{x,y}$ is not a constant, as required by the definition of a contraction mapping. The two definitions coincide if and only if $\sup\limits_{x,y \in X_1} k_{x,y} \equiv \bar{k} \in (0,1)$, but this condition is not guaranteed to be verified. The sup necessarily exists, since $\{k_{x,y},\ x, y \in X_1,\ x \neq y\}$ is a bounded subset of $\mathbb{R}$, but it can take the value 1, meaning that it is not strictly less than 1 as required by the definition of a contraction mapping.

Contraction mappings with a domain and image in the same complete metric space have a remarkable property, described in the classic Theorem 4.13.

THEOREM 4.13 (Banach fixed-point theorem).– Let (X, d) be a complete metric space and $f : X \to X$ a contraction mapping in X of coefficient $k \in (0,1)$: then f admits a single fixed point, that is there exists a single $\bar{x} \in X$ such that $f(\bar{x}) = \bar{x}$.

PROOF.– Let $a \in X$ be an arbitrary element. We define the sequence $(x_n)_{n\in\mathbb{N}} \subset X$ by recursion as:

$$\begin{cases} x_0 = a \\ x_n = f(x_{n-1}),\ n \geqslant 1 \end{cases}$$

The first step of the proof consists simply of showing that, if this sequence admits a limit in X, then this limit is a fixed point for f. The uniqueness of the fixed point will be a simple consequence of the definition of the contraction mapping. Instead, the convergence of the sequence $(x_n)_{n\in\mathbb{N}}$, which is harder to prove, will be verified later.

– *If there exists* $X \ni \bar{x} = \lim_{n\to+\infty} x_n$*, then* $\bar{x}$ *is a fixed point for* f: The proof of this statement relies on a simple continuity argument. Since we know that a contraction mapping is continuous, if we let n tend toward $+\infty$ in the definition of the sequence, that is, $x_n = f(x_{n-1})$, we obtain:

$$\lim_{n\to+\infty} x_n = \lim_{n\to+\infty} f(x_{n-1}) \iff \bar{x} = f(\lim_{n\to+\infty} x_{n-1}) \iff \bar{x} = f(\bar{x})$$

that is, $\bar{x}$ is a fixed point for f. This is the reason for considering the recursively defined sequence $(x_n)_{n\in\mathbb{N}}$ described above.

– *Uniqueness of the fixed point*: Let $\bar{x}, \bar{y} \in X$ be two fixed points for f, that is, $f(\bar{x}) = \bar{x}$, $f(\bar{y}) = \bar{y}$. We can show that their distance is null, that is, $\bar{x} = \bar{y}$, using the definite positiveness of the distance and the definition of contraction mapping:

$$d(\bar{x}, \bar{y}) = d(f(\bar{x}), f(\bar{y})) \leqslant kd(\bar{x}, \bar{y})$$

but since $k \in (0, 1)$, this inequality only holds if $d(\bar{x}, \bar{y}) = 0$, that is, $\bar{x} = \bar{y}$.

– *Convergence of the sequence*: Here, the hypothesis that (X, d) is complete will be crucial, because if we can show that $(x_n)_{n\in\mathbb{N}}$ is Cauchy, then, by completeness, it is convergent. We begin by noting that for all $n \geqslant 1$, using the definition of the sequence and the hypothesis that f is a contraction, we can write:

$$d(x_{n+1}, x_n) = d(f(x_n), f(x_{n-1})) \leqslant kd(x_n, x_{n-1})$$

hence, by iteration:

$$\begin{aligned} d(x_{n+1}, x_n) &\leqslant kd(x_n, x_{n-1}) \\ &\leqslant k^2 d(x_{n-1}, x_{n-2}) \\ &\leqslant \ldots \leqslant k^n d(x_1, x_0) \end{aligned} \qquad [4.15]$$

that is the distance between consecutive elements, x_{n+1} and x_n, in sequence $(x_n)_{n\in\mathbb{N}}$ is majorized by $k^n d(x_1, x_0)$; note that *the power of k is equal to the smallest index value.*

Now, let us take two arbitrary but different natural indices $n, m \in \mathbb{N}$. Without loss of generality, we may consider that $n < m$, hence $m - n = p \in \mathbb{N}$, or $m = n + p$, and so $d(x_m, x_n) = d(x_{n+p}, x_n)$. Iterating the triangular property of the distance, we obtain:

$$d(x_{n+p}, x_n) \leqslant d(x_{n+p}, x_{n+p-1}) + d(x_{n+p-1}, x_{n+p-2}) + \ldots + d(x_{n+1}, x_n)$$

We see that all terms on the right side of the inequality are distances between two consecutive elements of the sequence $(x_n)_{n\in\mathbb{N}}$; using this fact, we can apply the majorization given by [4.15] and write:

$$\begin{cases} d(x_{n+p}, x_{n+p-1}) \leqslant k^{n+p-1} d(x_1, x_0) \\ d(x_{n+p-1}, x_{n+p-2}) \leqslant k^{n+p-2} d(x_1, x_0) \\ \qquad \vdots \\ d(x_{n+1}, x_n) \leqslant k^n d(x_1, x_0) \end{cases}$$

that is:

$$\begin{aligned} d(x_{n+p}, x_n) &\leqslant (k^{n+p-1} + k^{n+p-2} + \ldots + k^n) d(x_1, x_0) \\ &= (k^{p-1} + k^{p-2} + \ldots + 1) k^n d(x_1, x_0) \\ &= \left(\sum_{j=0}^{p-1} k^j \right) k^n d(x_1, x_0) \\ &\underset{k^j > 0}{\leqslant} \left(\sum_{j=0}^{+\infty} k^j \right) k^n d(x_1, x_0), \end{aligned}$$

As $\sum_{j=0}^{+\infty} k^j$ is a geometric series in $k \in (0, 1)$, it converges to $\frac{1}{1-k}$, so we have:

$$d(x_{n+p}, x_n) \leqslant \frac{k^n}{1-k} d(x_1, x_0)$$

Remembering that $d(x_{n+p}, x_n) = d(x_m, x_n)$, $m > n \in \mathbb{N}$ of arbitrary value, we have:

$$d(x_m, x_n) \leqslant \frac{k^n}{1-k} d(x_1, x_0) \underset{n \to +\infty}{\longrightarrow} 0$$

This implies that $(x_n)_{n\in\mathbb{N}}$ is a Cauchy sequence, and thus converges to an element $\bar{x} \in X$ by the hypothesis of completeness of X. □

It is important to note that *the first element in the sequence* $(x_n)_{n\in\mathbb{N}}$ *is completely arbitrary*: even if this element is distant from the fixed point $\bar{x}$, the sequence will reach the fixed point by the limit. In some occasions, a starting point x_0 may be selected in such a way as to accelerate the speed at which the sequence convergences.

The following nice exercise is proposed by Sondaz (2010).

Exercise 4.1

1) Give an example of a metric space (X, d) and contraction mapping $f : X \to X$ with no fixed point.

2) Give an example of a complete metric space (X, d) and an application $f : X \to X$ which strictly reduces distances, that is such that $d(f(x), f(y)) < d(x, y)$ $\forall x, y \in X, x \neq y$, and which admits no fixed point.

3) Show that the Cauchy problem:

$$\begin{cases} x'(t) = \frac{1}{2}\sin x(t) \\ x(0) = 1 \end{cases} \qquad [4.16]$$

has a unique solution $\varphi : [-1, 1] \to \mathbb{R}$.

Solution to Exercise 4.1

For points 1 and 2, the answer evidently involves undermining the fixed point theorem by removing a hypothesis. For point (1), we consider a non-complete metric space. For point (2), we consider an application which strictly reduces distances; as we have seen, this hypothesis is less strict than requiring the application to be a contraction mapping.

1) We need to consider a non-complete metric space. We have already seen that $((0, 1), |\ |)$ is not complete. In this space, let us consider, for instance, the function $f : (0, 1) \to (0, 1)$, $f(x) = \frac{1}{2}x$. Then:

$$\forall x, y \in (0, 1), \quad |f(x) - f(y)| = \left|\frac{1}{2}x - \frac{1}{2}y\right| = \frac{1}{2}|x - y| \leqslant \frac{1}{2}|x - y|$$

so f is a contraction with coefficient $k = 1/2$. The fixed point equation for f, that is, $f(x) = x$, evidently has no solutions in $(0, 1)$ since $\frac{1}{2}x = x$ if and only if $x = 0 \notin (0, 1)$.

2) Consider the metric space $(X, d) = ([0, +\infty), |\ |)$ and the application $f : [0, +\infty) \to [0, +\infty)$ defined by $f(x) = \sqrt{x^2 + 1}$. Taking two arbitrary fixed elements $x, y \in [0, +\infty)$, due to Lagrange's mean value theorem, there exists an element $\xi \subset [0, +\infty)$ which is strictly included in the interval between x and y, such that:

$$f(x) - f(y) = (x - y)f'(\xi) = (x - y)\frac{\xi}{\sqrt{\xi^2 + 1}}$$

Since $\sqrt{\xi^2+1} > \sqrt{\xi^2} = \xi \in [0,+\infty)$, then $\frac{\xi}{\sqrt{\xi^2+1}} < 1$, so $|f(x) - f(y)| < |x-y|$, that is f strictly reduces the distances.

Nevertheless, in $[0,+\infty)$ the fixed point equation for f, that is, $x = \sqrt{x^2+1}$, can be written as $x^2 = x^2+1$, meaning $1 = 0$, which is obviously a contradiction; thus, f does not admit a fixed point.

3) We know from differential equation theory that solving the Cauchy problem [4.16] is equivalent to determining a function $\varphi \in C([-1,1])$ which satisfies the following Volterra integral equation:

$$\varphi(t) = \frac{1}{2}\int_0^t \sin\varphi(s)ds + 1 \qquad [4.17]$$

Let us verify this statement. On one side, if φ is a solution of [4.16], by definition, φ is differentiable and thus continuous. Integrating both sides of the differential equation from 0 to t, we obtain $\int_0^t \varphi'(s)ds = \frac{1}{2}\int_0^t \sin\varphi(s)ds$, that is, $\varphi(t) - \varphi(0) = \frac{1}{2}\int_0^t \sin\varphi(s)ds$. The initial condition [4.16] gives us $\varphi(0) = 1$, thus φ satisfies $\varphi(t) = \frac{1}{2}\int_0^t \sin\varphi(s)ds + 1$.

On the other side, supposing that φ satisfies [4.17], the integral function $\frac{1}{2}\int_0^t \sin\varphi(s)ds + 1$ is derivable $\forall t \in [-1,1]$, since $\sin\circ\varphi$ is continuous and the integration operation makes any continuous function derivable. Deriving [4.17] gives us $\varphi'(t) = \frac{1}{2}\sin\varphi(t)$ with $\varphi(0) = 1$, that is, φ satisfies [4.16].

These considerations highlight the interest of the space $C([-1,1])$, which is a Banach space when it is endowed with the norm $\|f\| = \sup_{t\in[-1,1]} |f(t)|$. Consider the following application:

$$\begin{aligned} F : C([-1,1]) &\longrightarrow C([-1,1]) \\ f &\longmapsto F(f) \end{aligned}$$

where $F(f)$ is the real-value continuous function on $[-1,1]$ defined by the analytical expression $F(f)(t) = \frac{1}{2}\int_0^t \sin\varphi(s)ds + 1$, for all $t \in [-1,1]$. Clearly, if we can show that F is a contraction, by invoking the fixed point theorem we will complete the proof that there is only one solution to the Cauchy problem [4.16].

To do this, let us consider any two functions $f, g \in C([-1,1])$ and an arbitrary $t \in [-1,1]$; then:

$$|F(f)(t) - F(g)(t)| = \frac{1}{2}\left|\int_0^t [\sin f(s) - \sin g(s)]ds\right|$$

$$\leqslant \frac{1}{2}\left|\int_0^t |\sin f(s) - \sin g(s)|ds\right|$$

(using the formula $\sin p - \sin q = 2\sin\left(\frac{p-q}{2}\right)\cos\left(\frac{p+q}{2}\right)$) :

$$= \left|\int_0^t \left|\sin\frac{f(s)-g(s)}{2}\cos\frac{f(s)+g(s)}{2}\right| ds\right|$$

$(|\cos(\alpha)| \leqslant 1)$

$$\leqslant \left|\int_0^t \left|\sin\frac{f(s)-g(s)}{2}\right| ds\right|$$

$(|\sin(\alpha)| \leqslant |\alpha|)$

$$\leqslant \left|\int_0^t \left|\frac{f(s)-g(s)}{2}\right| ds\right|$$

$$\leqslant \frac{1}{2}\left|\int_0^t \|f-g\|ds\right|$$

$$\leqslant \frac{\|f-g\|}{2}\left|\int_0^t ds\right| = \frac{\|f-g\|}{2}|t|$$

$(t \in [-1,1] \implies |t| \leqslant 1)$

$$\leqslant \frac{\|f-g\|}{2}$$

In summary: $|F(f)(t) - F(g)(t)| \leqslant \frac{\|f-g\|}{2}$ $\forall t \in [-1,1]$, hence:

$$\|F(f) - F(g)\| = \sup_{t\in[-1,1]} |F(f)(t) - F(g)(t)| \leqslant \frac{1}{2}\|f-g\|$$

that is F is a contraction. □

4.4. Remarkable examples of Banach and Hilbert spaces

In this section, we shall introduce function spaces which are of crucial importance in mathematics. We shall demonstrate that some of these spaces are Banach spaces, while others are Hilbert spaces; we shall then present density theorems related to these spaces.

4.4.1. L^p and ℓ^p spaces: presentation and completeness

In the following definitions, $\mathbb{K}$ will be either $\mathbb{R}$ or $\mathbb{C}$. Let $(X, \mathcal{A}, \mu)$ be a measure space. For all $1 \leqslant p < +\infty$, we define:

$$\boxed{\mathcal{L}^p(X, \mathcal{A}, \mu) = \left\{ f : X \to \mathbb{K},\ f \text{ measurable } : \int_X |f|^p d\mu < +\infty \right\}}$$

The set $\mathcal{L}^p(X, \mathcal{A}, \mu)$ becomes a vector space if we define the *pointwise vector structure*, that is $\forall \alpha, \beta \in \mathbb{K}, \forall f, g \in \mathcal{L}^p(X, \mathcal{A}, \mu)$:

$$\begin{array}{rcl} \alpha f + \beta g : X & \Longrightarrow & \mathbb{K} \\ x & \mapsto & (\alpha f + \beta g)(x) = \alpha f(x) + \beta g(x) \end{array}$$

This linear combination operation is well defined thanks to the famous *Minkowski inequality*[6] for integrals (which we will not prove):

$$\boxed{\left(\int_X |f+g|^p d\mu\right)^{1/p} \leqslant \left(\int_X |f|^p d\mu\right)^{1/p} + \left(\int_X |g|^p d\mu\right)^{1/p}} \quad [4.18]$$

Since multiplication by the scalars α, β has no effect on the integrability of f, g, the definition is coherent.

Writing:

$$\boxed{\|f\|_p = \left(\int_X |f|^p d\mu\right)^{1/p}}$$

the properties of the Lebesgue integral give us:

– positiveness (non-definite) and homogeneity:

$$\|f\|_p \geqslant 0, \ \|\lambda f\|_p = |\lambda| \|f\|_p \quad \forall f \in \mathcal{L}^p(X, A, \mu), \ \lambda \in \mathbb{K}$$

– *the Minkowski inequality [4.18] becomes the triangular inequality*[7] *for* $\| \ \|_p$:

$$\|f+g\|_p \leqslant \|f\|_p + \|g\|_p, \quad \forall f, g \in \mathcal{L}^p(X, A, \mu)$$

6 By iteration, we can write the *generalized Minkowski inequality*, which we shall use later: $\left(\int_X \left|\sum_{k=1}^n f_k\right|^p d\mu\right)^{1/p} \leqslant \sum_{k=1}^n \left(\int_X |f_k|^p d\mu\right)^{1/p}$.

7 By iteration: $\left\|\sum_{k=1}^n f_k\right\|_p \leqslant \sum_{k=1}^n \|f_k\|_p$. [4.18]

– but:

$$\|f\|_p = 0 \not\Longrightarrow f = 0 \text{ (the null function)}$$

so any function $g \in \mathcal{L}^p(X, A, \mu)$ which is null a.e. is such that $\|g\|_p = 0$. Thus, the fact that $\|f\|_p = 0$ does not imply that f is the null function, that is, that $f(x) = 0$ $\forall x \in X$.

Hence, $\| \ \|_p$ *is a semi-norm (or pseudo-norm) on* $\mathcal{L}^p(X, \mathcal{A}, \mu)$. Unfortunately, the presence of a semi-norm makes it impossible to use the (highly useful) property that, if the norm of the difference between two elements in a normed space is null, then the two elements coincide, since they can differ over a set of measure zero. This norm feature is used to show the uniqueness of a mathematical object in cases where it is difficult to prove it directly, this is why it is important to preserve it.

The solution to the problem is to apply the quotient of $\mathcal{L}^p(X, \mathcal{A}, \mu)$ w.r.t. a suitable subspace that allows us to get rid of the redundant functions. It should be clear that this subspace is:

$$\mathcal{N} = \{f : X \to \mathbb{K},\ f \text{ measurable} \ : \ f = 0 \ \text{ a.e}\}$$

The quotient space

$$\boxed{L^p(X, \mathcal{A}, \mu) = \mathcal{L}^p(X, \mathcal{A}, \mu)/\mathcal{N}}$$

formed by the equivalence classes of functions which are measurable on X, absolutely integrable in power p and equal a.e, is thus a *normed vector space with norm* $\| \ \|_p$.

Using the considerations presented in Appendix 1, it becomes apparent that, fixed a representative f of an equivalence class of $L^p(X, \mathcal{A}, \mu)$, all other functions g belonging to the same class can be written as $g = f + h$, where $h : X \to \mathbb{K}$ is null a.e.

For simplicity's sake, a representative function and the equivalence class to which it belongs are generally noted using the same symbol. Furthermore, in cases where X, $\mathcal{A}$ and μ do not need to be specified, we may simply write L^p.

REMARK.– Take $X \subseteq \mathbb{K}^n$ with the Lebesgue measure. Let us consider two functions $f, g \in \mathcal{L}^p(X, \mathcal{A}, \mu)$ which are *continuous* on X and which differ, at least, at the point $x_0 \in X$: $f(x_0) \neq g(x_0)$. By definition of continuity:

$$\forall \varepsilon > 0 \ \exists \delta_\varepsilon > 0 \ : \ x \in U_{\delta_\varepsilon}(x_0) \implies f(x) \in U_\varepsilon(f(x_0)) \text{ and } g(x) \in U_\varepsilon(g(x_0))$$

but, by the separability property of $\mathbb{K}^n$, $\exists \varepsilon > 0$ such that $U_\varepsilon(f(x_0)) \cap U_\varepsilon(g(x_0)) = \emptyset$, that is, $\exists \delta_\varepsilon > 0$ such that $x \in U_{\delta_\varepsilon}(x_0)$ implies $f(x) \neq g(x)$, that is, if two continuous functions f and g are different at a point x_0, they must also be different on a neighborhood $U_{\delta_\varepsilon}(x_0)$ of radius $\delta_\varepsilon > 0$. This neighborhood has a *non-null Lebesgue measure*, so the two functions are not equal a.e.

In other words: *two functions which are continuous on $X \subset \mathbb{K}^n$ cannot be equal a.e.*: either they are the same function, or they are different on a non-null Lebesgue measure set. Thus, *two continuous functions of $\mathcal{L}^p(X, \mathcal{A}, \mu)$ which are different in at least one point are two different elements of $L^p(X, \mathcal{A}, \mu)$*, as they are representatives of two different equivalence classes.

If $p = 2$, then we can define an inner product on $L^2(X, \mathcal{A}, \mu)$:

$$\boxed{\langle f, g\rangle = \int_X fg \, d\mu} \quad \text{if } \mathbb{K} = \mathbb{R} \quad \text{and} \quad \boxed{\langle f, g\rangle = \int_X f\bar{g} \, d\mu} \quad \text{if } \mathbb{K} = \mathbb{C}$$

These inner products are well defined thanks to *Hölder's inequality for integrals* (which we shall not prove here): if $p, q > 0$ are *conjugate exponents*, that is, $\frac{1}{p} + \frac{1}{q} = 1$, then it holds that:

$$\boxed{\int_X |fg| \, d\mu \leqslant \left(\int_X |f|^p \, d\mu\right)^{1/p} \left(\int_X |g|^q \, d\mu\right)^{1/q}} \qquad [4.19]$$

Evidently, $p = q = 2$ are conjugate exponents and thus the inner product introduced above is well defined. The proof that this verifies the axioms of the inner product is left to the reader; here we note simply that Hölder's inequality for $p = q = 2$ implies the validity of *the Cauchy-Schwarz inequality for the space $L^2(X, \mathcal{A}, \mu)$*. In fact, for all $f, g \in L^2(X, \mathcal{A}, \mu)$:

$$|\langle f, g\rangle_{L^2}| = \left|\int_X f\bar{g} \, d\mu\right| \leqslant \int_X |fg| \, d\mu \underset{[4.19]}{\leqslant} \left(\int_X |f|^2 \, d\mu\right)^{1/2} \left(\int_X |g|^2 \, d\mu\right)^{1/2}$$

$$= \|f\|_2 \|g\|_2$$

One notable instance of L^p spaces is represented by the ℓ^p spaces, which are defined through the following choices:

– X is taken to be a countable set, typically $X = \mathbb{N}$ or $X = \mathbb{Z}$;

– $\mathcal{A} = \mathcal{P}(X)$, the set of parts of X;

– μ is the *counting measure*, that is $\mu : \mathcal{P}(X) \to [0, +\infty]$, $\mu(A) = \text{card}(A)$ $\forall A \in \mathcal{P}(X)$ which has a finite cardinal and $\mu(A) = +\infty$ if card(A) is not finite.

Using these choices, any function $f : X \Longrightarrow \mathbb{K}$ is measurable and it can be identified with a sequence of elements in $\mathbb{K}$, written $(x_n)_{n\in\mathbb{N}}$. Thus, explicitly[8]:

$$\boxed{\ell^p(\mathbb{N},\mathbb{K}) = \left\{(x_n)_{n\in\mathbb{N}},\ \sum_{n\in\mathbb{N}} |x_n|^p < +\infty\right\}}$$

the same considerations hold if we exchange $\mathbb{N}$ for $\mathbb{Z}$.

In cases where there is no need to specify $\mathbb{N}$, $\mathbb{Z}$ or any other countable set, we simply write ℓ^p. The linear structure of these spaces is the same as that of the L^p spaces, that is pointwise defined, and the norm of $(x_n)_{n\in\mathbb{N}} \in \ell^p(\mathbb{N},\mathbb{K})$ is:

$$\boxed{\|(x_n)_{n\in\mathbb{N}}\|_p = \left(\sum_{n\in\mathbb{N}} |x_n|^p\right)^{1/p}}$$

The same holds if we exchange $\mathbb{N}$ for $\mathbb{Z}$. The triangular inequality for this norm follows from *the Minkowski inequality for series*:

$$\boxed{\left(\sum_{n\in\mathbb{N}} |x_n + y_n|^p\right)^{1/p} \leqslant \left(\sum_{n\in\mathbb{N}} |x_n|^p\right)^{1/p} + \left(\sum_{n\in\mathbb{N}} |y_n|^p\right)^{1/p}} \qquad [4.20]$$

As in the case of L^p spaces, if $p = 2$, an inner product can be defined on ℓ^2:

$$\boxed{\langle x_n, y_n\rangle = \sum_{n\in\mathbb{N}} x_n y_n} \quad \text{if } \mathbb{K} = \mathbb{R} \quad \text{and} \quad \boxed{\langle x_n, y_n\rangle = \sum_{n\in\mathbb{N}} x_n \overline{y_n}} \quad \text{if } \mathbb{K} = \mathbb{C}$$

The same holds if we exchange $\mathbb{N}$ for $\mathbb{Z}$, or any other countable set.

The inner product is well defined thanks to *Hölder's inequality for series*: if $p, q > 0$, $\frac{1}{p} + \frac{1}{q} = 1$, then it holds that:

$$\boxed{\sum_{n\in\mathbb{N}} |x_n y_n| \leqslant \left(\sum_{n\in\mathbb{N}} |x_n|^p\right)^{1/p} \left(\sum_{n\in\mathbb{N}} |y_n|^q\right)^{1/q}}$$

REMARK.–

– The inner product of $\ell^2(\mathbb{N},\mathbb{K})$ is the infinite-dimensional generalization of the inner product of $\ell^2(\mathbb{Z}_N)$.

8 The spaces $\ell^p(\mathbb{N},\mathbb{K})$ are vector subspaces of the vector space $\mathbb{K}^{\mathbb{N}} := \{(x_n)_{n\in\mathbb{N}},\ x_n \in \mathbb{K}\ \forall n \in \mathbb{N}\}$ of sequences with values in $\mathbb{K}$ possessing a pointwise defined linear structure. The same holds if $\mathbb{N}$ and $\mathbb{Z}$ are switched, in which case we speak of bilateral sequences.

– The role of the Minkowski and Hölder inequalities in defining L^p and ℓ^p spaces should be clear: the Minkowski inequality guarantees the existence of a linear structure, and Hölder's inequality ensures that the inner product is well defined in the case where $p = 2$.

– $\| \ \|_p$ norms with $p \neq 2$ are not Hilbert norms, in fact it is possible to provide examples of elements in all the L^p spaces, with $p \neq 2$, for which the parallelogram law is not verified.

Now, let us demonstrate that *L^p and ℓ^p spaces with $1 \leqslant p < +\infty$, $p \neq 2$ are Banach spaces, and for $p = 2$, Hilbert spaces.*

The completeness of $L^2([0,1])$ spaces was demonstrated independently by the Austrian mathematician Ernst Sigismund Fischer (1875-1954) and by Frigyes Riesz[9] in 1907. In 1910, Riesz demonstrated that all $L^p[0,1]$ spaces are complete.

THEOREM 4.14 (Riesz-Fischer theorem).– For all $1 \leqslant p < +\infty$, the spaces $(L^p(X, \mathcal{A}, \mu), \| \ \|_p)$ and $(\ell^p, \| \ \|_p)$ are complete.

PROOF.– We will report Riesz's demonstration, who brought out the heavy artillery to prove these results, using the characterization theorem for complete normed vector spaces, Fatou's lemma, the generalized Minkowski inequality, the monotone convergence theorem and the dominated convergence theorem to construct his proof.

Let us consider any series $\sum_{k=0}^{\infty} f_k$ in $L^p(X, \mathcal{A}, \mu)$, $1 \leqslant p \leqslant +\infty$, which is absolutely convergent, that is:

$$\sum_{k=0}^{\infty} \|f_k\|_p = M < +\infty$$

then, by Theorem 4.12, to show that $L^p(X, \mathcal{A}, \mu)$ is complete, we must simply prove that $\sum_{k \in \mathbb{N}} f_k$ is convergent in norm, that is that $\exists S \in L^p(X, \mathcal{A}, \mu)$ such that:

$$\left\| \sum_{k=0}^{n} f_k - S \right\|_p \underset{n \Longrightarrow +\infty}{\longrightarrow} 0 \qquad [4.21]$$

The first step in determining the function S is to define the sequence:

$$(g_n)_{n \in \mathbb{N}}, \ g_n = \sum_{k=0}^{n} |f_k|, \quad \forall n \in \mathbb{N}$$

9 Frigyes Riesz (1880-1956) was a Hungarian mathematician who made many hugely important contributions to the development of functional analysis, among other areas.

Using the generalized Minkowski theorem in equation [4.18], we know that:

$$\left(\int_X (g_n)^p d\mu\right)^{1/p} = \|g_n\|_p = \left\|\sum_{k=0}^{n} |f_k|\right\|_p \leqslant \sum_{k=0}^{n} \|f_k\|_p$$

$$\leqslant \sum_{k=0}^{\infty} \|f_k\|_p \underset{\text{(by hypothesis)}}{=} M < +\infty$$

hence:

$$\int_X (g_n)^p d\mu \leqslant M^p, \qquad \forall n \in \mathbb{N} \qquad [4.22]$$

that is, $(g_n)^p$ is a sequence of monotonic increasing functions of integrable functions, and the sequence of integrals is bounded.

The monotone convergence Theorem 3.3 tells us that the pointwise limit function $\lim\limits_{n\to+\infty}(g_n)^p(x)$ is finite a.e. on X, that is, $\forall x \in E \subseteq X$ and $\mu(X\backslash E) = 0$. This implies the existence $\forall x \in E$ of a finite pointwise limit:

$$g(x) \equiv \lim_{n\to+\infty} g_n(x) \left(= \lim_{n\to+\infty}[(g_n)^p(x)]^{1/p}\right)$$

Since $\forall x \in E$, $\sum\limits_{k=0}^{\infty} f_k(x) \leqslant \sum\limits_{k=0}^{\infty} |f_k(x)| = g(x)$, the series $\sum\limits_{k=0}^{\infty} f_k(x)$ converges a.e. on X.

Now, let us construct the required function $S : X \to \mathbb{K}$:

$$S(x) = \begin{cases} \sum\limits_{k=0}^{\infty} f_k(x) & x \in E \\ 0 & x \in X\backslash E \end{cases}$$

This definition ensures that S is measurable. The fact that $S \in L^p(X, \mathcal{A}, \mu)$, that is, $\int_X S^p d\mu$ exists and is finite a.e., is a consequence of the dominated convergence theorem (Theorem 3.5) and Fatou's lemma (Theorem 3.4). This can be proved by considering the sequence of partial sums for S^p,

that is:

$$(S_n)^p = \left(\sum_{k=0}^{n} f_k\right)^p \leqslant \left(\sum_{k=0}^{n} |f_k|\right)^p = (g_n)^p.$$

$(g_n)^p$ is an increasing positive sequence, thus:

$$(S_n)^p(x) \leqslant (g_n)^p(x) \leqslant \lim_{n\to+\infty}(g_n)^p(x) = g^p(x), \qquad \forall x \in E \qquad [4.23]$$

By monotony, $\lim\limits_{n\to+\infty}(g_n)^p(x) = \liminf\limits_{n\to+\infty}(g_n)^p(x)$ and thus, by Fatou's lemma, we have:

$$\int_X g^p d\mu \leqslant \lim_{n\to+\infty}\int_X (g_n)^p d\mu \underset{[4.22]}{\leqslant} \lim_{n\to+\infty} M^p = M^p < +\infty$$

The positive measurable function g^p is therefore integrable a.e. on X. Using this information and equation [4.23], that is $(S_n)^p \leqslant g^p\ \forall n \in \mathbb{N}$, the dominated convergence theorem can be used to guarantee that S^p, the a.e. limit of $(S_n)^p$, converges on X, that is $S \in L^p(X, \mathcal{A}, \mu)$.

To complete our proof, we must demonstrate that function S verifies equation [4.21], that is:

$$\left\|\sum_{k=0}^{n} f_k - S\right\|_p \underset{n\Longrightarrow+\infty}{\longrightarrow} 0 \Longleftrightarrow \lim_{n\to+\infty}\int_E \left|\sum_{k=0}^{n} f_k - \sum_{k=0}^{\infty} f_k\right|^p d\mu = 0$$

note that we do not need to write the integration on $X\backslash E$ since $\mu(X\backslash E) = 0$. With our notation, the condition of convergence in norm $\| \ \|_p$ for the series $\sum\limits_{k=0}^{\infty} f_k$ to S can be rewritten in a simpler way as follows:

$$\lim_{n\to+\infty}\|S_n - S\|_p = 0 \Longleftrightarrow \lim_{n\to+\infty}\int_E |S_n - S|^p\, d\mu = 0$$

Evidently, if we can show that the integral and the limit can switch places, then the result will be proved, since, in this case:

$$\lim_{n\to+\infty}\int_E |S_n - S|^p\, d\mu = \int_E \lim_{n\to+\infty}|S_n - S|^p\, d\mu$$

$$\underset{(S \text{ is independent of } n)}{=} \int_E \left|\lim_{n\to+\infty} S_n - S\right|^p d\mu = 0$$

To make this exchange possible, we can write the following majorization:

$$|S_n(x) - S(x)|^p \leqslant (|S_n(x)| + |S(x)|)^p \leqslant (g(x) + g(x))^p = (2g(x))^p$$

$$= 2^p(g(x))^p \quad \forall x \in E$$

As $\int_X g^p d\mu \leqslant M^p < +\infty$, this majorization ensures that the sequence $(|S_n(x) - S(x)|^p)_{n\in\mathbb{N}}$ verifies the conditions of the dominated convergence theorem, meaning that the limit and integral can be exchanged.

As we saw previously, this ensures that the series $\sum\limits_{k=0}^{\infty} f_k$ in $L^p(X, \mathcal{A}, \mu)$, which we presumed to be absolutely convergent, is also simply convergent. Hence, all $L^p(X, \mathcal{A}, \mu)$ spaces with $1 \leqslant p < \infty$ are complete.

Since ℓ^p spaces are special cases of L^p spaces, this result also holds for these spaces $\forall 1 \leqslant p < \infty$. □

Exercise 4.2

Let $a = (a_n)_{n\in\mathbb{N}}$ be a sequence of strictly positive real numbers, and let $\ell_a^2(\mathbb{N}, \mathbb{C})$ be the vector space formed by the sequences of complex numbers $(u_n)_{n\in\mathbb{N}}$ which verify $\sum_{n\in\mathbb{N}} a_n|u_n|^2 < +\infty$. Show that the application defined by:

$$\langle u, v\rangle_{\ell_a^2} = \sum_{n\in\mathbb{N}} a_n u_n \overline{v_n}$$

is well defined on $\ell_a^2(\mathbb{N}, \mathbb{C}) \times \ell_a^2(\mathbb{N}, \mathbb{C})$ (i.e. $\langle u, v\rangle$ exists for all $u, v \in \ell_a^2(\mathbb{N}, \mathbb{C})$), and deduce that this is an inner product.

Solution to Exercise 4.2

Since $u, v \in \ell_a^2(\mathbb{N}, \mathbb{C})$, $\sqrt{a}u$ and $\sqrt{a}\overline{v}$ belong to $\ell^2(\mathbb{N}, \mathbb{C})$, then:

$$\langle u, v\rangle_{\ell_a^2} = \sum_{n\in\mathbb{N}} \sqrt{a_n}u_n\overline{\sqrt{a_n}v_n} = \langle\sqrt{a_n}u_n, \sqrt{a_n}v_n\rangle_{\ell^2} < +\infty$$

The sesquilinearity and conjugate symmetry of $\langle u, v\rangle_{\ell_a^2}$ follow directly from the analogous properties of the inner product of $\ell^2(\mathbb{N}, \mathbb{C})$. The only element to verify explicitly is definite positiveness. If $u \in \ell_a^2(\mathbb{N}, \mathbb{C})$, then $\langle u, u\rangle_{\ell_a^2} = \sum_{n\in\mathbb{N}} a_n|u_n|^2 \geqslant 0$ as it is a sum of positive terms. This formula also shows that $\langle u, u\rangle_{\ell_a^2} = 0 \iff a_n|u_n|^2 = 0$ for all $n \in \mathbb{N}$, but $a_n > 0$ for all $n \in \mathbb{N}$ by hypothesis, thus $|u_n|^2 = 0 \iff u_n = 0\ \forall n \in \mathbb{N}$, that is $u = 0_{\ell_a^2}$. □

Exercise 4.3

Take $s \in \mathbb{R}$, $s > 0$ and:

$$H^s = \left\{u = (u_n)_{n\in\mathbb{N}} \subset \mathbb{C}\ \forall n \in \mathbb{N} \ :\ \sum_{n\in\mathbb{N}}(1 + n^2)^s|u_n|^2 < +\infty\right\}$$

H^s is a Hilbert space which is often encountered when solving differential equations using the Fourier transform.

1) Show that H^s is a vector subspace of $\ell^2(\mathbb{N}, \mathbb{C})$.

2) Let $\phi : H^s \times H^s \to \mathbb{C}$ be the application defined by:

$$\phi(u, v) := \sum_{n\in\mathbb{N}}(1 + n^2)^s u_n\overline{v_n} \qquad \forall u, v \in H^s$$

presuming, for the moment, that the application is well defined, that is, the series converges. For any sequence $w = (w_n)_{n\in\mathbb{N}} \in H^s$, define the sequence $\tilde{w}$ as follows:

$$\tilde{w}_n = (1+n^2)^{s/2} w_n \qquad \forall n \in \mathbb{N}$$

a) Show that $\tilde{w} \in \ell^2(\mathbb{N},\mathbb{C})$ and it holds that:

$$\phi(u,v) = \langle \tilde{u}, \tilde{v} \rangle_{\ell^2} \qquad \forall u, v \in H^s$$

where $\langle\ ,\ \rangle_{\ell^2}$ is the usual inner product of $\ell^2(\mathbb{N},\mathbb{C})$.

b) Deduce that ϕ is well defined on $H^s \times H^s$, then that it constitutes an inner product, noted $\phi = \langle\ ,\ \rangle_{H^s}$.

3) We wish to show that $(H^s, \langle\ ,\ \rangle_{H^s})$ is a Hilbert space. To do this, let us fix an arbitrary Cauchy sequence $(u_m)_{m\in\mathbb{N}}$ in H^s.

a) Show that $(\tilde{u}_m)_{m\in\mathbb{N}}$ is Cauchy in $\ell^2(\mathbb{N},\mathbb{C})$.

b) Deduce that $(\tilde{u}_m)_{m\in\mathbb{N}}$ converges in $\ell^2(\mathbb{N},\mathbb{C})$ to a limit, which we shall note $\tilde{l}$.

c) Define the sequence $l = (l_n)_{n\in\mathbb{N}}$ by:

$$l_n = \frac{1}{(1+n^2)^{s/2}} \tilde{l}_n \qquad \forall n \in \mathbb{N}$$

Show that l belongs to H^s, that $(u_m)_{m\in\mathbb{N}}$ converges to l in H^s, and conclude your proof.

Solution to Exercise 4.3

1) To show that $H^s \subset \ell^2(\mathbb{N},\mathbb{C})$ we shall demonstrate, in order, that $u \in H^s \implies u \in \ell^2(\mathbb{N},\mathbb{C})$, that $H^s \neq \varnothing$, and that H^s is stable with respect to linear combinations of its elements.

For any sequence $u = (u_n)_{n\in\mathbb{N}} \subset \mathbb{C}$ it holds that $0 \leqslant |u_n|^2 \leqslant (1+n^2)|u_n|^2$ for all $n \in \mathbb{N}$, hence:

$$\sum_{n\in\mathbb{N}} |u_n|^2 \leqslant \sum_{n\in\mathbb{N}} (1+n^2)^s |u_n|^2 \underset{\text{def. of } H^s}{<} +\infty \implies u \in \ell^2(N,\mathbb{C})$$

Evidently, $0_{\ell^2} \in H^s$, thus $H^s \neq \varnothing$. Finally, taking $\lambda \in \mathbb{C}$ and $u, v \in H^s$, then:

$$0 \leqslant |u_n + \lambda v_n|^2 \leqslant (|u_n| + |\lambda||v_n|)^2 \leqslant 2(|u_n|^2 + |\lambda|^2|v_n|^2)$$

where the final inequality draws on the fact that the moduli are real numbers and that, for all $a, b \in \mathbb{R}$, $0 \leqslant (a-b)^2 = a^2 + b^2 - 2ab = 2a^2 - a^2 + 2b^2 - b^2 - 2ab$, so $a^2 + b^2 + 2ab \leqslant 2a^2 + 2b^2$, that is $(a+b)^2 \leqslant 2(a^2+b^2)$; writing $a = |u_n|$ and $b = |\lambda||v_n|$, we obtain the final inequality from the previous formula. Now, with respect to the series, we can write:

$$\sum_{n\in\mathbb{N}}(1+n^2)^s|u_n+\lambda v_n|^2 \leqslant 2\left(\sum_{n\in\mathbb{N}}(1+n^2)^s|u_n|^2 + |\lambda|^2\sum_{n\in\mathbb{N}}(1+n^2)^s|v_n|^2)\right) < +\infty$$

as $u, v \in H^s$, thus $u + \lambda v \in H^s$ and so H^s is a vector subspace of $\ell^2(\mathbb{N},\mathbb{C})$.

2) *a)* $\tilde{w} \in \ell^2(\mathbb{N},\mathbb{C})$ if $\sum_{n\in\mathbb{N}} |\tilde{w}_n|^2 < +\infty$, but:

$$\sum_{n\in\mathbb{N}} |\tilde{w}_n|^2 = \sum_{n\in\mathbb{N}}(1+n^2)^s|w_n|^2 < +\infty$$

since $w \in H^s$, so $\tilde{w} \in \ell^2(\mathbb{N},\mathbb{C})$. Now, taking $u, v \in H^s$:

$$\phi(u,v) = \sum_{n\in\mathbb{N}}(1+n^2)^s u_n\overline{v_n} = \sum_{n\in\mathbb{N}}(1+n^2)^{s/2}u_n\overline{(1+n^2)^{s/2}v_n} = \sum_{n\in\mathbb{N}} \tilde{u}_n\overline{\tilde{v}_n} = \langle \tilde{u},\tilde{v}\rangle_{\ell^2}$$

b) We have:

$$\begin{aligned}\phi(u,v) &= \textstyle\sum_{n\in\mathbb{N}}(1+n^2)^s u_n v_n \leqslant \sum_{n\in\mathbb{N}} |(1+n^2)^s u_n\overline{v}_n| \\ &= \sum_{n\in\mathbb{N}} |(1+n^2)^{s/2}u_n(1+n^2)^{s/2}\overline{v}_n| \\ &= \sum_{n\in\mathbb{N}} |\tilde{u}_n\overline{\tilde{v}_n}| = \langle \tilde{u},\tilde{v}\rangle_{\ell^2} \underset{\text{Cauchy-Schwarz}}{\leqslant} \|\tilde{u}\|_{\ell^2}\|\tilde{v}\|_{\ell^2} < +\infty\end{aligned}$$

thus $\phi(u,v)$ is well defined for all $u, v \in H^2$. By the fact that $\phi(u,v) = \langle \tilde{u},\tilde{v}\rangle_{\ell^2}$, we know that ϕ is an inner product: it is Hermitian and sesquilinear, since $\langle\ ,\ \rangle_{\ell^2}$ possesses these properties. Regarding the definite positiveness, we simply note that for all $u \in H^s$, $\phi(u) = 0$ implies $\sum_{n\in\mathbb{N}}(1+n^2)^{s/2}u_n(1+n^2)^{s/2}\overline{u_n} = \langle \tilde{u},\tilde{u}\rangle_{\ell^2} = \|\tilde{u}\|_{\ell^2} = 0$, that is, $\tilde{u} = 0$, that is, $(1+n^2)^{s/2}u_n = 0 \iff u_n = 0\ \forall n \in \mathbb{N}$. Hence ϕ is a complex inner product on H^2, and this is noted $\phi(u,v) = \langle u, v\rangle_{H^s}$.

3) *a)* To prove that if $u = (u_m)_{m\in\mathbb{N}}$ is an arbitrary Cauchy sequence in H^s then $(\tilde{u}_m)_{m\in\mathbb{N}}$ is a Cauchy sequence in $\ell^2(\mathbb{N},\mathbb{C})$, we write the Cauchy condition in its squared form for u:

$$\forall\varepsilon > 0\ \exists N_\varepsilon \in \mathbb{N}\ :\ m, k \leqslant N_\varepsilon \implies \|u_m - u_k\|^2_{H^s} < \varepsilon^2$$

but $\|u_m - u_k\|^2_{H^s} = \langle u_m - u_k, u_m - u_k\rangle_{H^s} \underset{(2.(\mathrm{a}))}{=} \langle \widetilde{u_m - u_k}, \widetilde{u_m - u_k}\rangle_{\ell^2}$, and:

$$\widetilde{u_m - u_k} = (1+n^2)^{s/2}(u_m - u_k) = (1+n^2)^{s/2}u_m - (1+n^2)^{s/2}u_k = \tilde{u}_m - \tilde{u}_k$$

hence $\|u_m - u_k\|^2_{H^s} = \langle \widetilde{u_m - u_k}, \widetilde{u_m - u_k}\rangle_{\ell^2} = \langle \tilde{u}_m - \tilde{u}_k, \tilde{u}_m - \tilde{u}_k\rangle_{\ell^2} = \|\tilde{u}_m - \tilde{u}_k\|^2_{\ell^2}$, which implies that $(\tilde{u}_m)_{m\in\mathbb{N}}$ is a Cauchy sequence in $\ell^2(\mathbb{N},\mathbb{C})$.

b) Given that $\ell^2(\mathbb{N}, \mathbb{C})$ is complete, the Cauchy sequence $(\tilde{u}_m)_{m\in\mathbb{N}}$ converges to an element in $\ell^2(\mathbb{N}, \mathbb{C})$ which we note $\tilde{l}$.

c) Let us consider the sequence $l = \tilde{l}/(1+n^2)^{s/2}$ and show that it belongs to H^s by calculating the square of its norm in H^s:

$$\|l\|^2_{H^s} = \sum_{n\in\mathbb{N}} (1+n^2)^s |l_n|^2 = \sum_{n\in\mathbb{N}} \cancel{(1+n^2)^s} \frac{|\tilde{l}_n|^2}{\cancel{((1+n^2)^s}} = \sum_{n\in\mathbb{N}} |\tilde{l}_n|^2 < +\infty$$

so $l \in H^2$.

Now, let us show that $(u_m)_{m\in\mathbb{N}}$ converges to ℓ: using the result from point (2a), we have $\langle u_m - l, u_m - l\rangle_{H^s} = \langle \widetilde{u_m - l}, \widetilde{u_m - l}\rangle_{\ell^2}$. Since we have also seen that $\widetilde{u_m - l} = \tilde{u}_m - \tilde{l}$, it holds that $\|u_m - l\|^2_{H^s} = \|\tilde{u}_m - \tilde{l}\|^2_{\ell^2} \underset{m\to+\infty}{\longrightarrow} 0$, by (3b), that is, $(u_m)_{m\in\mathbb{N}}$ converges to l in H^s. We have thus demonstrated that the arbitrary Cauchy sequence $(u_m)_{m\in\mathbb{N}}$ converges inside H^s, that is, H^s constitutes a Hilbert space. □

4.4.2. L^∞ *and* ℓ^∞ *spaces*

The case where $p = \infty$ has been deliberately excluded up to this point, and will be examined separately here. Let $(X, \mathcal{A}, \mu)$ be a measure space, as before, and let $\mathbb{K} = \mathbb{R}$ or $\mathbb{C}$. We begin by defining the space:

$$\boxed{\mathcal{L}^\infty(X, \mathcal{A}, \mu) = \{f : X \implies \mathbb{K} \ : \ \exists M \in \mathbb{R}, M \geqslant 0, \text{ such that } |f(x)| \leqslant M \text{ a.e.}\}}$$

The elements of $\mathcal{L}^\infty(X, \mathcal{A}, \mu)$ are known as *essentially bounded functions*, that is, functions which are bounded on the complement of a null measure set w.r.t. μ.

As in the case of L^p spaces, we need to introduce the equivalence relation:

$$f, g \in \mathcal{L}^\infty(X, \mathcal{A}, \mu), \ f \sim g \text{ if } f = g \text{ a.e.}$$

to make the quotient space:

$$L^\infty(X, \mathcal{A}, \mu) = \mathcal{L}^\infty(X, \mathcal{A}, \mu)/_\sim$$

a normed vector space with norm given by:

$$\boxed{\|f\|_\infty = \inf\{M \geqslant 0 \ : \ |f(x)| \leqslant M \text{ a.e.}\}}$$

which we shall call ess sup(f), read as the *essential supremum* of f, which, by definition, satisfies:

$$|f(x)| \leqslant ||f||_\infty \text{ a.e.}$$

for all $f \in L^\infty(X, \mathcal{A}, \mu)$.

The symbol $\| \ \|_\infty$ has its origins in the fact that if $1 \leqslant p < +\infty$ and $f \in L^p \cap L^\infty$, then:

$$\|f\|_\infty = \lim_{p \Longrightarrow +\infty} \|f\|_p$$

As in the case of L^p spaces, the case of continuous functions requires further clarification. If a continuous function is such that $|f(x)| > M$, then, by completeness, there exists a neighborhood of positive radius in which f is not bounded by M. Thus, *a continuous and essentially bounded function is actually a bounded function in the usual sense.*

We also define:

$$\boxed{\ell^\infty(\mathbb{N}, \mathbb{K}) = L^\infty(\mathbb{N}, \mathcal{P}(\mathbb{N}), \mu_{\text{counting}}) = \{(x_n)_{n\in\mathbb{N}} \ : \ x_n \in \mathbb{K} \ \ \forall n \in \mathbb{N}, \ \exists M \geqslant 0 \ : \ |x_n| \leqslant M\}}$$

that is, ℓ^∞ is *the space of bounded sequences* (a similar definition is obtained if we exchange $\mathbb{N}$ for $\mathbb{Z}$). $\ell^\infty(\mathbb{N}, \mathbb{K})$ is a normed space with:

$$\boxed{\|(x_n)_{n\in\mathbb{N}}\|_\infty = \sup_{n\in\mathbb{N}} |x_n|}$$

THEOREM 4.15.– $(L^\infty(X, \mathcal{A}, \mu), \| \ \|_\infty)$ and $(\ell^\infty(\mathbb{N}, \mathbb{K}), \| \ \|_\infty)$ are Banach spaces.

PROOF.– Let us set out the proof for $L^\infty(X, \mathcal{A}, \mu)$, then the fact that $\ell^\infty(\mathbb{N}, \mathbb{K})$ is a Banach space will be an automatic implication.

We must show that, if $(f_n)_{n\in\mathbb{N}}$ is a Cauchy sequence of elements of $L^\infty(X, \mathcal{A}, \mu)$, then it converges to an element in $L^\infty(X, \mathcal{A}, \mu)$.

By the definition of a Cauchy sequence, we have:

$$\forall \varepsilon > 0 \ \exists N_\varepsilon > 0 \ : \ n, m \geqslant N_\varepsilon \implies \|f_n - f_m\|_\infty < \varepsilon \qquad [4.24]$$

This will be used later.

Now, let us consider the sets of points where the functions in the sequence behave in a "peculiar" manner:

$$A_k = \{x \in X : |f_k(x)| > ||f_k||_\infty\},$$

$$B_{n,m} = \{x \in X : |f_n(x) - f_m(x)| > ||f_n - f_m||_\infty\}$$

by the definition of $L^\infty(X, \mathcal{A}, \mu)$, $\mu(A_k) = \mu(B_{n,m}) = 0$ and:

$$\forall x \in A_k^c : \ |f_k(x)| \leqslant ||f_k||_\infty, \ \forall x \in B_{n,m}^c : \ |f_n(x) - f_m(x)| \leqslant ||f_n - f_m||_\infty$$

To eliminate the dependency of the indices k, n, m, we construct the set:

$$E = \bigcup_{k \in \mathbb{N}} A_k \cup \bigcup_{n,m \in \mathbb{N}} B_{n,m}$$

which has a null measure, $\mu(E) = 0$, as a countable union of null measure sets.

Now, we observe that:

$$\forall x \in E^c,\ \forall n, m \geqslant N_\varepsilon\ : | f_n(x) - f_m(x) | \leqslant \|f_n - f_m\|_\infty \underset{4.24}{<} \varepsilon \qquad [4.25]$$

so $(f_n(x))_{n \in \mathbb{N}}$ is a Cauchy sequence of elements of $\mathbb{K}$, which is complete; thus, there exists a pointwise limit $f(x) = \lim_{n \Longrightarrow +\infty} f_n(x)$.

Equation [4.25] of course holds if $n \Longrightarrow +\infty$, thus $\forall \varepsilon > 0$ we have:

$$\forall x \in E^c,\ \forall m \geqslant N_\varepsilon\ : | \lim_{n \to +\infty} f_n(x) - f_m(x) | = | f(x) - f_m(x) | < \varepsilon$$

which is the definition of uniform convergence of the sequence $(f_n)_{n \in \mathbb{N}} \subset L^\infty$ to f on E^c. A standard result of calculus guarantees that if a sequence of bounded functions converges uniformly to a function, then even the limit function is bounded; in our case, this implies that f is essentially bounded on E^c.

The final step is to extend the definition of f to a function $\tilde{f}$ defined on all X (since the elements of $L^\infty(X, \mathcal{A}, \mu)$ are defined on all X) while retaining the property of essential boundedness. This is trivial, as we simply take:

$$\tilde{f}(x) = \begin{cases} f(x) & \text{if } x \in E^c \\ 0 & \text{if } x \in E \end{cases}$$

Since $\mu(E) = 0$, $\tilde{f} : X \to \mathbb{K}$ is the representative of an equivalence class of $L^\infty(X, \mathcal{A}, \mu)$ to which the Cauchy sequence $(f_n)_{n \in \mathbb{N}}$ converges, this concludes our proof. □

Exercise 4.4

Consider a sequence $a = (a_k)_{k \in \mathbb{Z}}$ and, for all $u \in \ell^\infty(\mathbb{Z}, \mathbb{C})$, let $a * u$ be the bilateral sequence defined for $k \in \mathbb{Z}$ by:

$$(a * u)_k = \sum_{m \in \mathbb{Z}} a_m u_{k-m}$$

Let us take, for all $f \in \ell^\infty(\mathbb{Z}, \mathbb{C})$, $T(u) := a * u + f$.

1) For the purposes of this question, we take $a = \delta_1$, that is, the sequence defined by $a_1 = 1$ and $a_j = 0$ if $j \neq 1$. Calculate $a * u$ as a function of u.

2) First, suppose that $a = (a_k)_{k\in\mathbb{Z}} \in \ell^1$ verifying $\|a\|_1 = \sum_{k\in\mathbb{Z}} |a_k| < 1$.

a) Show that $(a * u)_k$ is well defined for all $k \in \mathbb{Z}$ and that $a * u \in \ell^\infty$.

b) Show that $T : \ell^\infty(\mathbb{Z}, \mathbb{C}) \to \ell^\infty(\mathbb{Z}, \mathbb{C})$ is a contraction.

c) Deduce that there exists a single unique solution $u \in \ell^\infty(\mathbb{Z}, \mathbb{C})$ to the equation $T(u) = u$.

3) Now, let us suppose that $a = (a_k)_{k\in\mathbb{Z}} \in \ell^2$ verifies $\|a\|_2 := \sum_{k\in\mathbb{Z}} |a_k|^2 < 1$.

a) Using an example, show that we can have $a \notin \ell^1(\mathbb{Z}, \mathbb{C})$.

b) Show that $(a * u)_k$ is well defined for all $k \in \mathbb{Z}$ and that $a * u \in \ell^\infty(\mathbb{Z}, \mathbb{C})$.

c) Deduce that, for all $u \in \ell^2(\mathbb{Z}, \mathbb{C})$, $T(u) \in \ell^\infty(\mathbb{Z}, \mathbb{C})$ and that if $u, v \in \ell^2(\mathbb{Z}, \mathbb{C})$, then $\|T(u) - T(v)\|_\infty < \|u - v\|_2$.

d) Now, take $a = \frac{1}{2}\delta_1$ and let $f = \mathbf{1}$ be the constant sequence $f_j = 1$ for all $j \in \mathbb{Z}$. Calculate $T(u)$ as a function of u and determine $\lim_{k\to+\infty} (T(u))_k$.

e) Deduce that there is no $u \in \ell^2(\mathbb{Z}, \mathbb{C})$ such that $T(u) = u$. Does this contradict the fixed-point theorem?

Hint: There is no need to determine u to answer this question.

f) Determine $u \in \ell^\infty(\mathbb{Z}, \mathbb{C})$ such that $T(u) = u$.

Solution to Exercise 4.4

1) By definition:

$$(\delta_1 * u)_k = \sum_{m\in\mathbb{Z}} \delta_{m,1} u_{k-m} = u_{k-1}$$

2) a) By direct calculation:

$$(a * u)_k = \sum_{m\in\mathbb{Z}} a_m u_{k-m} \leqslant \sum_{m\in\mathbb{Z}} |a_m u_{k-m}| \leqslant \|u\|_\infty \sum_{m\in\mathbb{Z}} |a_m| = \|u\|_\infty \|a\|_1 < +\infty$$

since $a \subset \ell^1(\mathbb{Z}, \mathbb{C})$ and $u \in \ell^\infty(\mathbb{Z}, \mathbb{C})$. Furthermore, as the majorization is independent of k, $\|a * u\|_\infty = \sup_{k\in\mathbb{Z}}\{(a * u)_k\} \leqslant \|u\|_\infty \|a\|_1 < +\infty$.

b) Once again, by direct calculation, we have $\|T(u) - T(v)\|_\infty = \|a * u + f - a * v - f\|_\infty = \|a * u - a * v\|_\infty$, but $u \mapsto a * u$ is linear, so from what we saw in the previous question: $\|T(u) - T(v)\|_\infty = \|a * (u - v)\|_\infty \leqslant \|a\|_1 \|u - v\|_\infty$. Since $\|a\|_1 < 1$ by hypothesis, T is a contraction.

c) Since $(\ell^\infty(\mathbb{Z}, \mathbb{C}), \| \ \|_\infty)$ is a complete normed (and therefore metric) space, the fixed-point theorem gives us the existence of a single element $\bar{u} \in \ell^\infty(\mathbb{Z}, \mathbb{C})$ such that $T(\bar{u}) = \bar{u}$, that is, $\bar{u} = a * \bar{u} + f$.

3) a) The simplest example of a sequence $a \in \ell^2(\mathbb{Z}, \mathbb{C})$ such that $a \notin \ell^1(\mathbb{Z}, \mathbb{C})$ is probably $a_k = \begin{cases} 0 & k \leqslant 0 \\ \frac{1}{k} & \text{otherwise.} \end{cases}$ In this case, $\sum\limits_{k \in \mathbb{Z}} |a_k| = \sum\limits_{k=1}^{\infty} \frac{1}{k}$, the harmonic series, which we know to be divergent, so $a \notin \ell^1(\mathbb{Z}, \mathbb{C})$. On the other hand, $\sum\limits_{k \in \mathbb{Z}} |a_k|^2 = \sum\limits_{k=1}^{\infty} \frac{1}{k^2}$, which is convergent, that is, $a \in \ell^2(\mathbb{Z}, \mathbb{C})$.

b) Using the given hypotheses, for all fixed $k \in \mathbb{Z}$, the Cauchy-Schwarz inequality can be applied to give:

$$\sum_{m \in \mathbb{Z}} |a_m u_{k-m}| \leqslant \left(\sum_{m \in \mathbb{Z}} |a_m| \right)^{1/2} \left(\sum_{m \in \mathbb{Z}} |u_{k-m}| \right)^{1/2}$$

$$\leqslant \|a\|_2^2 \left(\sum_{n \in \mathbb{Z}} |u_n| \right)^{1/2} = \|a\|_2 \|u\|_2$$

using the change of variable $n = k - m$, with fixed $k \in \mathbb{Z}$ and $m \in \mathbb{Z}$, thus $n \in \mathbb{Z}$. Since $(a * u)_k = \sum\limits_{m \in \mathbb{Z}} a_m u_{k-m} \leqslant \sum\limits_{m \in \mathbb{Z}} |a_m u_{k-m}| < +\infty$, for all fixed $k \in \mathbb{Z}$, the sequence $a * u$ is well defined. Furthermore, as in question 2a, since the majorization does not depend on k, $\|a * u\|_\infty = \sup\limits_{k \in \mathbb{Z}} \{(a * u)_k\} \leqslant \|a\|_2 \|u\|_2 < +\infty$.

c) $T(u) = a * u + f$ is the sum of two elements of $\ell^\infty(\mathbb{Z}, \mathbb{C})$ (f by hypothesis, and $a * u$ as demonstrated above), so $T(u) \in \ell^\infty(\mathbb{Z}, \mathbb{C})$. Once again, $u \mapsto a * u$ is clearly linear, so, using the result from the previous question:

$$\|T(u) - T(v)\|_\infty = \|a * u - a * v\|_\infty = \|a * (u - v)\|_\infty \leqslant \|a\|_2 \|u - v\|_2 < \|u - v\|_2$$

since, by hypothesis, $\|a\|_2 < 1$.

d) Using the result from question 1, we have $(T(u))_k = \frac{u_{k-1}}{2} + 1$. Moreover, as $u \in \ell^2(\mathbb{Z}, \mathbb{C})$, $\sum\limits_{k \in \mathbb{Z}} |u_k|^2$ converges, we necessarily have $u_k \underset{k \to +\infty}{\longrightarrow} 0$, which implies $(T(u))_k \underset{k \to +\infty}{\longrightarrow} 1$.

e) Taking $T(u) = u$, we would have $u_k = \frac{u_{k-1}}{2} + 1$, and, taking the limit for k which tends to infinity on both sides, we would obtain the absurd result $0 = 1$. There is no contradiction with the fixed-point theorem, since the inequality $\|T(u) - T(v)\|_\infty <$

$\|u-v\|_2$ does not involve $\|T(u)-T(v)\|_2 \dots$ Evidently, as there is no fixed point, T cannot be a contraction on $\ell^2(\mathbb{Z}, \mathbb{C})$.

f) A sequence $u \in \ell^\infty(\mathbb{Z}, \mathbb{C})$ such that $T(u) = u$ is a bounded sequence $u_k = \frac{u_{k-1}}{2} + 1$ (this is an "arithmetico-geometric" sequence). Taking $u_k = v_k + \alpha$, with unknown v_k and α, then $v_k + \alpha = \frac{u_{k-1}}{2} + 1 = \frac{v_{k-1}+\alpha}{2} + 1$, that is, $v_k = \frac{v_{k-1}}{2} + 1 - \frac{\alpha}{2}$ thus, if we take $\alpha = 2$, we obtain a geometric sequence $v_k = \frac{v_{k-1}}{2}$; by a standard result for geometric sequences, $v_k = 2^{-k} v_0$. Furthermore, $v_0 = u_0 - \alpha$ and $\alpha = 2$, hence $v_0 = u_0 - 2$, implying that $u_k = 2^{-k}(u_0-2)+2$. For all $k \geqslant 0$, $2^{-k} < 1$, but for $k < 0$, 2^{-k} is not bounded, so to obtain a bounded u_k, we need to eliminate its factor, that is, to impose $u_0 - 2 = 0$. Finally, we see that the only sequence $u \in \ell^\infty(\mathbb{Z}, \mathbb{C})$ such that $T(u) = u$, that is, the only fixed point for the contraction $T : \ell^\infty(\mathbb{Z}, \mathbb{C}) \to \ell^\infty(\mathbb{Z}, \mathbb{C})$, is the constant sequence of 2, $u_k = 2$ for all $k \in \mathbb{Z}$. □

4.4.3. *Inclusion relationships between ℓ^p spaces*

Let us introduce the following functional space:

$$\boxed{\ell^0(\mathbb{N}, \mathbb{K}) \equiv \ell_{\text{fin}}(\mathbb{N}, \mathbb{K}) = \{(x_n)_{n\in\mathbb{N}} \subset \mathbb{K},\ \exists N \in \mathbb{N} \ :\ x_n = 0\ \forall n \geqslant N\}} \quad [4.26]$$

that is, the space of sequences with a finite number of elements $\neq 0$. Clearly, $\ell^0(\mathbb{N}, \mathbb{K}) \subset \ell^p(\mathbb{N}, \mathbb{K})\ \forall p \geqslant 1$.

THEOREM 4.16.– Taking $p, q \in \mathbb{R}$, $1 \leqslant p \leqslant q < \infty$, then:

$$\boxed{\ell^0(\mathbb{N}, \mathbb{K}) \subset \ell^1(\mathbb{N}, \mathbb{K}) \subset \dots \subset \ell^p(\mathbb{N}, \mathbb{K}) \subset \dots \subset \ell^q(\mathbb{N}, \mathbb{K}) \subset \dots \subset \ell^\infty(\mathbb{N}, \mathbb{K})}$$

PROOF.– Given that $\ell^0(\mathbb{N}, \mathbb{K}) \subset \ell^1(\mathbb{N}, \mathbb{K})$, the demonstration that $\ell^p(\mathbb{N}, \mathbb{K}) \subset \ell^\infty(\mathbb{N}, \mathbb{K})\ \forall 1 \leqslant p < \infty$ is almost trivial since:

$$(x_n)_{n\in\mathbb{N}} \in \ell^p(\mathbb{N}, \mathbb{K}) \iff \sum_{n\in\mathbb{N}} |x_n|^p < +\infty$$

which gives us $|x_n| \underset{n \Longrightarrow +\infty}{\longrightarrow} 0$, that is, $|x_n|$ is bounded and thus $(x_n)_{n\in\mathbb{N}} \in \ell^\infty(\mathbb{N}, \mathbb{K})$.

It only remains to prove that $\ell^p(\mathbb{N}, \mathbb{K}) \subset \ell^q(\mathbb{N}, \mathbb{K})$ if $1 \leqslant p \leqslant q$: as $|x_n| \underset{n \Longrightarrow +\infty}{\longrightarrow} 0$, then, in particular, $\exists N \in \mathbb{N}$ such that $|x_n| \leqslant 1$, $\forall n \geqslant N$ thus $|x_n|^q \leqslant |x_n|^p$ $\forall n \geqslant N$, which implies that:

$$\sum_{n\in\eth\mathbb{N}} ||x_n|^q \leqslant \sum_{n\in\eth\mathbb{N}} ||x_n|^p$$

The convergence of $\sum_{n\in\mathbb{N}} |x_n|^p$ therefore implies the convergence of $\sum_{n\in\mathbb{N}} |x_n|^q$, that is, $\ell^p \subset \ell^q$. $\square$

REMARK.– The completeness of an infinite-dimensional metric space depends on the metric selected for the space. To verify this statement, let us examine the completeness of $(\ell^1, \| \|_\infty)$, that is, ℓ^1 interpreted as a subspace of ℓ^∞ and equipped with the norm of this latter space.

Exercise 4.5

Show that $(\ell^1, \| \|_\infty)$ is not complete.

Solution to Exercise 4.5

Since $\ell^1 \subset \ell^\infty$, to solve this problem we must prove that ℓ^1 is not a closed subset of ℓ^∞ with respect to the norm $\| \|_\infty$, that is, there exists at least one sequence that converges (and so it is Cauchy) outside $(\ell^1, \| \|_\infty)$.

The elements of ℓ^1 are sequences $x \equiv (x_n)_{n\in\mathbb{N}}$, so a sequence of elements of ℓ^1 is a sequence of sequences. For all fixed $m \in \mathbb{N}$, we shall note this sequence $x^m \equiv (x_n^m)_{n\in\mathbb{N}}$.

Now, let us verify that the sequence of elements of ℓ^1 defined by:

$$x_n^m = \begin{cases} 0 & \text{if } n = 0 \\ \frac{1}{n} & \text{if } 1 \leqslant n \leqslant m \\ 0 & \text{if } n > m \end{cases}$$

converges in $\ell^\infty \setminus \ell^1$. For all fixed $m \in \mathbb{N}$, the sequence x^m is explicitly defined as follows:

$$\left(0, 1, \frac{1}{2}, \ldots, \frac{1}{m}, 0, 0, \ldots\right)$$

which shows that $x^m \in \ell^1$ for all fixed $m \in \mathbb{N}$.

Now, consider the sequence $x^* \equiv (x_n^*)_{n\in\mathbb{N}}$ defined by:

$$x_n^* = \begin{cases} 0 & \text{if } n = 0 \\ \frac{1}{n} & \text{if } n \geqslant 1 \end{cases}$$

Clearly, $(x_n^*)_{n\in\mathbb{N}}$ is bounded, and thus belongs to ℓ^∞, but $\|(x_n^*)_{n\in\mathbb{N}}\|_1 = \sum_{n=1}^{\infty} \frac{1}{n} = +\infty$, so $(x_n^*)_{n\in\mathbb{N}} \notin \ell^1$. If we can show that $(x^m)_{m\in\mathbb{N}}$ converges to x^* in norm $\| \|_\infty$, this will complete our proof.

To do this, we calculate:

$$\|x^m - x^*\|_\infty = \sup_{n\in\mathbb{N}} |x_n^m - x_n^*| = \sup_{n>m} \frac{1}{n}$$

Up to $n = m$, the difference $x_n^m - x_n^*$ is null, but when $n > m$, the difference becomes $|0 - \frac{1}{n}| = \frac{1}{n}$. By the definition of sup, $\sup_{n>m} \frac{1}{n} = \sup\{\frac{1}{m+1}, \frac{1}{m+2}, \dots\} = \frac{1}{m+1}$ and thus:

$$\|x^m - x^*\|_\infty = \frac{1}{m+1} \underset{m\to+\infty}{\longrightarrow} 0$$

□

4.4.4. *Inclusion relationships between L^p spaces*

In general, there are no inclusion relationships between $L^p(X, \mathcal{A}, \mu)$ spaces. For instance, consider $L^1(\mathbb{R})$, $L^2(\mathbb{R})$ and the following functions:

$$f(x) = \begin{cases} x^{-2/3} & \text{if } 0 < x < 1 \\ 0 & \text{otherwise} \end{cases}, \quad g(x) = \begin{cases} x^{-2/3} & \text{if } x > 1 \\ 0 & \text{otherwise} \end{cases}$$

Clearly, $f \in L^1(\mathbb{R})$, but $f \notin L^2(\mathbb{R})$, since[10]:

$$\int_{\mathbb{R}} |f(x)|dx = \int_0^1 \frac{1}{x^{2/3}}dx < +\infty, \quad \int_{\mathbb{R}} |f(x)|^2 dx = \int_0^1 \frac{1}{x^{4/3}}dx = +\infty$$

and $g \in L^2(\mathbb{R})$, but $g \notin L^1(\mathbb{R})$, because:

$$\int_{\mathbb{R}} |g(x)|dx = \int_1^{+\infty} \frac{1}{x^{2/3}}dx = +\infty, \quad \int_{\mathbb{R}} |g(x)|^2 dx = \int_1^{+\infty} \frac{1}{x^{4/3}}dx < +\infty$$

Inclusions among L^p spaces can be obtained by imposing additional conditions. Since spaces $L^1(\mathbb{R})$ and $L^2(\mathbb{R})$ are particularly important, we shall examine the conditions used for these spaces – which are often verified in practical applications – in Theorem 4.17.

THEOREM 4.17.– The following statements are true:

1) if $f \in L^1(\mathbb{R})$, with f bounded, then $f \in L^2(\mathbb{R})$;

2) if $f \in L^2(\mathbb{R})$, with f null outside of a finite interval, then $f \in L^1(\mathbb{R})$.

10 Recall that if $a > 0$ and $b \in \mathbb{R}$, $\int_0^a \frac{1}{x^\alpha}dx < +\infty$ and $\int_b^{+\infty} \frac{1}{x^\beta}dx < +\infty$ if and only if $\alpha < 1$ and $\beta > 1$.

PROOF.–

1) If f is in $L^1(\mathbb{R})$ and is bounded, say $|f(x)| \leqslant M \ \forall x \in \mathbb{R}$, $M \geqslant 0$, then:

$$\int_{\mathbb{R}} |f(x)|^2 dx = \int_{\mathbb{R}} |f(x)| \cdot |f(x)| dx \leqslant \int_{\mathbb{R}} M|f(x)|dx = M\|f\|_1 < +\infty$$

thus $f \in L^2(\mathbb{R})$.

2) If f is in $L^2(\mathbb{R})$ and is null outside of a finite interval, say $f(x) = 0 \ \forall x \notin [a,b]$, then:

$$\int_{\mathbb{R}} |f(x)|dx = \int_{[a,b]} |f(x)|dx \underset{\mathbf{1}(x)=1 \ \forall x \in [a,b]}{=} \int_{[a,b]} \mathbf{1}(x) \cdot |f(x)|dx = \langle \mathbf{1}, |f| \rangle_{L^2[a,b]}$$

$$\underset{\text{(Cauchy-Schwarz)}}{\leqslant} \left(\int_{[a,b]} dx \right)^{1/2} \left(\int_{[a,b]} |f(x)|^2 dx \right)^{1/2} = \sqrt{b-a}\,\|f\|_2 < +\infty$$

so $f \in L^1(\mathbb{R})$. □

Statement 1 remains valid for all $f \in L^1(\mathbb{R}^n)$, $n \geqslant 1$, while statement 2 remains valid if we replace an interval with a finite-measure part of $\mathbb{R}^n$.

More generally, in the case where $\mu(X) < +\infty$, it is possible to create a highly useful string of inclusions.

THEOREM 4.18.– If $(X, \mathcal{A}, \mu)$ is a measure space with a *finite measure*, $\mu(X) < +\infty$, and if $q > p > 1$, then:

$$\boxed{L^\infty(X, \mathcal{A}, \mu) \subset \ldots \subset L^q(X, \mathcal{A}, \mu) \subset \ldots \subset L^p(X, \mathcal{A}, \mu) \subset \ldots \subset L^1(X, \mathcal{A}, \mu)}$$

PROOF.– First, let us verify the thesis for L^∞, then for L^1 and L^2 (which provide a clearer illustration of the approach used), and finally for L^p and L^q.

If $f \in L^\infty(X, \mathcal{A}, \mu)$, then $\int_X |f|^p d\mu \leqslant \int_X \|f\|_\infty^p d\mu = \|f\|_\infty^p \mu(X) < +\infty$, hence $f \in L^p(X, \mathcal{A}, \mu)$.

If $f \in L^2(X, \mathcal{A}, \mu)$, then:

$$\int_X |f|d\mu = \int_X |\mathbf{1} \cdot f|d\mu \underset{\text{Hölder inequ. [4.19]}}{\leqslant} \left(\int_X 1^2 d\mu \right)^{\frac{1}{2}} \left(\int_X |f|^2 d\mu \right)^{\frac{1}{2}}$$

$$= \sqrt{\mu(X)}\|f\|_2 < +\infty$$

hence $f \in L^1(X, \mathcal{A}, \mu)$.

Taking $E = \{x \in X : |f(x)| \geqslant 1\}$ and $F = \{x \in X : |f(x)| \leqslant 1\}$, then $X = E \cup F$, and let $p < q$. Then $|f(x)|^p \leqslant |f(x)|^q \ \forall x \in E$ and $|f(x)|^p \leqslant 1 \ \forall x \in F$. Thus, if $f \in L^q$:

$$\int_X |f(x)|^p d\mu \leqslant \int_E |f(x)|^q d\mu + \int_F 1\, d\mu \leqslant \int_X |f(x)|^q d\mu + \int_X 1\, d\mu$$
$$= \|f\|_q^q + \mu(X) < +\infty$$

that is, $f \in L^p$. □

4.4.5. *Density theorems in $L^p(X,A,\mu)$*

We shall begin our examination of dense varieties in L^p by considering step functions.

4.4.5.1. *Step functions*

Let $(X, \mathcal{A}, \mu)$ be any measure space and $\mathbb{K} = \mathbb{R}$ or $\mathbb{C}$. A piecewise constant function on X with values in $\mathbb{K}$ is known as a step or *simple* function. For all $N \in \mathbb{N}$, the rigorous definition of the space of these functions is:

$$\Sigma = \left\{ s : X \to \mathbb{K} \ : \ \exists (\alpha_i)_{i=1}^N \in \mathbb{K} \ : \ s = \textstyle\sum_{i=1}^N \alpha_i \chi_{E_i}, \ E_i \text{ measurable and } \mu(E_i) \right.$$
$$\left. < +\infty \text{ if } \alpha_i \neq 0 \right\}$$

The function $\chi_{E_i} = \begin{cases} 1 \text{ if } x \in E_i \\ 0 \text{ if } x \notin E_i \end{cases}$ is the indicator function of E_i.

THEOREM 4.19.– $\boxed{\overline{\Sigma} = L^p(X, A, \mu)}$ $\forall 1 \leqslant p < \infty$, where the closure should be interpreted with respect to the topology of $L^p(X, A, \mu)$ taking $\Sigma \subset L^p(X, A, \mu)$.

4.4.5.2. *Intersections: $L^p \cap L^q$ and $\ell^p \cap \ell^q$*

THEOREM 4.20.– Let $(X, \mathcal{A}, \mu)$ be any measure space and $\mathbb{K} = \mathbb{R}$ or $\mathbb{C}$, then:

$$\boxed{\overline{L^p(X, \mathcal{A}, \mu) \cap L^q(X, \mathcal{A}, \mu)} = \begin{cases} L^p(X, \mathcal{A}, \mu) \\ L^q(X, \mathcal{A}, \mu) \end{cases}} \qquad \forall 1 \leqslant p, q \leqslant \infty$$

In the first case, the intersection should be interpreted as a subset of $L^p(X, A, \mu)$ and the closure with respect to the metric topology generated by the norm $\| \ \|_p$. In the second case, the intersection should be interpreted as a subset of $L^q(X, A, \mu)$ and the closure with respect to the topology relative to the norm $\| \ \|_q$.

Notably, as ℓ^p spaces are nested, it holds that:

$$\boxed{\overline{\ell^p(\mathbb{N}, \mathbb{K})} = \ell^q(\mathbb{N}, \mathbb{K})} \qquad \forall 1 \leqslant p < q < \infty$$

As before, for all fixed q, ℓ^p should be interpreted as a subspace of ℓ^q and the closure should be interpreted with respect to the norm $\| \ \|_q$.

THEOREM 4.21.– For all $p \in \mathbb{R}$, $1 \leqslant p < +\infty$:

$$\boxed{\overline{\ell^0(\mathbb{N}, \mathbb{K})} = \ell^p(\mathbb{N}, \mathbb{K})}$$

that is, $\ell^0(\mathbb{N}, \mathbb{K})$ is dense in $\ell^p(\mathbb{N}, \mathbb{K})$ with respect to the topology generated by the norm $\| \ \|_p$.

PROOF.– Let $(x_n)_{n \in \mathbb{N}}$ be an arbitrary sequence in $\ell^p(\mathbb{N}, \mathbb{K})$. Consider the sequence:

$$x_n^N := \begin{cases} x_n & \text{if } n < N \\ 0 & \text{otherwise} \end{cases}$$

then:

$$\|x_n - x_n^N\|_p = \sum_{n \in \mathbb{N}} |x_n - x_n^N|^p = \sum_{n=N}^{+\infty} |x_n|^p \underset{N \to +\infty}{\longrightarrow} 0$$

as this is the remainder of a convergent series (since $(x_n)_{n \in \mathbb{N}}$ belongs to $\ell^p(\mathbb{N}, \mathbb{K})$), which proves the density of $\ell^0(\mathbb{N}, \mathbb{K})$ in $\ell^p(\mathbb{N}, \mathbb{K})$. □

4.4.5.3. *Test functions*

Let $\Omega \subseteq \mathbb{R}^n$ be an open set.

DEFINITION 4.14.–

$$\begin{aligned} \mathcal{C}_c^\infty(\Omega) = *\{ & f : \Omega \implies \mathbb{K}, f \text{ indefinitely derivable on } \Omega \text{ and} \\ & \text{supp}(f) \text{ compact in } \mathbb{R}^n \} \end{aligned}$$

where $\text{supp}(f) = \overline{\{x \in \Omega : f(x) \neq 0\}}$ *is said to be the support of* f.

The functions in $\mathcal{C}_c^\infty(\Omega)$ are known as *test functions*, as they are so regular that they are often used to test the action and properties of certain "wild" operators. Test functions play a crucial role in distribution theory and in analyzing differential equations. The identically null function is obviously a test function; other explicit examples are much harder to find. The canonical example of a test function on $\mathbb{R}$ for any value of $\varepsilon > 0$ is given by:

$$f(x) = \begin{cases} \exp\left(-\frac{1}{1-\left(\frac{x}{\varepsilon}\right)^2}\right) & \text{if } |x| < \varepsilon \\ 0 & \text{if } |x| \geqslant \varepsilon. \end{cases}$$

For the purposes of our discussion, we need a simple symbol to denote the partial derivative of a function with n variables with respect to a *multi-index* $\mathbf{l} = (l_1, l_2, \ldots, l_d) \in \mathbb{N}^d$ of length $|\mathbf{l}| = l_1 + l_2 + \ldots + l_d$. The canonical notation is:

$$D^{\mathbf{l}} f(x) = \frac{\partial^{|\mathbf{l}|} f}{\partial x_1^{l_1} \partial x_2^{l_2} \ldots \partial x_d^{l_d}}(x) \qquad \forall x \in \mathbb{R}^n$$

hence $D^{\mathbf{l}} f(x)$ is the partial derivative of f in x l_1 times with respect to x_1, l_2 with respect to x_2, etc. This symbol appears in the (non-trivial) definition of a topology on $\mathcal{C}_c^\infty(\Omega)$ with respect to which the convergence of a sequence of test functions $(f_n)_{n\in\mathbb{N}}$ to a test function f is equivalent to fulfilling the following two conditions:

– there exists a compact set $K \subset \Omega$ such that $\text{supp}(f_n) \subseteq K$ for all $n \in \mathbb{N}$;

– $\forall x \in \mathbb{R}^n$, $\forall \mathbf{l} \in \mathbb{N}^d$: $D^{\mathbf{l}} f_n(x) \underset{n\to+\infty}{\longrightarrow} D^{\mathbf{l}} f(x)$, uniformly.

The space $\mathcal{C}_c^\infty(\Omega)$ with this topology is usually written as $\mathcal{D}(\Omega)$ and is *complete*. The following density result holds true.

THEOREM 4.22.– Considering the Borel σ algebra and the Lebesgue measure, then:

$$\boxed{\overline{\mathcal{C}_c^\infty(\Omega)} = L^p(\Omega)} \qquad \forall 1 \leqslant p < \infty$$

where $\mathcal{C}_c^\infty(\Omega)$ should be interpreted as a subspace of $L^p(\Omega)$ and interpret the closure with respect to the topology generated by the norm $\| \ \|_p$.

By the definition of closure, $\mathcal{C}_c^\infty(\Omega)$ *is not complete with respect to the topology generated by the norm* $\| \ \|_p$, since there are sequences of elements of $\mathcal{C}_c^\infty(\Omega)$ which converge to elements in $L^p(\Omega) \backslash \mathcal{C}_c^\infty(\Omega)$.

4.4.5.4. *Schwartz space*

For simplicity's sake – particularly in terms of notation – we shall start by examining the case of a function with a single variable.

Taking $k, l \in \mathbb{N}$ and $f \in \mathcal{C}^\infty(\mathbb{R})$, for all $x \in \mathbb{R}$, we write:

$$f^{k,l}(x) = x^k \frac{d^l f}{dx^l}(x)$$

DEFINITION 4.15 (Schwartz space, $n = 1$).– *The function space of functions $f \in C^\infty(\mathbb{R})$ such that:*

$$\lim_{|x|\to+\infty} |f^{k,l}(x)| = 0 \qquad \forall k, l \in \mathbb{N}$$

is known as the Schwartz space, or the space of rapidly decreasing functions. The canonical notation for this space is $\mathcal{S}(\mathbb{R})$.

Any element in $\mathcal{S}(\mathbb{R})$ is thus an infinitely derivable function everywhere such that, if we consider its derivative of any order and multiply this value by any power of its variable, it converges to 0 as the variable tends to $\pm\infty$. To satisfy this characteristic, a function must decrease very rapidly to zero at infinity, hence the alternative name for the functions of $\mathcal{S}(\mathbb{R})$.

Evidently, $\mathcal{D}(\mathbb{R}) \subset \mathcal{S}(\mathbb{R})$, since test functions are null at infinity, but the inclusion is strict, as we see from the most important example of a rapidly-decreasing function: the Gaussian $f(x) = e^{-x^2}$, which does not belong to $\mathcal{D}(\mathbb{R})$, as its support is not compact.

Now, let us consider a function with n real variables $f \in C^\infty(\mathbb{R}^n)$. In this case, given two multi-indices $\mathbf{l}, \mathbf{k} \in \mathbb{N}^n$, we write:

$$f^{\mathbf{k},\mathbf{l}}(x) = x_1^{k_1} x_2^{k_2} \cdots x_n^{k_n} D^{\mathbf{l}} f(x) \qquad \forall x \in \mathbb{R}^n$$

DEFINITION 4.16 (Schwartz space, arbitrary n).– *The function space of functions $f \in C^\infty(\mathbb{R}^n)$ such that:*

$$\lim_{\|x\|\to+\infty} |f^{\boldsymbol{k},\boldsymbol{l}}(x)| = 0 \qquad \forall \boldsymbol{k}, \boldsymbol{l} \in \mathbb{N}^n$$

is the Schwartz space, or rapidly decreasing function space. The canonical notation for this space is $\mathcal{S}(\mathbb{R}^n)$.

By construction, $\mathcal{S}(\mathbb{R}^n)$ is stable with respect to partial derivation and to multiplication by a polynomial. Functions of $\mathcal{S}(\mathbb{R}^n)$ (and their derivatives) decay at infinity faster than the reciprocal of a polynomial.

As in the case where $n = 1$, $\mathcal{D}(\mathbb{R}^n) \subset \mathcal{S}(\mathbb{R}^n)$ and the inclusion is strict, as the Gaussian $f(x) = e^{-\|x\|^2}$ belongs to $\mathcal{S}(\mathbb{R}^n)$, but not to $\mathcal{D}(\mathbb{R}^n)$.

It is possible to define a topology on $\mathcal{S}(\mathbb{R}^n)$ in which a sequence $(f_n)_{n\in\mathbb{N}}$ of functions of $\mathcal{S}(\mathbb{R}^n)$ converges to $f \in \mathcal{S}(\mathbb{R}^n)$ if $f_n^{\mathbf{k},\mathbf{l}} \underset{n\to+\infty}{\longrightarrow} f$ uniformly $\forall \mathbf{k}, \mathbf{l} \in \mathbb{N}^n$. *With respect to this topology, the Schwartz space is complete.*

Just as we saw for the test function space, Schwartz space plays an important role in distribution theory (which was formalized by Laurent Schwartz himself) and in the context of partially-derived differential equations.

The fact that $\mathcal{D}(\mathbb{R}^n) \subset \mathcal{S}(\mathbb{R}^n)$ and that $\mathcal{D}(\mathbb{R}^n)$ is $\| \ \|_p$-dense in $L^p(\mathbb{R}^n)$ implies the following result (Theorem 4.23).

THEOREM 4.23.– Considering the Borel σ-algebra and the Lebesgue measure, then:

$$\boxed{\overline{\mathcal{S}(\mathbb{R}^n)} = L^p(\mathbb{R}^n)} \qquad \forall 1 \leqslant p < \infty$$

interpreting $\mathcal{S}(\mathbb{R}^n)$ as a subspace of $L^p(\mathbb{R}^n)$ and considering the closure with respect to the topology generated by the norm of $\| \ \|_p$.

From the definition of closure, $\mathcal{S}(\mathbb{R}^n)$ *is not complete with respect to the topology generated by the norm* $\| \ \|_p$: there exist sequences of $\mathcal{S}(\mathbb{R}^n)$ which converge to elements of $L^p(\mathbb{R}^n) \backslash \mathcal{S}(\mathbb{R}^n)$.

4.5. Summary

In this chapter, we have examined the compatibility of the topological structure of inner product vector spaces with the linear structure: the sum and product by a scalar operations are continuous in the topology generated by the inner product, as is the inner product itself, and the canonically induced norm. This compatibility is essential, as it implies that the limit operation commutes with the operations cited above; this result is fundamental in both theoretical and practical contexts.

We also saw that all finite-dimensional vector spaces possess the same Euclidean topological structure up to a homeomorphism.

Hilbert and Banach spaces were introduced as special cases of inner product or normed vector spaces, respectively, such that all Cauchy sequences converge within the space (completeness property). Any finite-dimensional inner product space is a Hilbert space, while any finite-dimensional normed vector space is a Banach space. All Hilbert spaces are Banach spaces, but the reverse is not usually true.

Complete normed vector spaces can be characterized in a simple but very useful way: they are all, and only, spaces in which absolutely convergent series are also simply convergent.

Any contraction defined on a complete metric space possesses a unique fixed point.

We presented the Hilbert spaces L^2 and ℓ^2, along with examples of non-Hilbert, but Banach, spaces, L^p and ℓ^p, with $1 \leqslant p \leqslant \infty$, $p \neq 2$. The Minkowski inequality can be used to define a linear structure on all of these spaces, while Hölder's inequality is used to define an inner product when $p = 2$.

ℓ^p spaces are nested with increasing p; on the other hand, there is generally no inclusion relationship in L^p spaces, with the notable exception of finite measure spaces, for which L^p spaces are nested, but in the opposite way to ℓ^p spaces, that is, with decreasing p.

Finally, we demonstrated that L^p spaces coincide with the closure of many widely used function spaces, such as the test function space and the Schwartz space.

5

The Geometric Structure of Hilbert Spaces

Among the infinite-dimensional vector spaces, Hilbert spaces are the closest to the Euclidean spaces $\mathbb{K}^n$ presented in Chapter 1 with respect to their geometric structure, which is the focus of the present chapter.

Infinite-dimensional Banach spaces do not share this characteristic, with structural properties that can be far more complicated than those of Hilbert spaces.

The rich geometric structure of Hilbert spaces makes it possible to extend the discrete Fourier transform (DFT) to spaces in infinite dimensions, using the concepts of series and the continuous Fourier transform.

Suggested reading for those wishing to go further into the subjects discussed in this chapter and in Chapter 6 includes Berberian (1961), Abbati and Cirelli (1997), Saxe (2000), Debnath and Mikusinski (2005) and Moretti (2013).

The first step in analyzing the geometric structure of Hilbert spaces is to consider the concept of orthogonal complement.

5.1. The orthogonal complement in a Hilbert space and its properties

The set of all vectors which are orthogonal to the vectors of a subset in a Hilbert space is of crucial importance in understanding the geometric properties of these spaces.

The definition and properties of this set are given below.

DEFINITION 5.1.– *Let $\mathcal{H}$ be a Hilbert space and $M \subseteq \mathcal{H}$ any subset. The orthogonal complement of M in $\mathcal{H}$ is:*

$$M^\perp = \{x \in \mathcal{H} : \langle x, y \rangle = 0 \ \forall\, y \in M\}$$

that is $M^\perp$ contains all of the elements of $\mathcal{H}$ which are orthogonal to elements in M.

We denote with span(M) the vector subspace of $\mathcal{H}$ generated by M, that is, the set of (finite) linear combinations of vectors in M. In Theorem 5.1, we shall write $(M^\perp)^\perp = M^{\perp\perp}$ and $(M^{\perp\perp})^\perp = M^{\perp\perp\perp}$.

THEOREM 5.1 (Properties of the orthogonal complement).– Let $\mathcal{H}$ be a Hilbert space and $M \subseteq \mathcal{H}$ an arbitrary subset. Then:

1) $\{0_\mathcal{H}\}^\perp = \mathcal{H}$ and $\mathcal{H}^\perp = \{0_\mathcal{H}\}$;

2) $M \cap M^\perp = \begin{cases} \{0_\mathcal{H}\} & \text{if } 0_\mathcal{H} \in M \\ \varnothing & \text{if } 0_\mathcal{H} \notin M \end{cases}$;

3) $M^\perp$ is a closed vector subspace of $\mathcal{H}$;

4) if $N \subseteq \mathcal{H}$, then $M \subseteq N \Rightarrow N^\perp \subseteq M^\perp$ ($\perp$ reverses the set inclusion relationships);

5) $M \subseteq M^{\perp\perp}$ (difference with respect to finite dimensions);

6) $(\overline{M})^\perp = M^\perp$;

7) $M^{\perp\perp\perp} = M^\perp$;

8) $M^\perp = (\mathit{span}\, M)^\perp = (\overline{\mathit{span}\, M})^\perp$;

9) if $\overline{M} = \mathcal{H} \Rightarrow M^\perp = \{0_\mathcal{H}\}$ (the orthogonal complement of a dense subset is the zero vector).

The proof is given below. First, however, we note that the fact that $M^\perp$ is always closed is a very useful property for demonstrating that a linear variety of $\mathcal{H}$ is closed: we must simply demonstrate that this variety coincides with the orthogonal complement of a subset of $\mathcal{H}$. We also remark how noticeable it is that, thanks to the orthogonal complement, we pass from the category of sets, in which M belongs, to that of topological vector spaces, where $M^\perp$ belongs to moreover, the property of closure for $M^\perp$.

PROOF.–

1) The property follows from the fact that $0_\mathcal{H}$ is the only vector in $\mathcal{H}$ which is orthogonal to all the others.

2) $0_\mathcal{H}$ is the only vector which is orthogonal to itself.

3) $M^\perp$ is a *vector subspace*: if $x, x' \in M^\perp$, then $\langle x, y\rangle = \langle x', y\rangle = 0\ \forall y \in M$, hence $\langle \alpha x + \beta x', y\rangle = \alpha\langle x, y\rangle + \beta\langle x', y\rangle = 0\ \forall y \in M$, i.e. $M^\perp$ is a vector subspace, as it is stable with respect to linear combinations of its elements.

$M^\perp$ is *closed*: We must show that $M^\perp$ contains all the limit points of sequences in $M^\perp$. Let $(x_n)_{n\in\mathbb{N}} \subset M^\perp$ be a sequence which is convergent (and thus Cauchy) to a limit x; then, since $M^\perp \subseteq \mathcal{H}$ and $\mathcal{H}$ is complete, $x \in \mathcal{H}$. For all $y \in M$, $\langle x_n, y\rangle = 0$ $\forall n \in \mathbb{N}$, so, from the continuity of the inner product, we can write:

$$0 = \lim_{n\to+\infty} \langle x_n, y\rangle = \langle \lim_{n\to+\infty} x_n, y\rangle = \langle x, y\rangle$$

thus $x \perp y\ \forall y \in M$, that is $x \in M^\perp$.

4) Since $M \subseteq N$, the vectors of $\mathcal{H}$ which are orthogonal to the vectors of N are also orthogonal to the vectors of M (although the contrary is not necessarily true). Thus, $y \in \mathcal{H}$, $y \in N^\perp$ implies $y \in M^\perp$, that is $N^\perp \subseteq M^\perp$.

5) Every vector in M is orthogonal to every vector in $M^\perp$ by definition, but there may also be other vectors in $\mathcal{H}$ which are orthogonal to $M^\perp$, hence $M \subseteq M^{\perp\perp}$.

6) The equality of the sets can be demonstrated by demonstrating the two inclusions in the opposite direction:

– $(\overline{M})^\perp \subseteq M^\perp$: this follows from $M \subseteq \overline{M}$ and property 4;

– $M^\perp \subseteq (\overline{M})^\perp$: we must show that $y \in M^\perp \implies y \in (\overline{M})^\perp$. The elements of $\overline{M}$ are the union of all elements in M with the limits of the sequences in M, so we must show that, if $y \in M$ is orthogonal to all of the elements $(x_n)_{n\in\mathbb{N}} \subset M^\perp$ of an arbitrary convergent sequence in $M^\perp$, then y is also orthogonal to the limit of this sequence. This can be proved using the continuity of the inner product: by hypothesis, $\langle x_n, y\rangle = 0\ \forall n \in \mathbb{N}$, thus:

$$0 = \lim_{n\to+\infty} \langle x_n, y\rangle = \langle \lim_{n\to+\infty} x_n, y\rangle$$

that is, $y \perp \lim_{n\to+\infty} x_n$.

7) From property 5, $M \subseteq M^{\perp\perp}$, and from property 4, $M^{\perp\perp\perp} \subseteq M^\perp$ for any subset M of $\mathcal{H}$. Now, writing $N = M^\perp$, the final inclusion can be rewritten as $N^{\perp\perp} \subseteq N$, implying, by property 4, $N^\perp \subseteq N^{\perp\perp\perp}$. M, and thus N, are arbitrary subsets of $\mathcal{H}$, hence the inclusions $M^{\perp\perp\perp} \subseteq M^\perp$, $N^\perp \subseteq N^{\perp\perp\perp}$ imply equality between a subset of $\mathcal{H}$ and its triorthogonal complement.

8) Consider an arbitrary element in span(M): $y_0 = \sum_{i=1}^{n} \alpha_i y_i$, $y_i \in M$ and $\alpha_i \in \mathbb{K}$ $\forall i = 1, \ldots, n$. Taking any fixed $x \in M^\perp$, by the sesquilinearity (or bilinearity) of $\langle \, , \, \rangle$ (for $\mathbb{K} = \mathbb{C}$ or $\mathbb{K} = \mathbb{R}$), we can write:

$$\langle x, y_0 \rangle = \langle x, \sum_{i=1}^{n} \alpha_i y_i \rangle = \sum_{i=1}^{n} \overline{\alpha_i} \langle x, y_i \rangle \underset{x \perp y_i}{=} 0$$

hence $x \in (\text{span}(M))^\perp$, and since x is arbitrary, this implies $M^\perp \subseteq (\text{span}(M))^\perp$. Given that $M \subseteq \text{span}(M)$, by (4), $(\overline{\text{span}(M)})^\perp \subseteq M^\perp$, that is $(\text{span}(M))^\perp = M^\perp$; furthermore, by (6), $(\text{span}(M))^\perp = (\overline{\text{span}(M)})^\perp = M^\perp$.

9) $M^\perp = (\overline{M})^\perp = \mathcal{H}^\perp = \{0_\mathcal{H}\}$, as the only vector which is orthogonal to all vectors in $\mathcal{H}$ is the zero vector. □

5.2. Projection onto closed convex sets: theorem and consequences

In this section, we shall consider a particularly important geometric property which has already been covered in the context of Euclidean spaces: the orthogonal projection minimizes the distance between a given vector and those of the subset on which it projects.

This result will be presented and proved for the case of a closed convex subset and then used for a closed vector subspace.

DEFINITION 5.2.– *A subset S of a vector space is* convex *if:*

$$\forall x, y \in S, \ \forall \lambda \in [0, 1] \ : \ \lambda x + (1 - \lambda) y \in S$$

that is, if S is stable with respect to convex combinations, that is, for linear combinations where the coefficients sum to 1.

In geometric terms, a convex subspace can be characterized by the fact that any pair of points may be connected by a line segment which remains within the subspace. Evidently, *all vector subspaces are convex*, as they are stable with respect to all linear combinations, including convex combinations.

Note that the *half sum* of x and y (i.e. $\frac{x+y}{2}$) is a convex combination with $\lambda = 1/2$.

THEOREM 5.2.– Let $\mathcal{H}$ be a Hilbert space and S a closed, convex and proper subset[1] of $\mathcal{H}$. Then, $\forall x \in \mathcal{H}$ (fixed) there exists a single point $y_0 \in S$ such that:

$$\|x - y_0\| = \inf_{y \in S} \|x - y\|$$

1 If $S = \mathcal{H}$, then the theorem may be verified trivially with $y_0 = x$.

that is such that y_0 minimizes the distance between x and the points in S.

Before presenting the proof of this theorem, we should note that *this result also holds for any closed vector subspace of* $\mathcal{H}$: the theorem of projection onto a closed convex space generalizes property 3 from Theorem 1.12 to infinite-dimensional Hilbert spaces.

DEFINITION 5.3.– *The vector* y_0 *in the previous theorem is the* orthogonal projection *of* x *on* S, *noted* $y_0 = P_S(x)$. *The non-negative real quantity* $d(x, S) = \|x - P_S(x)\|$ *is the* distance between x and the closed, convex and proper subset S.

It is evident that if $x \in S$ then $P_S(x) = x$ and $d(x, S) = 0$, so the information provided by the theorem is interesting when $x \notin S$.

PROOF.–

$\boxed{\exists}$: For simplicity's sake, let us note[2], $\delta \equiv \inf_{y \in S} \|x - y\|$. We shall demonstrate the existence of y_0 using a non-constructive technique typical of the Hilbert school. Consider a sequence $(y_n)_{n\in\mathbb{N}} \subset S$ which satisfies the equation[3]:

$$\lim_{n\to+\infty} \|x - y_n\| = \delta \qquad [5.1]$$

The interest of such a sequence is that, by the continuity of the norm, [5.1] can be rewritten as:

$$\delta = \left\| \lim_{n\to+\infty} (x - y_n) \right\| = \left\| \lim_{n\to+\infty} x - \lim_{n\to+\infty} y_n \right\| = \left\| x - \lim_{n\to+\infty} y_n \right\|$$

hence to prove the existence of y_0, we can simply take $y_0 := \lim_{n\to+\infty} y_n$.

We begin by noting that S is closed and is thus itself *complete*; to demonstrate the existence of the limit of y_n, we must show that $(y_n)_{n\in\mathbb{N}}$ is a Cauchy sequence in S.

To show that $(y_n)_{n\in\mathbb{N}}$ is a Cauchy sequence we will use the parallelogram law [1.6] (which holds since the norm is Hilbertian, see Theorem 4.3) on the elements $x - y_n$ and $x - y_m$:

$$\|(x-y_n)+(x-y_m)\|^2+\|(x-y_n)-(x-y_m)\|^2 = 2(\|(x-y_n)\|^2+\|(x-y_m)\|^2)$$

which can be rewritten as:

$$\|2x - y_n - y_m\|^2 + \|y_n - y_m\|^2 = 2(\|(x - y_n)\|^2 + \|(x - y_m)\|^2)$$

2 $\delta(x, S)$ would be more "correct", since δ generally changes with x and S.

3 For example, $\delta^2 \leqslant \|x - y_n\|^2 \leqslant \delta^2 + \frac{1}{n}$, $\forall n \in \mathbb{N}$.

that is:

$$\|y_n - y_m\|^2 = 2(\|(x - y_n\|^2 + \|x - y_m\|^2) - \|2x - y_n - y_m\|^2 \qquad [5.2]$$

and $2x - y_n - y_m = 2\left(x - \frac{1}{2}(y_n + y_m)\right)$, so [5.2] becomes:

$$\|y_n - y_m\|^2 = 2(\|(x - y_n\|^2 + \|x - y_m\|^2) - 4\left\|x - \frac{1}{2}(y_n + y_m)\right\|^2 \qquad [5.3]$$

Note that $\frac{1}{2}(y_n + y_m) \in S$ by convexity, then $\left\|x - \frac{1}{2}(y_n + y_m)\right\| \geqslant \delta$ by definition of δ, thus $-4\left\|x - \frac{1}{2}(y_n + y_m)\right\|^2 \leqslant -4\delta^2$, so equation [5.3] gives us:

$$\|y_n - y_m\|^2 \leqslant 2(\|x - y_n\|^2 + \|x - y_m\|^2) - 4\delta^2$$

Since $\lim_{n\to+\infty} \|x - y_n\|^2 = \lim_{m\to+\infty} \|x - y_m\|^2 = \delta^2$, the right-hand side of the previous inequality tends to 0 for sufficiently high values of n and m, therefore $(y_n)_{n\in\mathbb{N}}$ is a Cauchy sequence.

[!]: Let us now prove that only one y_0 exists which satisfies equation [5.1]. Let y_1 be another element in S which verifies $\|x - y_1\| = \delta$. Writing the parallelogram formula once again, but this time using $x - y_0$ and $x - y_1$, we obtain:

$$\|(x - y_0) + (x - y_1)\|^2 + \|(x - y_0) - (x - y_1)\|^2 = 2(\|x - y_0\|^2 + \|x - y_1\|^2)$$

that is:

$$\|2x - y_0 - y_1\|^2 + \|y_1 - y_0\|^2 = 4\delta^2$$

thus:

$$\begin{aligned} 0 \leqslant \|y_1 - y_0\|^2 &= 4\delta^2 - \|2x - y_0 - y_1\|^2 = 4\delta^2 - \left\|2\left(x - \frac{y_0 + y_1}{2}\right)\right\|^2 \\ &= 4\delta^2 - 4\left\|x - \frac{y_0 + y_1}{2}\right\|^2 = 4\left(\delta^2 - \left\|x - \frac{y_0 + y_1}{2}\right\|^2\right) \end{aligned}$$

We observe that $\frac{y_0+y_1}{2} \in S$ by convexity, and, since $\delta^2 = \inf_{y\in S} \|x - y\|^2$, it must hold that $\delta^2 \leqslant \left\|x - \frac{y_0+y_1}{2}\right\|^2$ and thus $\delta^2 - \left\|x - \frac{y_0+y_1}{2}\right\|^2 \leqslant 0$, that is:

$$0 \leqslant \|y_1 - y_0\|^2 = 4\left(\delta^2 - \left\|x - \frac{y_0 + y_1}{2}\right\|^2\right) \leqslant 0,$$

hence $y_1 = y_0$. □

As we have seen, the parallelogram formula is essential to the proof of this theorem. Since, by Theorem 4.3, only Hilbert norms verify this formula, the proof given above cannot be applied to Banach spaces which are not Hilbert spaces and, in fact, counter-examples show that *the theorem of projection onto a closed, convex and proper subset does not hold for any infinite-dimensional Banach space.*

The theorem of projection onto a closed convex space has very important consequences, which will be described in detail later.

For now, note that this theorem guarantees the existence and uniqueness of the orthogonal projection y_0, but it does not provide any information regarding the explicit expression of elements of the sequence $(y_n)_{n\in\mathbb{N}}$ in S which converges to y_0.

A geometric characterization of y_0, shown in Theorem 5.3, is therefore very useful. A remarkable application of this result will be presented in section 5.3.

THEOREM 5.3.– Let $\mathcal{H}$ be a real Hilbert space, S a closed, convex and proper subset of $\mathcal{H}$ and x a fixed element in $\mathcal{H}$. Then y_0 is the orthogonal projection of x onto S, that is $\|x - y_0\| = \inf_{y\in S} \|x - y\|$, if and only if:

$$\forall y \in S, \quad \langle x - y_0, y - y_0\rangle \leqslant 0$$

that is[4], if and only if the angle ϑ between vectors $x - y_0$ and $y - y_0$ is obtuse, as shown in Figure 5.1.

If $\mathcal{H}$ is *complex*, then we replace $\langle x - y_0, y - y_0\rangle \leqslant 0$ with $\Re(\langle x - y_0, y - y_0\rangle) \leqslant 0$.

PROOF.– This proof concerns the real case. Proof of the complex case is left to the reader.

$\boxed{\Rightarrow}$: we wish to show that, if $\|x - y_0\| = \inf_{y\in S} \|x - y\|$, then $\langle x - y_0, y - y_0\rangle \leqslant 0$ $\forall y \in S$. To do this, let us consider any $y \in S$, using the convexity of S to guarantee that $\lambda y + (1 - \lambda)y_0 \in S$ $\forall \lambda \in [0, 1]$. Thus, by hypothesis, and using the bilinearity and symmetry properties of the real inner product, we obtain:

$$\begin{aligned}\|x - y_0\|^2 &\leqslant \|x - [\lambda y + (1 - \lambda)y_0]\|^2 = \|x - y_0 - \lambda(y - y_0)\|^2 \\ &= \langle x - y_0 - \lambda(y - y_0), x - y_0 - \lambda(y - y_0)\rangle\end{aligned}$$

4 Since $\langle x - y_0, y - y_0\rangle = \|x - y_0\|\|y - y_0\| \cos(\vartheta)$, with ϑ the angle between vectors $x - y_0$ and $y - y_0$.

$$= \langle x - y_0, x - y_0 \rangle - \lambda \langle x - y_0, y - y_0 \rangle$$
$$- \underbrace{\lambda \langle y - y_0, x - y_0 \rangle}_{=\lambda \langle x - y_0, y - y_0 \rangle} + \lambda^2 \langle y - y_0, y - y_0 \rangle$$
$$= \|x - y_0\|^2 + \lambda^2 \|y - y_0\|^2 - 2\lambda \langle x - y_0, y - y_0 \rangle$$

Thus:

$$\|x - y_0\|^2 \leqslant \|x - y_0\|^2 + \lambda^2 \|y - y_0\|^2 - 2\lambda \langle x - y_0, y - y_0 \rangle$$

Simplifying and dividing by $\lambda \in (0, 1]$, we obtain:

$$0 \leqslant \lambda \|y - y_0\|^2 - 2\langle x - y_0, y - y_0 \rangle$$

that is:

$$\langle x - y_0, y - y_0 \rangle \leqslant \frac{\lambda}{2} \|y - y_0\|^2$$

for all λ in $(0, 1]$. Now, taking the limit by $\lambda \to 0$ to both members of the inequality, we obtain: $\lim_{\lambda \to 0} \langle x - y_0, y - y_0 \rangle = \langle x - y_0, y - y_0 \rangle \leqslant \lim_{\lambda \to 0} \frac{\lambda}{2} \|y - y_0\|^2 = 0$, completing the proof of the direct implication.

$\boxed{\Leftarrow}$: Let $x \in \mathcal{H}$ be fixed, and take $y_0 \in S$ such that $\forall y \in S, \langle x - x_0, y - y_0 \rangle \leqslant 0$. Then, we wish to show that $\|x - y_0\| = \inf_{y \in S} \|x - y\|$. We have:

$$\|x - y\|^2 = \|x - y_0 + y_0 - y\|^2 = \|x - y_0 - (y - y_0)\|^2$$
$$= \langle x - y_0 - (y - y_0), x - y_0 - (y - y_0) \rangle$$
$$= \langle x - y_0, x - y_0 \rangle - \langle x - y_0, y - y_0 \rangle - \langle y - y_0, x - y_0 \rangle + \langle y - y_0, y - y_0 \rangle$$
$$= \|x - y_0\|^2 + \|y - y_0\|^2 - 2\langle x - y_0, y - y_0 \rangle,$$

having used the symmetry of the real inner product. We thus have:

$$\|x - y_0\|^2 - \|x - y\|^2 = 2 \underbrace{\langle x - y_0, y - y_0 \rangle}_{\leqslant 0 \text{ by hypothesis}} - \underbrace{\|y - y_0\|^2}_{\geqslant 0} \leqslant 0$$

that is, $\|x - y_0\|^2 \leqslant \|x - y\|^2 \ \forall y \in S$, that is: $\|x - y_0\| = \inf_{y \in S} \|x - y\|$. □

The following corollary shows that any complement of a closed proper vector subspace of a Hilbert space is not trivial, and generalizes property 2 from Theorem 1.12 to infinite-dimensional Hilbert spaces.

THEOREM 5.4.– Let $\mathcal{H}$ be a Hilbert space and S a *closed*, *proper* vector subspace of $\mathcal{H}$, that is $S \neq \varnothing$ and $S \neq \mathcal{H}$. Then $S^{\perp}$ is not reduced to $\{0_{\mathcal{H}}\}$; in fact, for all fixed $x \in \mathcal{H} \backslash S$, the vector $u = x - P_S(x)$ is non-zero and belongs to $S^{\perp}$:

$$\forall x \in \mathcal{H} \backslash S, \quad u = x - P_S(x) \neq 0_{\mathcal{H}} \text{ and } u \in S^{\perp}$$

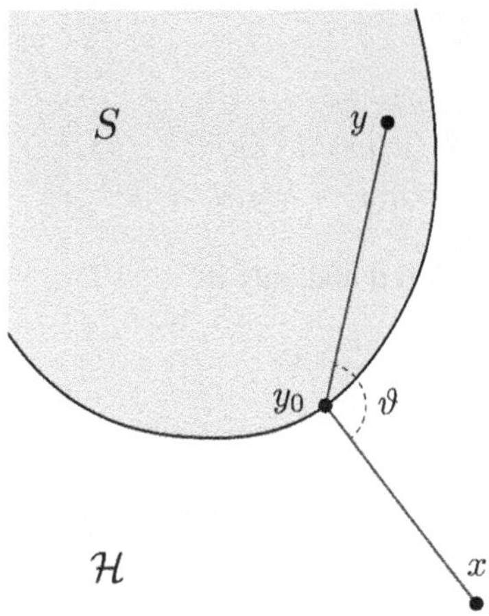

Figure 5.1. *Two-dimensional geometric visualization of the property verified by the projection onto a closed, convex and proper subset of $\mathcal{H}$. For a color version of this figure, see www.iste.co.uk/provenzi/spaces.zip*

The vector $\boxed{u = x - P_S(x)}$ is known as the *residual vector*, and the fact that $u \perp S$ fully justifies the use of the term "orthogonal projection" for $P_S(x)$.

PROOF.– Vector subspaces are convex, so the theorem of projection onto a closed convex subset holds, and thus $\exists \ P_S(x) \in S$, such that $\|u\| = \|x - P_S(x)\| = \inf_{y \in S} \|x - y\| \equiv \delta$.

Since $x \notin S$ and $P_S(x) \in S$, $u \neq 0_{\mathcal{H}}$. If we can show that $u \in S^{\perp}$, that is that $\langle u, s \rangle = 0 \ \forall s \in S$, this will prove the theorem.

To obtain this result, we start by noting that $\forall k \in \mathbb{K}, \forall s \in S$:

$$\|u + ks\|^2 = \|x - P_S(x) + ks\|^2 = \|x - \underbrace{(P_S(x) - ks)}_{\in S}\|^2 \geqslant \delta^2 = \|u\|^2$$

by the definition of δ and the fact that S is a vector subspace.

Hence:

$$\|u + ks\|^2 - \|u\|^2 \geqslant 0, \qquad \forall k \in \mathbb{K}, \forall s \in S$$

Furthermore:

$$\begin{aligned}\|u+ks\|^2 &= \langle u+ks, u+ks\rangle \\ &= \langle u,u\rangle + \langle u,ks\rangle + \langle ks,u\rangle + \langle ks,ks\rangle \\ &= \|u\|^2 + \bar{k}\langle u,s\rangle + k\langle s,u\rangle + |k|^2\|s\|^2\end{aligned}$$

Thus, $\|u+ks\|^2 - \|u\|^2 \geqslant 0$ if and only if:

$$\bar{k}\langle u,s\rangle + k\langle s,u\rangle + |k|^2\|s\|^2 \geqslant 0$$

As k is arbitrary, we can take $k = \langle u,s\rangle t$ with *any* $t \in \mathbb{R}$. Thus, the equation above becomes:

$$\begin{aligned}&\overline{\langle u,s\rangle} t\langle u,s\rangle + \langle u,s\rangle t \underbrace{\langle s,u\rangle}_{=\overline{\langle u,s\rangle}} + |\langle u,s\rangle|^2 t^2\|s\|^2 \geqslant 0 \\ \iff &|\langle u,s\rangle|^2 t + |\langle u,s\rangle|^2 t + |\langle u,s\rangle|^2 t^2 \|s\|^2 \geqslant 0 \\ \iff &t^2\left(|\langle u,s\rangle|^2\|s\|^2\right) + 2t|\langle u,s\rangle|^2 \geqslant 0 \quad \forall t \in \mathbb{R}\end{aligned}$$

It would be pointless to simplify $|\langle u,s\rangle|^2$, since we wish to calculate $\langle u,s\rangle$. The strategy to complete our proof consists of interpreting $t^2\left(|\langle u,s\rangle|^2\|s\|^2\right) + 2t|\langle u,s\rangle|^2$ as a second-degree polynomial function of the form $P(t) = at^2 + bt + c$ with:

$$\begin{cases} a = |\langle u,s\rangle|^2\|s\|^2 \\ b = 2|\langle u,s\rangle|^2 \\ c = 0 \end{cases}$$

The discriminant is equal to $\Delta = b^2 - 4ac = 4|\langle u,s\rangle|^4 \geqslant 0$. Thus, for $P(t)$ to be positive $\underline{\forall t \in \mathbb{R}}$, we must have $\Delta \leqslant 0$; this is only possible if $\Delta = 0$, but:

$$\Delta = 0 \iff 4|\langle u,s\rangle|^4 = 0 \ \forall s \in S \ \ (u \text{ being fixed since } x \text{ is fixed})$$

that is, $\langle u,s\rangle = 0 \ \forall s \in S$, hence $u \in S^\perp$. □

5.2.1. *Characterization of closed vector subspaces in Hilbert spaces*

Theorem 5.4 can be used to deduce a highly useful characterization of closed vector subspaces in Hilbert spaces.

We shall begin by considering an intermediate result.

LEMMA 5.1.– Let $\mathcal{H}$ be a Hilbert space and M any subset of $\mathcal{H}$, then:

$$\overline{span(M)} = \overline{span(M)}^{\perp\perp}$$

PROOF.– Taking:

$$\begin{cases} S = \overline{\text{span}(M)} \\ T = \overline{\text{span}(M)}^{\perp\perp} \end{cases}$$

Theorem 5.1 guarantees that $S \subseteq T$; if we can prove that $S \subset T$ is an impossible condition, then we will be left with $S = T$.

We begin by observing that S is a closed vector subspace of T and that T, as the orthogonal complement of a subset of $\mathcal{H}$, is a closed vector subspace of $\mathcal{H}$; T itself is thus a Hilbert space.

Reasoning by the absurd, if we assume $S \subset T$, then Theorem 5.4 can be applied to the pair S and T to ensure the existence of $u \in T$, $u \neq 0_{\mathcal{H}}$ and $u \in S^{\perp}$. However, this implies that $u \in S^{\perp} \cap T = \overline{\text{span}(M)}^{\perp} \cap \overline{\text{span}(M)}^{\perp\perp} = \{0_{\mathcal{H}}\}$, which contradicts the fact that $u \neq 0_{\mathcal{H}}$. Thence, $\overline{\text{span}(M)} = \overline{\text{span}(M)}^{\perp\perp}$. □

We now have all of the information needed to prove a helpful characterization of closed vector subspaces in Hilbert spaces: *closed vector subspaces are precisely those which coincide with their biorthogonal complement.*

This characterization is particularly powerful, as it creates a bridge between different structures that coexist in a Hilbert space: in fact, on one side, closure is related to the topological structure induced by the presence of a Hilbert norm, while, on the other side, the concept of orthogonality is related to the geometry of the Hilbert space by the presence of an inner produce. This bridge can be used, for instance, to verify the closure of a vector subspace by showing explicitly its biorthogonal complement, if this computation is easier than the direct verification of the closure.

THEOREM 5.5.– Let $\mathcal{H}$ be a Hilbert space and M a *vector subspace* of $\mathcal{H}$.

1) $M^{\perp\perp} = \overline{span(M)}$.

2) $\boxed{\overline{M} = M^{\perp\perp}}$.

3) M is a closed vector subspace of $\mathcal{H}$ if and only if $M = M^{\perp\perp}$.

PROOF.–

1) Given that M is a vector subspace, $M \equiv \text{span}(M)$. Furthermore, by property 8 from Theorem 5.1, $M^{\perp} = \text{span}(M)^{\perp} = \overline{\text{span}(M)}^{\perp}$. Hence, the previous lemma implies $M^{\perp\perp} = \overline{\text{span}(M)}^{\perp\perp} = \overline{\text{span}(M)}$.

2) We have shown that $M^{\perp\perp} = \overline{\text{span}(M)}$ and we know that $M \equiv \text{span}(M)$, thus $M = M^{\perp\perp}$.

3) Let us demonstrate the double implication:

$\boxed{\Rightarrow}$: we know from point 1) that $\overline{M} = M^{\perp\perp}$, but if M is closed then $M = \overline{M}$, and thus $M = M^{\perp\perp}$;

$\boxed{\Leftarrow}$: if $M = M^{\perp\perp}$, then M is automatically a closed vector subspace by the fact that it is the orthogonal complement of $M^{\perp}$. □

COROLLARY 5.1.– Let $\mathcal{H}$ be a Hilbert space and M, N any two parts of $\mathcal{H}$.

1) It holds that:

$$\boxed{(M \cup N)^{\perp} = M^{\perp} \cap N^{\perp}}$$

2) Additionally, if M and N are two *closed vector subspaces* of $\mathcal{H}$, then:

$$\boxed{(M \cap N)^{\perp} = \overline{\text{span}\left(M^{\perp} \cup N^{\perp}\right)}} \quad [5.4]$$

PROOF.–

1) Let us prove the two inclusions:

– $(M \cup N)^{\perp} \subseteq M^{\perp} \cap N^{\perp}$: taking $x \in (M \cup N)^{\perp}$ and $y \in M$, then y also belongs to $M \cup N$, thus $\langle x, y \rangle = 0$, that is $x \in M^{\perp}$. Now, taking $y \in N$, the same argument tells us that $x \in N^{\perp}$. Thus $x \in M^{\perp}$ and $x \in N^{\perp}$, that is, $x \in M^{\perp} \cap N^{\perp}$;

– $M^{\perp} \cap N^{\perp} \subseteq (M \cup N)^{\perp}$: taking $x \in M^{\perp} \cap N^{\perp}$, then $x \in M^{\perp}$ and $x \in N^{\perp}$. If $y \in M \cup N$, then $y \in M$ or $y \in N$, but in both cases $\langle x, y \rangle = 0$, that is, $x \in (M \cup N)^{\perp}$;

2) The relationship determined above holds for all parts of $\mathcal{H}$ and thus also holds for $M^{\perp}$ and $N^{\perp}$. In this case, point 1 becomes $\left(M^{\perp} \cup N^{\perp}\right)^{\perp} \subseteq M^{\perp\perp} \cap N^{\perp\perp} \underset{\text{th. [5.5]}}{=} \overline{\text{span}(M)} \cap \overline{\text{span}(N)} = M \cap N$ since M and N are presumed to be closed vector subspaces. Now, taking the orthogonal, we obtain:

$$(M \cap N)^{\perp} = \left(M^{\perp} \cup N^{\perp}\right)^{\perp\perp} \underset{\text{th. [5.5]}}{=} \overline{\text{span}(M^{\perp} \cup N^{\perp})}$$ □

5.3. Polar and bipolar subsets of a Hilbert space

In this section, we shall use a different approach to obtain the same result concerning the characterization of a closed part of a Hilbert space. In this case, we shall use the concept of polar sets, which is particularly important in the context of convex optimization theory.

DEFINITION 5.4 (Polar and bipolar).– *Let $\mathcal{H}$ be a Hilbert space and M any non-empty part of $\mathcal{H}$. The polar set of M, noted M°, is the subset of $\mathcal{H}$ defined by*[5]*:*

$$M^\circ := \{x \in \mathcal{H} : \forall y \in M,\ \Re(\langle x, y\rangle) \leqslant 1\} \equiv \{x \in \mathcal{H} : \sup_{y \in M} \Re(\langle x, y\rangle) \leqslant 1\}$$

The bipolar of M is the polar of the polar, that is:

$$M^{\circ\circ} := (M^\circ)^\circ = \{h \in \mathcal{H} : \forall x \in M^\circ,\ \Re(\langle h, x\rangle) \leqslant 1\} \equiv \{h \in \mathcal{H} : \sup_{x \in M^\circ} \Re(\langle h, x\rangle) \leqslant 1\}$$

Let us also recall the concept of convex hull.

DEFINITION 5.5 (Closed convex hull).– *The closed convex hull of a part M of $\mathcal{H}$ is the closure of the intersection of all convex parts of $\mathcal{H}$ containing M. It is the smallest closed convex subset of $\mathcal{H}$ which contains M.*

The following result contains remarkable properties of both the polar and bipolar.

THEOREM 5.6.– Let $\mathcal{H}$ be a Hilbert space and M any non-empty part of $\mathcal{H}$.

1) M° is a closed convex subset of $\mathcal{H}$ which contains the zero vector $0_{\mathcal{H}}$.

2) $M^{\circ\circ}$ coincides with the closed convex hull C of $M \cup \{0_{\mathcal{H}}\}$.

3) If M is a convex part of $\mathcal{H}$ which contains $0_{\mathcal{H}}$, then $\overline{M} = M^{\circ\circ}$.

4) If M is a vector subspace of $\mathcal{H}$, then $M^\circ = M^\perp$.

PROOF.–

1) The fact that M° contains $0_{\mathcal{H}}$ is an obvious consequence of the fact that $\langle 0_{\mathcal{H}}, y\rangle = 0 < 1$ for all $y \in M$. To verify convexity, let us consider $\lambda \in [0, 1]$ and $x_1, x_2 \in M^\circ$; by the left linearity of the inner product:

$$\Re(\langle x_1 + (1-\lambda)x_2, y\rangle) = \lambda\Re(\langle x_1, y\rangle) + (1-\lambda)\Re(\langle x_2, y\rangle) \leqslant \lambda + (1-\lambda) = 1$$

showing that M° is convex. All that remains is to prove the closure; to do this, we first remark that, for all fixed $y \in \mathcal{H}$, the application $\phi_y : \mathcal{H} \to \mathbb{R}$, $x \mapsto \phi_y(x) := \Re(\langle x, y\rangle)$ is continuous. Writing:

$$\mathcal{H}_y := \phi_y^{-1}\{[-\infty, +1]\} = \{x \in \mathcal{H} : \Re(\langle x, y\rangle) \leqslant 1\}$$

5 Evidently, the real part $\Re$ of the inner product can be eliminated if $\mathcal{H}$ is a real Hilbert space.

then $\mathcal{H}_y$ is a closed subset of $\mathcal{H}$ as it is the reciprocal image of a closed subset of $\mathbb{R}$ *via* the continuous map ϕ_y (remember that $(-\infty, +1]$ is closed, since its complement set in $\mathbb{R}$ is $(1, +\infty)$, which is open). By definition, the elements of M° must belong to $\mathcal{H}_y$ for all $y \in M$, that is $M^\circ = \bigcap_{y \in M} \mathcal{H}_y$, and thus it is closed in $\mathcal{H}$ as it is the intersection of closed parts of $\mathcal{H}$.

2) Now, let us demonstrate the opposite inclusions.

$\boxed{\mathcal{C} \subseteq M^{\circ\circ}}$: first, it is useful to show that $M \subseteq M^{\circ\circ}$. To do this, we note that $M^{\circ\circ} = \bigcap_{x \in M^\circ} \mathcal{H}_x$. Next, let us take arbitrary but fixed $y \in M$ and $x \in M^\circ$; then notably, $x \in \mathcal{H}_y$, that is, $\Re(\langle x, y\rangle) \leqslant 1$ and since $\Re(\langle x, y\rangle) = \Re(\langle y, x\rangle)$, we also have $\Re(\langle y, x\rangle) \leqslant 1$, that is, $y \in \mathcal{H}_x$. Since this holds for all $x \in M^\circ$, $y \in \bigcap_{x \in M^\circ} \mathcal{H}_x = M^{\circ\circ}$. Then: $y \in M \implies y \in M^{\circ\circ}$, that is, $M \subseteq M^{\circ\circ}$.

By (1), we know that $M^{\circ\circ}$, as a polar set, is convex, closed and contains $\{0_\mathcal{H}\}$. We have just seen that $M \subseteq M^{\circ\circ}$, thus $M^{\circ\circ}$ is a closed convex set which contains $M \cup \{0_\mathcal{H}\}$. Given that $\mathcal{C}$, the closed convex hull of $M \cup \{0_\mathcal{H}\}$, is the smallest convex subset of $\mathcal{H}$ which contains $M \cup \{0_\mathcal{H}\}$, it must be included in $M^{\circ\circ}$.

$\boxed{M^{\circ\circ} \subseteq \mathcal{C}}$: the fact that $0_\mathcal{H} \in \mathcal{C}$ comes into play at this stage of the proof. From Theorem 5.3, for all $x \in \mathcal{H}$ it holds that:

$$\begin{aligned}&\Re(\langle x - P_\mathcal{C}x, 0_\mathcal{H} - P_\mathcal{C}x\rangle) \leqslant 0 \iff \Re(\langle x - P_\mathcal{C}x, -P_\mathcal{C}x\rangle) \\ &\leqslant 0 \iff \Re(\langle x - P_\mathcal{C}x, P_\mathcal{C}x\rangle) \geqslant 0\end{aligned}$$

and for all $y \in M$, we also have $\Re(\langle x - P_\mathcal{C}x, y - P_\mathcal{C}x\rangle) \leqslant 0$, that is, $\Re(\langle x - P_\mathcal{C}x, y - P_\mathcal{C}x\rangle) \leqslant \varepsilon$ for all $\varepsilon > 0$, that is, by linearity of the inner product:

$$\Re(\langle x - P_\mathcal{C}x, y - P_\mathcal{C}x\rangle) = \Re(\langle x - P_\mathcal{C}x, y\rangle) - \Re(\langle x - P_\mathcal{C}x, P_\mathcal{C}x\rangle) \leqslant \varepsilon$$

that is, given that $\varepsilon + \Re(\langle x - P_\mathcal{C}x, P_\mathcal{C}x\rangle)$ is a real number > 0:

$$\Re(\langle x - P_\mathcal{C}x, y\rangle) \leqslant \varepsilon + \Re(\langle x - P_\mathcal{C}x, P_\mathcal{C}x\rangle) \iff \frac{\Re(\langle x - P_\mathcal{C}x, y\rangle)}{\varepsilon + \Re(\langle x - P_\mathcal{C}x, P_\mathcal{C}x\rangle)} \leqslant 1$$

which can be rewritten as:

$$\Re\left(\left\langle \frac{x - P_\mathcal{C}x}{\varepsilon + \Re(\langle x - P_\mathcal{C}x, P_\mathcal{C}x\rangle)}, y\right\rangle\right) \leqslant 1 \qquad \forall y \in M,\ \forall \varepsilon > 0$$

that is, the element $z(x) := \frac{x - P_\mathcal{C}x}{\varepsilon + \Re(\langle x - P_\mathcal{C}x, P_\mathcal{C}x\rangle)} \in M^\circ$ for all $x \in \mathcal{H}$.

As this result holds for any $x \in \mathcal{H}$, it can be applied when $x \in M^{\circ\circ}$; in this case, by definition, we have $\Re(\langle x, z(x)\rangle) \leqslant 1$, that is:

$$\Re(\langle x, z(x)\rangle) = \Re\left(\left\langle x, \frac{x - P_\mathcal{C}x}{\varepsilon + \Re(\langle x - P_\mathcal{C}x, P_\mathcal{C}x\rangle)}\right\rangle\right) \leqslant 1$$

hence:

$$\Re(\langle x, x - P_{\mathcal{C}}x \rangle) \leqslant \varepsilon + \Re(\langle x - P_{\mathcal{C}}x, P_{\mathcal{C}}x \rangle) = \varepsilon + \Re(\langle P_{\mathcal{C}}x, x - P_{\mathcal{C}}x \rangle)$$

which gives us:

$$\Re(\langle x - P_{\mathcal{C}}x, x - P_{\mathcal{C}}x \rangle) \leqslant \varepsilon \iff \|x - P_{\mathcal{C}}x\|^2 \leqslant \varepsilon \quad \forall \varepsilon > 0$$

but this is possible if and only if $x - P_{\mathcal{C}}x = 0_{\mathcal{H}}$, that is, $x = P_{\mathcal{C}}x$; however, since $P_{\mathcal{C}}x \in \mathcal{C}$, $x \in \mathcal{C}$. This completes our proof that $x \in M^{\circ\circ} \implies x \in \mathcal{C}$, that is, $M^{\circ\circ} \subseteq \mathcal{C}$.

3) By (2), if M is a convex part of $\mathcal{H}$ containing $0_{\mathcal{H}}$, then $M^{\circ\circ}$ is the smallest convex part which contains M. If M is convex, then $\overline{M}$ is also convex, and, furthermore, is the smallest closed set which contains M; consequently, $\overline{M} = M^{\circ\circ}$.

4) Let us now prove the opposite inclusions.

$\boxed{M^{\perp} \subseteq M^{\circ}}$: if $x \in M^{\perp}$, then, for all $y \in M$, $\langle x, y \rangle = 0 < 1$, therefore $x \in M^{\circ}$.

$\boxed{M^{\circ} \subseteq M^{\perp}}$: taking $x \in M^{\circ}$, we wish to prove that $x \in M^{\perp}$, that is $\langle x, y \rangle = 0$ $\forall y \in M$. This is done using the fact that M is taken to be a vector subspace of $\mathcal{H}$: if $y \in M$, then $ty \in M$ $\forall t \in \mathbb{R}\backslash\{0\}$. Since $x \in M^{\circ}$ and $ty \in M$, it must hold that:

$$\Re(\langle x, ty \rangle) \leqslant 1 \iff t\Re(\langle x, y \rangle) \leqslant 1 \; \forall t \in \mathbb{R}\backslash\{0\} \iff \Re(\langle x, y \rangle) = 0 \quad \forall y \in M$$

If $\mathcal{H}$ is a real Hilbert space, this concludes our proof. If $\mathcal{H}$ is complex, we also need to show that the imaginary part of the inner product is zero. We do this using Theorem 1.2, which tells us that $\Im(\langle x, y \rangle) = \Re(\langle x, iy \rangle)$, thus $\Im(\langle x, y \rangle) = \Re(\langle x, iy \rangle) = 0$ as we have previously proven that $\Re(\langle x, z \rangle) = 0$ $\forall z \in M$ and $z = iy \in M$ when $y \in M$ if M is a complex vector subspace. Finally, $\langle x, y \rangle = 0$ $\forall y \in M$ and thus $x \in M^{\perp}$. □

Properties 3 and 4 from Theorem 5.6 imply property 2 of Theorem 5.5, that is, $\overline{M} = M^{\perp\perp}$. In fact, on one side, $M^{\circ} = M^{\perp}$, so by repeating the polar operation twice we obtain $M^{\circ\circ} = M^{\perp\perp}$. Furthermore, $M^{\circ\circ} = \overline{M}$, thus $\overline{M} = M^{\perp\perp}$.

5.4. The (orthogonal) projection theorem in a Hilbert space

We shall now present and demonstrate the most important corollary of the theorem of orthogonal projection on a closed convex set.

THEOREM 5.7 (Orthogonal projection theorem).– Let $\mathcal{H}$ be a Hilbert space on $\mathbb{K} = \mathbb{R}$ or $\mathbb{C}$ and let S be a *closed* proper subspace of $\mathcal{H}$. Then:

$$\mathcal{H} = S \oplus S^{\perp}$$

that is, $\forall x \in \mathcal{H},\ \exists s \in S,\ \exists t \in S^{\perp}\ :\ x = s + t$, and this decomposition is unique, that is, if:

$$\begin{cases} x = s + t \\ x = s' + t' \end{cases} \quad \text{with } s, s' \in S,\ t, t' \in S^{\perp}, \quad \text{then: } \begin{cases} s = s' \\ t = t' \end{cases}$$

If S is not a proper subspace, then we have the trivial decomposition $\mathcal{H} = \mathcal{H} \oplus \{0_{\mathcal{H}}\}$.

PROOF.– Take a fixed $x \in \mathcal{H}$, the orthogonal projection $P_S(x) \in S$ of x onto S and the residual vector u: $u = x - P_S(x) \in S^{\perp}$. By Theorem 5.7, x can be decomposed as follows:

$$x = \underbrace{P_S(x)}_{\in S} + \underbrace{x - P_S(x)}_{\in S^{\perp}}$$

We must now show that a decomposition of this type is unique. Consider the decompositions $x = s + t$ and $x = s' + t'$, with $s, s' \in S,\ t, t' \in S^{\perp}$, then $s + t = s' + t'$, that is:

$$\underbrace{s - s'}_{\in S} = \underbrace{t' - t}_{\in S^{\perp}}$$

As S and $S^{\perp}$ are vector spaces, they are stable by subtraction, hence the inclusions shown in curly brackets. We have $S \ni s - s' = t' - t \in S^{\perp}$, thus $s - s' \in S \cap S^{\perp}$ and $t' - t \in S \cap S^{\perp}$. However $S \cap S^{\perp} = \{0_{\mathcal{H}}\}$, so we must have $s' = s$ and $t' = t$. □

IMPORTANT OBSERVATION.– Whenever we recognize the presence of a closed vector subspace S of a Hilbert space $\mathcal{H}$, the orthogonal projection theorem, gives a much profound meaning to the otherwise trivial decomposition $x = x - y + y$, with $x \notin S$ and $y \in S$: in fact, we know that $y = P_S(x)$ and y is orthogonal to $x - y$. This latter property guarantees the possibility to use the Pythagorean theorem to write $\|x\|^2 = \|x - y\|^2 + \|y\|^2$, which is often extremely useful in both theoretical and practical contexts, as we will see later in this chapter and the following one.

The results introduced above are applied in the exercise below.

Exercise 5.1

Let Ω be a bounded subset of $\mathbb{R}^n$, and consider the set M of functions $f : \Omega \to \mathbb{R}$, $f \in L^2(\Omega)$ which are constant a.e. Show that:

1) M is a closed vector subspace of $L^2(\Omega)$;

2) $\forall f \in L^2(\Omega)$, the projection of f onto M is the function which is constant a.e. and equal to the average of f on Ω, that is: $P_M f = \frac{1}{|\Omega|}\int_\Omega f(x)dx$, $|\Omega| = m(\Omega)$, with m the Lebesgue measure on $\mathbb{R}^n$;

3) The orthogonal complement of M in $L^2(\Omega)$ is given by the functions h in $L^2(\Omega)$ with zero average, that is $\int_\Omega h(x)dx = 0$.

Solution to Exercise 5.1

1) M can be characterized as the vector subspace of $L^2(\Omega)$ generated by the function $1(x) = 1$ which is constant a.e. on Ω. As there is only one generator, M is isomorphic to $\mathbb{R}$, which is closed.

2) Taking $f \in L^2(\Omega)$, then, by the projection theorem, if we write $f = f - P_M f + P_M f$, we have $f - P_M f \in M^\perp$. Let g be an element in M such that $g(x) = c \neq 0$ a.e. on Ω, and let us calculate the inner product between $f - P_M f$ and g:

$$
\begin{aligned}
0 &= \langle f - P_M f, g\rangle_{L^2(\Omega)} \\
&= \int_\Omega (f(x) - P_M f)g(x)dx \quad (P_M f \in M \\
&\implies \text{const. a.e., so we interpret } P_M f \in \mathbb{R}) \\
&= \int_\Omega f(x)g(x)dx - \int_\Omega (P_M f)g(x)dx = c\int_\Omega f(x)dx - c(P_M f)\int_\Omega dx \\
&= c\left(\int_\Omega f(x)dx - (P_M f)|\Omega|\right)
\end{aligned}
$$

that is, since $c \neq 0$:

$$P_M f = \frac{1}{|\Omega|}\int_\Omega f(x)dx$$

3) Taking any $h \in M^\perp$, then by definition: $\langle h, g\rangle_{L^2(\Omega)} = 0 \; \forall g \in M$. Now, taking $g(x) = c \neq 0$ a.e., we have:

$$0 = \langle h, g\rangle_{L^2(\Omega)} = \int_\Omega h(x)g(x)dx = c\int_\Omega h(x)dx, \qquad \forall k \in \mathbb{R}$$

hence $\int_\Omega h(x)dx = 0$.

What we have just proven and the orthogonal projection theorem imply that any function $f \in L^2(\Omega)$, $\Omega \subset \mathbb{R}^n$ such that $m(\Omega) < +\infty$ can be represented *in a unique manner* as:

$$f = \langle f\rangle_\Omega + h$$

where $\langle f \rangle_\Omega$ is the constant function on Ω and equal to the average of f on Ω and $h \in L^2(\Omega)$ such that $\langle h \rangle_\Omega = 0$. This implies that h must be an oscillating function, with oscillations that cancel out when we consider its average. This result is already remarkable by its own, but it will be further refined by the Fourier expansion of f, that will be described later in this chapter. □

5.5. Orthonormal systems and Hilbert bases

As we saw in Chapters 1 and 2, the presence of an orthonormal basis in a Euclidean space makes it easy to calculate vector components and to characterize orthogonal projections. Furthermore, using the Fourier basis, it is also possible to define Fourier coefficients and the DFT.

In this section, we shall describe the conditions which must be added in order to extend these considerations to infinite-dimensional Hilbert spaces. Let us begin with a definition.

DEFINITION 5.6 (Orthonormal system).– *An orthonormal family of elements in a Hilbert space is known as an orthonormal system.*

The properties of orthonormal systems will be analyzed in the context of separable Hilbert spaces, which are defined below.

DEFINITION 5.7.– *A Hilbert space $\mathcal{H}$ is said to be* separable *if there exists a subset $E \subseteq \mathcal{H}$ which is* countable *and* dense *in $\mathcal{H}$:* $\text{card}(E) = \aleph_0$, $\overline{E} = \mathcal{H}$.

As the vast majority of Hilbert spaces are, in fact, separable, we shall give a counter-example of a non-separable Hilbert space in section 5.5.3. The main advantage of working with separable Hilbert spaces is set out in Theorem 5.8.

THEOREM 5.8.– All orthonormal systems in a separable infinite-dimensional Hilbert space $\mathcal{H}$ are countable.

PROOF.– Let M be an infinite orthonormal system in $\mathcal{H}$. Given that $\mathcal{H}$ is separable, there exists a subset $E \subseteq \mathcal{H}$ which is countable and dense in $\mathcal{H}$: $\overline{E} = \mathcal{H}$.

From the characterization 2 of density given in Definition 4.4, we can guarantee that, for any element $x \in M$ and any arbitrary but fixed $\varepsilon > 0$, $\exists u_x \in E$ such that $\|x - u_x\| < \varepsilon$. If we can show that the correspondence defined by the function:

$$\begin{aligned} \imath : M &\longrightarrow E \\ x &\longmapsto \imath(x) = u_x \end{aligned}$$

is injective, then the theorem will be proven. In fact, if this is the case, M is in bijective correspondence with $\imath(M) \subseteq E$ which is an infinite part of a countable set, and is therefore itself countable.

To this aim, we take any $y \in M$ such that $y \neq x$, and $u_y \in E$ such that $\|x-u_y\| < \varepsilon$ for all arbitrary but fixed $\varepsilon > 0$. Since $x \neq y$ are two distinct points arbitrarily selected in M, the injectivity of $\imath$ corresponds to the fact that $\imath(x) \neq \imath(y)$, that is $u_x \neq u_y$. To prove this, we begin by noting that, since x and y belong to an orthonormal system, their distance is equal to $\sqrt{2}$ and we can write:

$$\begin{aligned}\sqrt{2} &= \|x-y\| = \|x-u_x+u_y-y+u_x-u_y\| \\ &\underset{\text{triang. ineq.}}{\leqslant} \|x-u_x\|+\|y-u_y\|+\|u_x-u_y\| \\ &< 2\varepsilon + \|u_x-u_y\|\end{aligned}$$

that is $\|u_x-u_y\| > \sqrt{2}-2\varepsilon$. $\sqrt{2}-2\varepsilon > 0 \iff \varepsilon < \sqrt{2}/2$, thus, we simply need to fix $\varepsilon \in (0, \sqrt{2}/2)$, $\|u_x-u_y\| > 0$ to obtain $u_x \neq u_y$. □

This theorem is the reason for selecting a discrete value, for example $n \in \mathbb{N}$ or $\mathbb{Z}$, to label the elements of an orthonormal system in a separable Hilbert space.

CONVENTION.– From now on, all Hilbert spaces $\mathcal{H}$ will be *assumed to be separable*, unless otherwise stated.

The two most important propositions related to orthonormal systems are Bessel's inequality and the Fischer-Riesz theorem.

5.5.1. *Bessel's inequality and Fourier coefficients*

The expansion of a vector $v \in \mathbb{K}^n$, $n < +\infty$, with respect to an orthonormal basis $(u_i)_{i=1}^n$ is written as $v = \sum_{i=1}^{n} \langle v, u_i\rangle u_i$, $\langle v, u_i\rangle$ being the components of v in this basis. Furthermore, the Plancherel identity holds true: $\|v\|^2 = \sum_{i=1}^{n} |\langle v, u_i\rangle|^2$. If we want to extend this property to an orthonormal system of an infinite-dimensional Hilbert space $\mathcal{H}$ we immediately see that a necessary condition must be verified: for any element $x \in \mathcal{H}$, the sequence $(\langle x, u_n\rangle)_{n\in\mathbb{N}}$ must decay toward 0 when $n \to +\infty$; otherwise, the series $\sum_{n\in\mathbb{N}} \langle x, u_n\rangle u_n$ would not converge. The following result guarantees that this necessary condition is always satisfied; the Plancherel identity, on the other hand, is not guaranteed to hold.

THEOREM 5.9 (Bessel's inequality).– Let $(u_n)_{n\in\mathbb{N}} \subset \mathcal{H}$ be an orthonormal system in a Hilbert space $\mathcal{H}$. Then, $\forall x \in \mathcal{H}$, it holds that:

$$\sum_{n\in\mathbb{N}} |\langle x, u_n\rangle|^2 \leqslant \|x\|^2 \qquad [5.5]$$

More precisely, the difference between the two sides of inequality [5.5] may be quantified as:

$$\|x\|^2 - \sum_{n\in\mathbb{N}} |\langle x, u_n\rangle|^2 = \left\| x - \sum_{n\in\mathbb{N}} \langle x, u_n\rangle u_n \right\|^2 \qquad [5.6]$$

PROOF.– Bessel's inequality can be proved by showing that the difference $\|x\|^2 - \sum_{n\in\mathbb{N}} |\langle x, u_n\rangle|^2$ is equal to the square of a norm, which is $\geqslant 0$.

For simplicity's sake, we shall write $\lambda_n = \langle x, u_n\rangle \iff \overline{\lambda_n} = \langle u_n, x\rangle \ \forall n \in \mathbb{N}$ and consider any $N \in \mathbb{N}$. By Carnot's theorem (Theorem 1.5) we have:

$$\left\| x - \sum_{n=0}^{N} \lambda_n u_n \right\|^2 = \|x\|^2 - \langle x, \sum_{n=0}^{N} \lambda_n u_n\rangle - \langle \sum_{n=0}^{N} \lambda_n u_n, x\rangle + \left\| \sum_{n=0}^{N} \lambda_n u_n \right\|^2$$

Applying sesquilinearity to the two intermediary terms, and the generalized Pythagorean theorem to the final term, the previous equality can be rewritten as follows:

$$\left\| x - \sum_{n=0}^{N} \lambda_n u_n \right\|^2 = \|x\|^2 - \sum_{n=0}^{N} \overline{\lambda_n}\langle x, u_n\rangle - \sum_{n=0}^{N} \lambda_n \langle u_n, x\rangle + \sum_{n=0}^{N} |\lambda_n|^2 \|u_n\|^2$$

From the definitions of λ_n and $\overline{\lambda_n}$, and using the fact that $\|u_n\|^2 = 1$ for all n, the final equality becomes:

$$\left\| x - \sum_{n=0}^{N} \lambda_n u_n \right\|^2 = \|x\|^2 - \cancel{\sum_{n=0}^{N} \overline{\lambda_n}\lambda_n} - \sum_{n=0}^{N} \lambda_n \overline{\lambda_n} + \cancel{\sum_{n=0}^{N} |\lambda_n|^2} = \|x\|^2 - \sum_{n=0}^{N} |\lambda_n|^2$$

that is:

$$\|x\|^2 - \sum_{n=0}^{N} |\langle x, u_n\rangle|^2 = \left\| x - \sum_{n=0}^{N} \langle x, u_n\rangle u_n \right\|^2$$

As we did not impose any restrictions on $N \in \mathbb{N}$, this equality holds true for an arbitrarily large value of N, that is:

$$\|x\|^2 - \sum_{n\in\mathbb{N}} |\langle x, u_n\rangle|^2 = \left\| x - \sum_{n\in\mathbb{N}} \langle x, u_n\rangle u_n \right\|^2 \qquad \square$$

Bessel's inequality allows us to generalize the definition of Fourier coefficients encountered in Chapter 2.

DEFINITION 5.8 (Generalized Fourier coefficients).– *The scalars* $\langle x, u_n\rangle \in \mathbb{K}$ *are said to be the generalized Fourier coefficients of* x *with respect to the orthonormal system* $(u_n)_{n\in\mathbb{N}}$*, and are written as:*

$$\boxed{\hat{x}(n) = \langle x, u_n\rangle} \quad \forall n \in \mathbb{N}$$

Bessel's inequality can be reformulated stating that, for all $x \in \mathcal{H}$, the sequence:

$$\hat{x} \equiv (\hat{x}_n)_{n\in\mathbb{N}}$$

belongs to $\ell^2(\mathbb{N}, \mathbb{K})$, and that:

$$\|\hat{x}\|_2 \leqslant \|x\| \quad \forall x \in \mathcal{H}$$

We see that *the sequence of generalized Fourier coefficients always decays toward* 0. For Hilbert spaces where x can be identified with a function, analyzing the speed of decay of Fourier coefficients provides interesting information concerning the regularity of the function itself.

Equation [5.6] gives an estimation of the difference between $\|x\|^2$ and $\|\hat{x}\|_2{}^2$ and, rewritten with the notation introduced above, immediately implies Corollary 5.2.

COROLLARY 5.2.– Let $\mathcal{H}$ be a Hilbert space and $(u_n)_{n\in\mathbb{N}}$ any orthonormal system in $\mathcal{H}$. Then:

$$\left\| x - \sum_{n\in\mathbb{N}} \hat{x}(n) u_n \right\|^2 = \|x\|^2 - \|\hat{x}\|_2{}^2$$

Specifically, $\sum_{n\in\mathbb{N}} \hat{x}(n) u_n$ converges to x if and only if $\|\hat{x}\|_2 = \|x\|$.

5.5.2. *The Fischer-Riesz theorem*

Theorem 5.10, which is fundamental in functional analysis, is sometimes referred to as the Fischer-Riesz theorem, for example in the classic Dunford and Schwartz (1958).

THEOREM 5.10 (Fischer-Riesz).– Let $\mathcal{H}$ be a Hilbert space, $(u_n)_{n\in\mathbb{N}}$ an orthonormal system in $\mathcal{H}$ and $(k_n)_{n\in\mathbb{N}}$ a sequence of scalars in $\mathbb{K} = \mathbb{R}$ or $\mathbb{C}$.

1) Then:

$$\sum_{n\in\mathbb{N}} k_n u_n \text{ converges (in norm } \| \ \| \text{ of } \mathcal{H}) \iff \sum_{n\in\mathbb{N}} |k_n|^2 \text{ converges (in } \mathbb{K})$$

that is, $\sum\limits_{n\in\mathbb{N}} k_n u_n$ converges $\iff (k_n)_{n\in\mathbb{N}} \in \ell^2(\mathbb{N}, \mathbb{K})$.

2) If $\sum\limits_{n\in\mathbb{N}} k_n u_n$ converges to the sum x, that is $x = \sum\limits_{n\in\mathbb{N}} k_n u_n$, then:

$$k_n = \langle x, u_n \rangle = \hat{x}(n)$$

and:

$$\|x\|^2 = \sum_{n\in\mathbb{N}} |k_n|^2$$

that is, Bessel's inequality becomes Plancherel's equality $\|x\|^2 = \sum\limits_{n\in\mathbb{N}} |\langle x, u_n \rangle|^2 = \|\hat{x}\|_2{}^2$.

PROOF.–

1) We wish to verify that studying the convergence of $\sum\limits_{n\in\mathbb{N}} k_n u_n$ is equivalent to studying the convergence of $\sum\limits_{n\in\mathbb{N}} |k_n|^2$. This will be done by using the fact that $\mathcal{H}$ and $\mathbb{K}$ are complete, so the Cauchy condition is necessary and sufficient for the sequences to converge, and by remembering that the series $\sum\limits_{n\in\mathbb{N}} k_n u_n$ is the sequence $(S_N)_{N\in\mathbb{N}} = \left(\sum\limits_{n=0}^{N} k_n u_n\right)_{N\in\mathbb{N}}$ of partial sums.

The Cauchy condition for $(S_N)_{N\in\mathbb{N}}$ is:

$$\forall \varepsilon > 0\ \exists K_\varepsilon > 0 \ : \ r > s \geqslant K_\varepsilon \implies \|S_r - S_s\| = \left\| \sum_{n=s+1}^{r} k_n u_n \right\| < \varepsilon$$

Since $\left\| \sum_{n=s+1}^{r} k_n u_n \right\| < \varepsilon \iff \left\| \sum_{n=s+1}^{r} k_n u_n \right\|^2 < \varepsilon^2 \equiv \delta$, as the inequality concerns two real positive numbers, the Cauchy condition for $(S_N)_{N\in\mathbb{N}}$ can be redefined as follows:

$$\forall\delta > 0\ \exists K_\delta > 0\ :\ r > s \geqslant K_\delta \implies \left\| \sum_{n=s+1}^{r} k_n u_n \right\|^2 < \delta$$

The usefulness of considering the squared norm is that, thanks to the orthogonality of u_n, we can use the generalized Pythagorean theorem to write:

$$\left\| \sum_{n=s+1}^{r} k_n u_n \right\|^2 = \sum_{n=s+1}^{r} \|k_n u_n\|^2 = \sum_{n=s+1}^{r} |k_n|^2 \cancelto{1}{\|u_n\|^2} = \sum_{n=s+1}^{r} |k_n|^2$$

The Cauchy condition for the sequence of partial sums of the series $\sum_{n\in\mathbb{N}} k_n u_n$ can then be rewritten as:

$$\forall\delta > 0\ \exists K_\delta > 0\ :\ r > s \geqslant K_\delta \implies \sum_{n=s+1}^{r} |k_n|^2 < \delta,$$

which is the Cauchy condition for the sequence of partial sums of the series $\sum_{n\in\mathbb{N}} |k_n|^2$. Hence, the study of the convergence of the two series is equivalent.

2) Assuming that the series $\sum_{m\in\mathbb{N}} k_m u_m$ converges toward the sum x, then, by continuity of the inner product:

$$\langle x, u_n\rangle = \langle \sum_{m\in\mathbb{N}} k_m u_m, u_n\rangle = \sum_{m\in\mathbb{N}} k_m \langle u_m, u_n\rangle = \sum_{m\in\mathbb{N}} k_m \delta_{m,n} = k_n, \quad \forall n \in \mathbb{N}.$$

Hence: $x = \sum_{n\in\mathbb{N}} \langle x, u_n\rangle u_n = \sum_{n\in\mathbb{N}} \hat{x}(n) u_n$. The fact that property 2 implies that Bessel's inequality becomes Plancherel's equality is a direct consequence of Corollary 5.2. An alternative proof is possible using the continuity of the norm:

$$\|x\|^2 = \| \sum_{n\in\mathbb{N}} \langle x, u_n\rangle u_n \|^2 = \sum_{n\in\mathbb{N}} |\langle x, u_n\rangle|^2 \cancelto{1}{\|u_n\|^2} = \sum_{n\in\mathbb{N}} |\langle x, u_n\rangle|^2 = \|\hat{x}\|_2{}^2 \qquad \square$$

COROLLARY 5.3.– *Let $\mathcal{H}$ be a Hilbert space, $x \in \mathcal{H}$ and let $(u_n)_{n\in\mathbb{N}}$ be an orthonormal system in $\mathcal{H}$. Then the series $\sum_{n\in\mathbb{N}} \hat{x}(n) u_n$ is always convergent (with respect to the norm $\|\ \|$ of $\mathcal{H}$).*

PROOF.– By Bessel's inequality, $(\hat{x}(n))_{n\in\mathbb{N}} \in \ell^2(\mathbb{N},\mathbb{K})$, that is, $\sum_{n\in\mathbb{N}} |\hat{x}(n)|^2$ is convergent in $\mathbb{K}$; by property 1 of the Fischer-Riesz theorem, the series $\sum_{n\in\mathbb{N}} \hat{x}(n)u_n$ is convergent in $\mathcal{H}$. □

NOTABLE EXAMPLE.–

The fact that $\sum_{n\in\mathbb{N}} \hat{x}(n)u_n$ is always convergent does not necessarily imply that it converges to x, as we show with the following counter example.

We take: $\mathcal{H} = L^2[-\pi,\pi]$, $u_n(t) = \frac{1}{\sqrt{\pi}}\sin(nt)$, $n \in \mathbb{N}$ and $t \in [-\pi,\pi]$. It is easy to verify that $(u_n)_{n\in\mathbb{N}}$ is an orthonormal system for $\mathcal{H}$. Taking $x(t) = \cos(t)$, by direct calculation, we obtain:

$$\sum_{n\in\mathbb{N}} \hat{x}(n)u_n = \sum_{n=1}^{\infty} \frac{1}{\pi}\left(\int_{-\pi}^{\pi} \cos(t)\sin(nt)dt\right)\sin(nt)$$

Furthermore, $\int_{-\pi}^{\pi} \cos(t)\sin(nt)dt = 0$ as it is the integral of an odd function on a symmetrical domain, thus:

$$\sum_{n\in\mathbb{N}} \hat{x}(n)u_n = \sum_{n=1}^{\infty} 0\cdot\sin(nt) = \mathbf{0}$$

where $\mathbf{0}$ is the identically null function on $[-\pi,\pi]$, which is clearly different from the function $\cos(t)$; thus, $\sum_{n\in\mathbb{N}} \hat{x}(n)u_n \neq x$.

5.5.3. *Characterizations of a Hilbert basis (or complete orthonormal system)*

The example above shows that an orthonormal system in a Hilbert space $\mathcal{H}$ does not necessarily guarantee that the series of Fourier coefficients of $x \in \mathcal{H}$ multiplied by the elements of this orthonormal system will converge in norm to x itself.

This fact naturally raises the question of whether a condition which ensures such a convergence exists. In this section, we shall prove that the answer to this question is affirmative.

In section 1.5, we saw that, in finite dimension, this condition is that the orthonormal system must be an orthonormal basis, that is, a maximal set of unitary vectors orthogonal to each other, where "maximal" means that no other unitary vector exists which is orthogonal to all of them.

Remarkably, this property also characterizes the bases of an infinite-dimensional Hilbert space, but the terminology used in this case is different.

DEFINITION 5.9 (Complete orthonormal system).– *Let* $(u_n)_{n\in\mathbb{N}} \subset \mathcal{H}$ *be an orthonormal system of a Hilbert space* $\mathcal{H}$*. If* $(u_n)_{n\in\mathbb{N}}$ *is not a proper set of another orthonormal system of* $\mathcal{H}$*, that is, if there are no other unitary vectors orthogonal to the vectors* $(u_n)_{n\in\mathbb{N}}$*, then this system is referred to as a* complete *(or* total*) orthonormal system, or as a* Hilbert basis.

The property of being a Hilbert basis, in the sense defined above, is equivalent to five other properties.

THEOREM 5.11.– Let $(u_n)_{n\in\mathbb{N}}$ be an orthonormal system of a Hilbert space $\mathcal{H}$. The following statements are equivalent:

1) $(u_n)_{n\in\mathbb{N}}$ is a Hilbert basis;

2) $\langle x, u_n\rangle \equiv \hat{x}(n) = 0 \; \forall n \in \mathbb{N} \iff x = 0_{\mathcal{H}}$, that is $0_{\mathcal{H}}$ is the only vector which is orthogonal to all vectors of a complete orthonormal system (or, equivalently, the only vector $x \in \mathcal{H}$ whose generalized Fourier coefficients are all zero is the null vector);

3) $\overline{span((u_n)_{n\in\mathbb{N}})} = \mathcal{H}$, that is $(u_n)_{n\in\mathbb{N}}$ generates a vector subspace which is dense in $\mathcal{H}$;

4) $\forall x \in \mathcal{H}$:

$$x = \sum_{n\in\mathbb{N}} \langle x, u_n\rangle u_n = \sum_{n\in\mathbb{N}} \hat{x}(n) u_n \qquad \textit{Generalized Fourier series expansion}$$

5) $\forall x, y \in \mathcal{H}$:

$$\langle x, y\rangle = \sum_{n\in\mathbb{N}} \langle x, u_n\rangle\langle u_n, y\rangle = \langle \hat{x}, \hat{y}\rangle_{\ell^2(\mathbb{N},\mathbb{K})} \qquad \textit{Parseval's identity}$$

6) $\forall x \in \mathcal{H}$:

$$\|x\|^2 = \sum_{n\in\mathbb{N}} |\langle x, u_n\rangle|^2 = \|\hat{x}\|_2{}^2 \qquad \textit{Plancherel's identity}$$

PROOF.– Our proof consists of the following steps: 1) $\Rightarrow$ 2) $\Rightarrow$ 3) $\Rightarrow$ 4) $\Rightarrow$ 5) $\Rightarrow$ 6) $\Rightarrow$ 1)

1) $\Rightarrow$ 2): reasoning by the absurd, if statement 1 is true and statement 2 is false, then $\exists x^* \in \mathcal{H}$, $x^* \neq 0_{\mathcal{H}}$ such that: $\langle x^*, u_n\rangle = 0 \; \forall n \in \mathbb{N}$, but then the vector $u^* = \dfrac{x^*}{\|x^*\|}$ is a unitary vector and orthogonal to all of the elements of $(u_n)_{n\in\mathbb{N}}$, thus $(u^*, (u_n)_{n\in\mathbb{N}})$ would be a larger orthonormal system than $(u_n)_{n\in\mathbb{N}}$, which contradicts the completeness of $(u_n)_{n\in\mathbb{N}}$.

2) $\Rightarrow$ 3): 2) $\Rightarrow ((u_n)_{n\in\mathbb{N}})^\perp = \{0_{\mathcal{H}}\} \iff \left(\overline{\text{span}((u_n)_{n\in\mathbb{N}})}\right)^\perp = \{0_{\mathcal{H}}\}$, by property 8 from Theorem 5.1, if we take the orthogonal complement of both sides,

we obtain $\left(\overline{\text{span}((u_n)_{n\in\mathbb{N}})}\right)^{\perp\perp} = \{0_{\mathcal{H}}\}^{\perp} = \mathcal{H}$, then $\mathcal{H} = \left(\overline{\text{span}((u_n)_{n\in\mathbb{N}})}\right)^{\perp\perp}$ $= \overline{\text{span}((u_n)_{n\in\mathbb{N}})}$, by Theorem 5.5.

3) $\Rightarrow$ 4): let us consider x, calculate the inner products with $(u_n)_{n\in\mathbb{N}}$ and write the series $\sum_{n\in\mathbb{N}} \langle x, u_n\rangle u_n$, which we know converges to a certain point $y \in \mathcal{H}$. We must show that, if statement 3 holds, then it follows that $y = x$. To this aim, note that the second part of the Fischer-Riesz theorem tells us that $\langle x, u_n\rangle = \langle y, u_n\rangle$ $\forall n \in \mathbb{N}$, that is $\langle x - y, u_n\rangle = 0$, $\forall n \in \mathbb{N}$, that is $x - y \in ((u_n)_{n\in\mathbb{N}})^{\perp} = \left(\overline{\text{span}((u_n)_{n\in\mathbb{N}})}\right)^{\perp} \underset{(3)}{=}$ $\mathcal{H}^{\perp} = \{0_{\mathcal{H}}\}$, that is, $y = x$.

4) $\Rightarrow$ 5): let us consider any $x, y \in \mathcal{H}$ and write their generalized Fourier series. By statement 4, we have:

$$x = \sum_{n\in\mathbb{N}} \langle x, u_n\rangle u_n \quad y = \sum_{m\in\mathbb{N}} \langle y, u_m\rangle u_m$$

thus:

$$\langle x, y\rangle = \langle \sum_{n\in\mathbb{N}} \langle x, u_n\rangle u_n, \sum_{m\in\mathbb{N}} \langle y, u_m\rangle u_m\rangle$$

By the continuity and linearity of the inner product, we have:

$$\langle x, y\rangle = \sum_{n\in\mathbb{N}} \langle x, u_n\rangle \langle u_n, \sum_{m\in\mathbb{N}} \langle y, u_m\rangle u_m\rangle$$

then, by the continuity and sesquilinearity of the inner product:

$$\langle x, y\rangle = \sum_{n\in\mathbb{N}} \sum_{m\in\mathbb{N}} \langle x, u_n\rangle \overline{\langle y, u_m\rangle} \langle u_n, u_m\rangle = \sum_{n\in\mathbb{N}} \sum_{m\in\mathbb{N}} \langle x, u_n\rangle \langle u_m, y\rangle\, \delta_{n,m}$$

that is:

$$\langle x, y\rangle = \sum_{n\in\mathbb{N}} \langle x, u_n\rangle \langle u_n, y\rangle$$

5) $\Rightarrow$ 6): consider $y = x$ in statement 5: $\|x\|^2 = \langle x, x\rangle = \sum_{n\in\mathbb{N}} \langle x, u_n\rangle\langle u_n, x\rangle$ $= \sum_{n\in\mathbb{N}} \langle x, u_n\rangle\overline{\langle x, u_n\rangle} = \sum_{n\in\mathbb{N}} |\langle x, u_n\rangle|^2$.

6) $\Rightarrow$ 1): reasoning by the absurd, if statement 6 is true and statement 1 is false, then $\exists u^* \in \mathcal{H}$, $\|u^*\| = 1$ and $\langle u^*, u_n\rangle = 0$ $\forall n \in \mathbb{N}$; this would give us $\sum_{n\in\mathbb{N}} |\langle u^*, u_n\rangle|^2 = 0$, which contradicts statement 4, since it states that $\sum_{n\in\mathbb{N}} |\langle u^*, u_n\rangle|^2 = \|u^*\|^2 = 1$. □

IMPORTANT NOTE CONCERNING PROPERTY 4.– The expansion into a generalized Fourier series on a Hilbert basis is an extension of the decomposition theorem for vectors on an orthonormal basis in a Euclidean space of finite dimension d, as shown in Table 5.1.

$\mathbb{K}^d$	Hilbert space $\mathcal{H}$
Orthonormal basis: $(u_i)_{i=1,\ldots,d}$	Hilbert basis: $(u_n)_{n\in\mathbb{N}}$
Expansion: $\forall x \in \mathbb{K}^d \quad x = \sum_{i=1}^{d} \langle x, u_i \rangle u_i$	Fourier series: $\forall x \in \mathcal{H} \quad x = \sum_{n\in\mathbb{N}} \langle x, u_n \rangle u_n$
Components: $\langle x, u_i \rangle$	Fourier coefficients: $\langle x, u_n \rangle$

Table 5.1. *Analogies between a finite-dimensional Euclidean space and an infinite-dimensional Hilbert space*

The generalization of the canonical basis of the space $\ell^2(\mathbb{Z}_N)$, introduced in section 2.1, is *the canonical Hilbert basis of* $\mathcal{H} = \ell^2(\mathbb{Z}, \mathbb{K})$ given by the vectors $(e_k)_{k\in\mathbb{Z}}$, $e_k(n) = \delta_{k,n}\ \forall n \in \mathbb{Z}$:

$$(e_1 = (1, 0, 0, \ldots),\ e_2 = (0, 1, 0, \ldots),\ \ldots)$$

The orthonormal property is obvious; completeness, for example, follows from the fact that the only vector which is orthogonal to $e_1, e_2, \ldots$ is the zero vector.

THEOREM 5.12.– All Hilbert spaces $\mathcal{H}$ admit a Hilbert basis.

PROOF.– Let O be the collection of all orthonormal families in $\mathcal{H}$. O is an ordered set by inclusion. If $\Phi \subset O$ is linearly ordered, then the union of all elements of Φ is a superior bound. Zorn's lemma (Moretti 2013) guarantees the existence of a maximal element in O. □

EXAMPLE OF A NON-SEPARABLE HILBERT SPACE.–

The Hilbert space in Theorem 5.11 was implicitly assumed to be separable. Any Hilbert space which does not verify any of the properties which characterize a Hilbert basis is non-separable. We shall use property 2 from Theorem 5.11 to illustrate an example of a non-separable Hilbert space. We begin by defining the following space:

$$\mathcal{H} = \{f : \mathbb{R} \to \mathbb{K} \ :\ \exists\, E_f \subset \mathbb{R},\ \mathrm{card}(E_f) \\ \leqslant \aleph_0 \ :\ f|_{E_f} \in \ell^2(\mathbb{N}, \mathbb{K}) \text{ et } f|_{\mathbb{R}\setminus E_f} = \mathbf{0}_{\mathbb{R}\setminus E_f}\}$$

This is the space made up of all functions f defined on $\mathbb{R}$ with a value in $\mathbb{K}$, which vanish everywhere except on a finite or countable subset E_f of $\mathbb{R}$, and such that the sequence $f : E_f \to \mathbb{K}$ is square summable.

$\mathcal{H}$ is a vector space, with respect to the pointwise-defined linear operations, which may be equipped with the following inner product:

$$\langle f, g\rangle = \sum_{x \in E_f \cap E_g} f(x)\overline{g(x)} \qquad f, g \in \mathcal{H}$$

This is well defined since, by definition of $\mathcal{H}$, the sum is either finite or a convergent series (evidently, if $\mathbb{K} = \mathbb{R}$, the conjugation operation becomes the identity). We can easily verify that $\mathcal{H}$ is a Hilbert space with respect to the topology induced by this inner product.

Reasoning by the absurd, let us suppose that $\mathcal{H}$ is separable, so that any Hilbert basis is be countable. Then let $u \equiv (u_n)_{n\in\mathbb{N}}$ be a Hilbert basis in $\mathcal{H}$, under the separability hypothesis, and take $U := \bigcup_{n\in\mathbb{N}} U_n$, where the sets $U_n \subset \mathbb{R}\ \forall n \in \mathbb{N}$ are such that $u_n|_{U_n} \in \ell^2(\mathbb{N}, \mathbb{K})$ and $u_n|_{\mathbb{R}\setminus U_n} = \mathbf{0}_{\mathbb{R}\setminus U_n}$. If we can show that there exists an element f_u in $\mathcal{H}$ which is orthogonal to all u_n and which is not the identically null function on $\mathbb{R}$, this would prove that property 2 of Theorem 5.11 does not hold: this contradiction implies that $\mathcal{H}$ cannot be separable.

To construct an element of this sort, we begin by noting that U is the union of countable or finite sets, and is thus, itself, either countable or finite. Considering any point $\bar{x} \in \mathbb{R}\setminus U$, we can therefore define $f_u : \mathbb{R} \to \mathbb{K}$ as:

$$f_u(x) = \begin{cases} 1 & \text{if } x = \bar{x} \\ 0 & \text{otherwise} \end{cases}$$

to obtain an element in $\mathcal{H}$ such that $\langle u_n, f_u\rangle = 0\ \forall n \in \mathbb{N}$, but $f \neq 0_{\mathbb{R}}$.

The fact that all complete orthonormal systems of a separable Hilbert space $\mathcal{H}$ of infinite dimension are countable should not lead us to think that $\mathcal{H}$ itself is of countable dimension *as a vector space*. In other words, if we consider $\mathcal{H}$ simply as a vector space, rather than a Hilbert space, then by definition its dimension is the cardinality of a basis in the algebraic sense, that is, a subset $B \subset \mathcal{H}$ of linearly independent elements in $\mathcal{H}$ such that any element in $\mathcal{H}$ can be obtained through a *finite* linear combination of elements in basis B. The following result, which we shall not prove, gives us quite surprising information about the difference between the cardinality of complete orthonormal systems and that of algebraic basis of an infinite dimensional Hilbert space.

THEOREM 5.13.– If the common cardinality of the Hilbert bases of a Hilbert space $\mathcal{H}$ (separable or otherwise) is $\aleph_0$, then the cardinality of the dimension of $\mathcal{H}$, as a vector space, cannot be less than $\aleph_1$.

It follows from this theorem that, as vector spaces, separable Hilbert spaces possess at least the cardinality of the continuum, that is a maximal system of linearly independent vectors possesses at least the cardinality of the continuum. The orthonormality requirement implies a further constraint, in the fact that the distance between the elements in the basis must be $\sqrt{2}$, this forces the cardinality of a complete orthonormal system to drop to that of the countable numbers.

Nevertheless, it is important to note – once again – that given a Hilbert basis, any element in an infinite-dimensional Hilbert space can be reconstructed *via* the generalized Fourier series in the sense of the Hilbert norm; this is by no means equivalent to the possibility of reconstructing elements by means of a finite linear combination.

This consideration shows that the concept of Hilbert basis is the most adequate to "parameterize" the elements of an infinite-dimensional Hilbert space *via* its generalized Fourier coefficients relative to the Hilbert basis, rather than a basis in the algebraic sense.

The reason for this lies in the fact that a Hilbert basis interacts with the rich geometric structure of the Hilbert space generated by the inner product *via* Fourier coefficients, while a mere algebraic basis only takes into account the linear structure.

The following definition establishes a specific terminology for the dimension of Hilbert spaces, adopted by certain authors, that we consider particularly adequate.

DEFINITION 5.10 (Orthogonal dimension).– *Let $\mathcal{H}$ be a Hilbert space. The orthogonal dimension of $\mathcal{H}$ is the common cardinality of all Hilbert bases in $\mathcal{H}$.*

Evidently, the orthogonal dimension coincides with the ordinary dimension for a finite-dimensional Hilbert space, but the same cannot be said in infinite dimensions.

5.5.4. *Isomorphisms between Hilbert spaces*

One final property which highlights the analogy between Hilbert spaces and finite-dimensional Euclidean spaces is the *existence of a prototype* for these spaces.

As we have seen, the dimension of a vector space V of finite dimension d is sufficient to characterize it up to an isomorphism. In fact, we know that, for any fixed basis of V, the correspondence $I : V \to \mathbb{K}^d$ which associates each vector v in V with its components (in $\mathbb{K}^d$) with respect to the chosen basis is an isomorphism. In this sense, $\mathbb{K}^d$ is the prototype of vector spaces on $\mathbb{K}$ of dimension $d < +\infty$. For (separable) infinite-dimensional Hilbert spaces, the prototype is $\ell^2(\mathbb{N}, \mathbb{K})$ and the generalized Fourier coefficients replace the vector components.

The concept of isomorphism between Hilbert spaces must be defined before we can establish a rigorous statement regarding this fact. The presence of the inner product implies that the canonical definition of isomorphism between vector spaces must be adapted to this situation.

DEFINITION 5.11 (Isomorphism between Hilbert spaces).– *Let $\mathcal{H}$ and $\mathcal{H}'$ be two Hilbert spaces on the same field $\mathbb{K}$. The transformation $U : \mathcal{H} \to \mathcal{H}'$ is an isomorphism of Hilbert spaces if:*

1) U is linear;

2) U is bijective;

3) U preserves the inner product, that is:

$$\langle U(x), U(y) \rangle_{\mathcal{H}'} = \langle x, y \rangle_{\mathcal{H}} \qquad \forall x, y \in \mathcal{H}$$

Condition 3 implies (in the specific case where $x = y$) that U preserves the norms, that is:

$$\|U(x)\|_{\mathcal{H}'} = \|x\|_{\mathcal{H}} \qquad \forall x \in \mathcal{H}$$

This also implies:

$$\|U(x) - U(y)\|_{\mathcal{H}'} = \|U(x - y)\|_{\mathcal{H}'} = \|x - y\|_{\mathcal{H}} \qquad \forall x, y \in \mathcal{H}$$

that is, U preserves the distances. In this case, we say that U is *isometric*.

The property of conservation of the norm implies $\|U(x)\|_{\mathcal{H}'} = 0 \iff \|x\|_{\mathcal{H}} = 0$; furthermore, by the definite positivity of the norm, it holds that $U(x) = 0_{\mathcal{H}'} \iff x = 0_{\mathcal{H}}$, that is $\ker(U) = \{0_{\mathcal{H}}\}$ and thus U is injective.

An isomorphism U between Hilbert spaces can thus be redefined as a surjective linear transformation which preserves the inner product. Actually, the linearity request is redundant, as we see from the following result.

THEOREM 5.14.– Let V, V' be two inner product spaces, of finite or infinite dimension, on the same field $\mathbb{K}$. If the transformation $U : V \to V'$ is surjective and preserves the inner product, then it is linear.

PROOF.– $\forall x, y, z \in V$ and $\forall \alpha, \beta \in \mathbb{K}$:

$$\begin{aligned} 0 = \langle 0, z \rangle &= \langle \alpha x + \beta y - \alpha x - \beta y, z \rangle = \langle \alpha x + \beta y, z \rangle - \alpha \langle x, z \rangle - \beta \langle y, z \rangle \\ &\underset{(U \text{ preserves } \langle\ \rangle)}{=} \langle U(\alpha x + \beta y), U(z) \rangle - \alpha \langle U(x), U(z) \rangle - \beta \langle U(y), U(z) \rangle \\ &\underset{(\text{linearity of } \langle\ \rangle)}{=} \langle U(\alpha x + \beta y) - \alpha U(x) - \beta U(y), U(z) \rangle \end{aligned}$$

Since, by hypothesis, U is surjective, as $z \in V$ varies, $U(z)$ represents any element of V', thus $U(\alpha x + \beta y) - \alpha U(x) - \beta U(y)$ is orthogonal to all of the elements of V', that is, $U(\alpha x + \beta y) - \alpha U(x) - \beta U(y) = 0_{\mathcal{H}'}$, hence:

$$U(\alpha x + \beta y) = \alpha U(x) + \beta U(y) \qquad \forall x, y \in V, \ \forall \alpha, \beta \in \mathbb{K}$$

and so U is linear. □

The definition of isomorphism between Hilbert spaces can thus be reformulated as follows.

DEFINITION 5.12 (Alternative definition of isomorphism between Hilbert spaces).– *Let $\mathcal{H}$ and $\mathcal{H}'$ be two Hilbert spaces on the same field $\mathbb{K}$. The transformation $U : \mathcal{H} \to \mathcal{H}'$ is an isomorphism of Hilbert spaces if:*

1) U is surjective;

2) U preserves the inner product.

The fact of being isomorphic is an equivalence relationship in the set of Hilbert spaces on the same field $\mathbb{K}$. The following result says that the orthogonal dimension plays, for a separable infinite-dimensional Hilbert space, the same role played by the dimension for a finite-dimensional vector space.

THEOREM 5.15.– $\mathcal{H}, \mathcal{H}'$: Hilbert spaces on the same field $\mathbb{K}$. $\mathcal{H}$ is isomorphic to $\mathcal{H}'$ if and only if the orthogonal dimension of $\mathcal{H}$ is the same as that of $\mathcal{H}'$.

5.5.5. $\ell^2(\mathbb{N}, \mathbb{K})$ *as the prototype of separable Hilbert spaces of infinite dimension*

LEMMA 5.2.– Let $(u_n)_{n\in\mathbb{N}}$ be a Hilbert basis of $\mathcal{H}$, then, for any sequence $(k_n)_{n\in\mathbb{N}}$ of $\ell^2(\mathbb{N}, \mathbb{K})$, there exists $x \in \mathcal{H}$ such that $(k_n)_{n\in\mathbb{N}} = (\langle x, u_n\rangle)_{n\in\mathbb{N}}$.

PROOF.– If $(k_n)_{n\in\mathbb{N}} \in \ell^2(\mathbb{N}, \mathbb{K})$, then, thanks to property 1 of Fischer-Riesz's theorem, $\sum\limits_{n\in\mathbb{N}} k_n u_n$ converges to a certain $x \in \mathcal{H}$. Then, property 2 of the same theorem guarantees that $(k_n)_{n\in\mathbb{N}} = (\langle x, u_n\rangle)_{n\in\mathbb{N}}$. □

THEOREM 5.16.– If the Hilbert space $\mathcal{H}$ has countable orthogonal dimension $\aleph_0$, then $\mathcal{H}$ is isomorphic to $\ell^2(\mathbb{N}, \mathbb{K})$.

PROOF.– Let $(u_n)_{n\in\mathbb{N}}$ be a countable Hilbert basis in $\mathcal{H}$ and consider the application:

$$\begin{aligned} U : \mathcal{H} &\longrightarrow \ell^2(\mathbb{N}, \mathbb{K}) \\ x &\longmapsto U(x) = (\langle x, u_n\rangle)_{n\in\mathbb{N}} \end{aligned}$$

U is surjective by Lemma 5.2 and it preserves the inner product by Parseval's identity:

$$\forall x, y \in \mathcal{H} \ : \ \langle x, y \rangle_{\mathcal{H}} = \sum_{n \in \mathbb{N}} \langle x, u_n \rangle \langle u_n, y \rangle = \sum_{n \in \mathbb{N}} \langle x, u_n \rangle \overline{\langle y, u_n \rangle}$$
$$\equiv \langle U(x), U(y) \rangle_{\ell^2(\mathbb{N},\mathbb{K})}$$

Hence, U is an isomorphism of Hilbert spaces. □

5.6. The Fourier Hilbert basis in L^2

The best-known example of a Hilbert basis, which is also the most important in terms of practical applications, is the Fourier basis. This basis is defined below in the context of the Hilbert space L^2.

5.6.1. $L^2[-\pi, \pi]$ *or* $L^2[0, 2\pi]$

Let us begin with $\mathcal{H} = L^2[-\pi, \pi]$ or $L^2[0, 2\pi]$ and $\mathbb{K} = \mathbb{C}$, then:

$$u_n(x) = \frac{1}{\sqrt{2\pi}} e^{inx}, \qquad n \in \mathbb{Z}$$

is a complete orthonormal system, called the Fourier basis of $L^2[-\pi, \pi]$ or $L^2[0, 2\pi]$. Note that this orthonormal system completes the orthonormal system $\frac{1}{\sqrt{\pi}} \sin(nx)$, $n \in \mathbb{N}$ which we used in section 5.5.2 as a counterexample to show that the convergence (in Hilbert norm) of the generalized Fourier series to the element defining the generalized Fourier coefficients is not guaranteed if we consider a *non-complete* orthonormal system.

Orthonormality is easy to prove. Considering $L^2[-\pi, \pi]$ (the proof for $L^2[0, 2\pi]$ is the same):

$$\langle u_n, u_m \rangle = \int_{-\pi}^{\pi} u_n(x) \overline{u_m(x)} dx = \frac{1}{2\pi} \int_{-\pi}^{\pi} e^{inx} e^{-imx} dx = \frac{1}{2\pi} \int_{-\pi}^{\pi} e^{i(n-m)x} dx$$

– if $n = m$, then $e^{i(n-m)x} = e^0 = 1$ and thus $\langle u_n, u_n \rangle = \|u_n\|^2 = 1$;

– If $n \neq m$, then, writing $y = i(n - m)$, the inner product can be written as:

$$\langle u_n, u_m \rangle = \frac{1}{2\pi} \int_{-\pi}^{\pi} e^{yx} dx$$
$$= \frac{1}{2\pi y} \left[e^{yx} \right]_{x=-\pi}^{x=\pi} = \frac{1}{2\pi i(n-m)} \left[e^{i(n-m)\pi} - e^{i(m-n)\pi} \right] = 0$$

In short, $\langle u_n, u_m \rangle = \delta_{n,m}$, proving orthonormality. The proof that the system is complete, instead, is much more complicated.

The Fourier expansion here is written as follows:

$$\forall f \in L^2[-\pi, \pi] \; : \; f = \sum_{n \in \mathbb{Z}} \langle f, u_n \rangle u_n$$

where:

$$\langle f, u_n \rangle \equiv \hat{f}(n) = \frac{1}{\sqrt{2\pi}} \int_{-\pi}^{\pi} f(x) e^{-inx} dx$$

is the n-th *Fourier coefficient* of f. Note that the convergence of the series should be interpreted as:

$$\int_{-\pi}^{\pi} \left| f(x) - \sum_{n=-N}^{N} \hat{f}(n) u_n(x) \right|^2 dx \underset{N \to +\infty}{\longrightarrow} 0$$

DEFINITION 5.13.– *Take* $\mathcal{H} = L^2[-\pi, \pi]$ *or* $L^2[0, 2\pi]$. *The application:*

$$\begin{aligned} F \equiv \hat{} : \mathcal{H} &\longrightarrow \ell^2(\mathbb{Z}, \mathbb{C}) \\ f &\longmapsto (\hat{f}(n))_{n \in \mathbb{Z}} \end{aligned}$$

is known as the Fourier transform of $\mathcal{H} = L^2[-\pi, \pi]$ *or* $L^2[0, 2\pi]$.

We see that F coincides with the transformation which implements the isomorphism between $L^2[-\pi, \pi]$ or $L^2[0, 2\pi]$ and its prototype $\ell^2(\mathbb{Z}, \mathbb{C})$!

The Fourier Hilbert basis of $L^2([-\pi, \pi])$ and $L^2([0, 2\pi])$ can be written in terms of real functions:

$$\begin{cases} u_0 \equiv \frac{1}{\sqrt{2\pi}} \\ \cos_n(x) \equiv \frac{1}{\sqrt{\pi}} \cos(nx), \quad n \in \mathbb{N} \\ \sin_n(x) \equiv \frac{1}{\sqrt{\pi}} \sin(nx), \quad n \in \mathbb{N} \end{cases}$$

It is important to note that the complex exponential of parameter $n \in \mathbb{Z}$ is replaced by two real sequences of parameter $n \in \mathbb{N}$; this is a consequence of Euler's formula, $e^{i\vartheta} = \cos \vartheta + i \sin \vartheta$, for all $\vartheta \in \mathbb{R}$.

The advantage of this basis is that it does not contain any imaginary parts; furthermore, the Fourier expansion in this case can be performed:

– for even functions, using u_0 and $\cos_n$;

– for odd functions, using $\sin_n$.

The reason for this result is easily explained: taking an even f, then $\hat{f}(n) = \frac{1}{\sqrt{\pi}} \int_{-\pi}^{\pi} f(x)\sin(nx)dx = 0 \; \forall n$, since $f(x)\sin(nx)$ is odd and $[-\pi, \pi]$ is a symmetrical domain. Similar arguments can be applied to odd functions to obtain the desired result.

5.6.2. $L^2(\mathbb{T})$

Our decision to consider the interval $[-\pi, \pi]$ or $[0, 2\pi]$ reflects the fact that the orthonormality of the system $(\frac{1}{\sqrt{2\pi}} e^{inx})_{n \in \mathbb{Z}}$ is very easy to prove. Actually, all of the properties stated for this system remain valid if $[-\pi, \pi]$ or $[0, 2\pi]$ is replaced by any other interval of size 2π.

Furthermore, these properties continue to hold if we consider functions defined on any real interval, that is, $f : \mathbb{R} \to \mathbb{C}$, on the condition that they are 2π-periodic. This can be formalized using a highly useful Hilbert space:

$$L^2(\mathbb{T}) = \Big\{ f : \mathbb{R} \to \mathbb{C} \,,\; f \text{ measurable} \,,\; f(x + 2\pi) = f(x),$$

$$\int_0^{2\pi} |f(x)|^2 \, dx < +\infty \Big\} / \sim$$

where $f \sim g$ if $f = g$ a.e., as usual. By periodicity, integration can be carried out on any interval of size 2π.

The symbol $\mathbb{T}$ represents the 1D *torus*, which may be identified with the unitary circumference. Any function $f : \mathbb{R} \to \mathbb{C}$ which is 2π-periodic may be identified with a function defined on $\mathbb{T}$ by means of the following diagram:

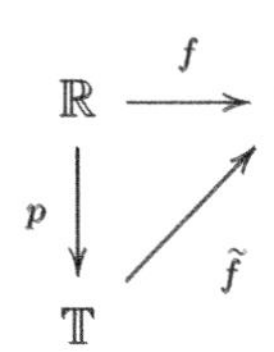

$$\begin{aligned} p : \mathbb{R} &\longrightarrow \mathbb{T} \\ x &\longmapsto (\cos x, \sin x) \end{aligned}$$

$$f : \mathbb{R} \to \mathbb{C} \text{ } 2\pi\text{-periodic}, \; \tilde{f} : \mathbb{T} \to \mathbb{C}, \; \tilde{f}(p(x)) = f(x)$$

$L^2(\mathbb{T})$ is isomorphic to $L^2[0, 2\pi]$ or $L^2[-\pi, \pi]$ *via* the application which restricts $f : \mathbb{R} \to \mathbb{C}$, $f \in L^2(\mathbb{T})$, to the interval $[0, 2\pi]$ or $[-\pi, \pi]$ (or any interval of size 2π):

$$\begin{aligned} I : L^2(\mathbb{T}) &\longrightarrow L^2[0, 2\pi] \\ f &\longmapsto If = f|_{[0,2\pi]} \text{ ou } f|_{[-\pi,\pi]} \end{aligned}$$

Using I, the complete orthonormal Fourier system can be transferred from $L^2[0, 2\pi]$ or $L^2[-\pi, \pi]$ onto $L^2(\mathbb{T})$:

$$\left(\frac{1}{\sqrt{2\pi}}e^{inx}\right)_{n\in\mathbb{Z}} \quad : \text{ Hilbert basis for } L^2(\mathbb{T})$$

and the definition of the Fourier transform can be extended on $L^2(\mathbb{T})$.

DEFINITION 5.14.– *The transformation:*

$$\begin{aligned} F \equiv \hat{}: L^2(\mathbb{T}) &\longrightarrow \ell^2(\mathbb{Z}, \mathbb{C}) \\ f &\longmapsto Ff = \widetilde{f} \end{aligned}$$

$Ff(n) = \hat{f}(n) = (\langle f, u_n \rangle)_{n\in\mathbb{Z}} = (\frac{1}{\sqrt{2\pi}} \int_0^{2\pi} f(x)e^{-inx}\, dx)_{n\in\mathbb{Z}}$. *is known as the Fourier transform on* $L^2(\mathbb{T})$.

We know that this transformation is an isomorphism between Hilbert spaces, and that $||\widehat{f}||_{\ell^2(\mathbb{Z},\mathbb{C})} = \sum\limits_{n\in\mathbb{Z}} |\langle f, u_n\rangle|^2 = ||f||_{L^2(\mathbb{T})}$.

5.6.3. $L^2[a, b]$

To handle elements of $f \in L^2[a, b]$, $a, b \in \mathbb{R}$, $a < b$, which are $(b-a)$-periodic, we must slightly modify the Fourier basis. The trick consists of multiplying the variable of f by an appropriate quantity – the *pulse* – which turns f into a $(b - a)$-periodic function. Formally, we define:

– $T = b - a$: the period;

– $\nu = \frac{1}{T}$: the frequency;

– $\omega = 2\pi\nu = \frac{2\pi}{T}$: the pulse.

We see that:

$$e^{i\omega n(x+T)} = \cos[\omega n(x + T)] + i \sin[\omega n(x + T)] = \cos[\omega nx$$

$$+\omega nT] + i \sin[\omega nx + \omega nT]$$

$$= \cos\left[\omega nx + \frac{2\pi}{T}nT\right] + i \sin[\omega nx + \frac{2\pi}{T}nT] = \cos[\omega nx$$
$$+2\pi n] + i \sin[\omega nx + 2\pi n]$$

$$= \cos(\omega nx) + i \sin(\omega nx) = e^{i\omega nx}$$

thus $x \mapsto e^{i\omega n x}$ is a T-periodic function. Using these considerations, we can show that a complete orthonormal system for $L^2[a,b]$ can be obtained using the following set of functions:

$$\begin{aligned} u_n : [a,b] &\longrightarrow \mathbb{C} \\ x &\longmapsto u_n(x) = \tfrac{1}{\sqrt{b-a}} e^{2\pi n i \frac{x-a}{b-a}}, \quad n \in \mathbb{Z} \end{aligned}$$

in the complex case, and:

$$\begin{cases} u_0 = \frac{1}{\sqrt{b-a}} \\ \cos_n(x) \equiv \sqrt{\frac{2}{b-a}} \cos\left(2\pi n \frac{x-a}{b-a}\right), & n \in \mathbb{N} \\ \sin_n(x) \equiv \sqrt{\frac{2}{b-a}} \sin\left(2\pi n \frac{x-a}{b-a}\right), & n \in \mathbb{N} \end{cases}$$

in the real case.

In the specific case of the Hilbert space $L^2[\ell,\ell]$, $\ell \in \mathbb{R}$, the Fourier basis can be written as:

$$u_n(x) = \frac{1}{\sqrt{2\ell}} e^{\pi i n \frac{x}{\ell}}, \qquad n \in \mathbb{Z}$$

in the complex case, and:

$$\begin{cases} u_0 = \frac{1}{\sqrt{2\ell}} \\ \cos_n(x) \equiv \sqrt{\frac{1}{\ell}} \cos\left(\pi n \frac{x}{\ell}\right), & n \in \mathbb{N} \\ \sin_n(x) \equiv \sqrt{\frac{1}{\ell}} \sin\left(\pi n \frac{x}{\ell}\right), & n \in \mathbb{N} \end{cases}$$

in the real case.

5.6.4. *Real Fourier series*

Using the real Hilbert basis of $L^2(\mathbb{T})$, that is:

$$\begin{cases} u_0 \equiv \frac{1}{\sqrt{2\pi}} \\ \cos_n(x) \equiv \frac{1}{\sqrt{\pi}} \cos(nx), & n \in \mathbb{N} \\ \sin_n(x) \equiv \frac{1}{\sqrt{\pi}} \sin(nx), & n \in \mathbb{N} \end{cases}$$

the real Fourier series expansion for any element $f \in L^2(\mathbb{T})$ is:

$$\boxed{f(t) \underset{L^2(\mathbb{T})}{=} \frac{a_0}{2} + \sum_{n=1}^{+\infty} a_n \cos(nt) + \sum_{n=1}^{+\infty} b_n \sin(nt)}$$

with:

$$a_0 = \frac{1}{\pi}\int_{\mathbb{T}} f(t)dt \quad \Longrightarrow \quad \frac{a_0}{2} = \frac{1}{2\pi}\int_{\mathbb{T}} f(t)dt = \langle f\rangle_{\mathbb{T}} \quad \text{(average of } f\text{)}$$

$$a_n = \frac{1}{\pi}\int_{\mathbb{T}} f(t)\cos(nt)dt \qquad \forall n = 1, 2, \ldots$$

$$b_n = \frac{1}{\pi}\int_{\mathbb{T}} f(t)\sin(nt)dt \qquad \forall n = 1, 2, \ldots$$

The coefficients a_0, a_n, b_n, $n = 1, 2, \ldots$ are known as the *real Fourier coefficients* of f.

Evidently, the equality must be interpreted in the sense of $L^2(\mathbb{T})$, that is:

$$\int_{\mathbb{T}} \left[f(t) - \left(\frac{a_0}{2} + \sum_{n=1}^{N} a_n \cos(nt) + \sum_{n=1}^{N} b_n \sin(nt) \right) \right]^2 dt \underset{N\to+\infty}{\longrightarrow} 0$$

The expression:

$$\boxed{S_N(t) = \frac{a_0}{2} + \sum_{n=1}^{N} a_n \cos(nt) + \sum_{n=1}^{N} b_n \sin(nt)}$$

is known as a *trigonometric polynomial of order* N. S_N is a 2π-periodic function, like the elements of $L^2(\mathbb{T})$.

To understand the presence of the constant $\frac{1}{\pi}$ in the real Fourier coefficients, consider the expansion of f with the respect to the system of cosine:

$$\sum_{n=1}^{+\infty} \langle f, \frac{1}{\sqrt{\pi}}\cos(nt)\rangle \frac{1}{\sqrt{\pi}}\cos(nt) = \sum_{n=1}^{+\infty} \left(\frac{1}{\pi}\int_{\mathbb{T}} f(t)\cos(nt)dt \right)\cos(nt)$$

the same holds true for the sine system and for the constant.

Incorporating the constant $\frac{1}{\pi}$ into the definition of the Fourier coefficients makes it possible to identify $\frac{a_0}{2}$ with the average value of f, so that *the real Fourier series can be interpreted as the superposition of the average value of f and combinations of harmonic waves of increasing frequency*. Notably:

– $t \mapsto a_1\cos(t) + b_1\sin(t)$ is known as the *fundamental harmonic*;

– $t \mapsto a_n\cos(nt) + b_n\sin(nt)$ is the *harmonic of order* n.

A *tuning fork* is able to produce a "pure" sound, that is one which consists exclusively of the fundamental harmonic; the vast majority of musical instruments, on the other hand, produce sounds which can be described by a Fourier series, that is a superposition of harmonics at frequencies which are multiples of the fundamental.

Using the orthogonal projection theorem and Plancherel's identity, we can say that the mean quadratic error (that is the norm L^2) between f and the trigonometric polynomial of order N is:

$$E_N = \int_{\mathbb{T}} [f(t) - S_N(t)]^2 \, dt = \int_{\mathbb{T}} f(t)^2 dt - \pi \left[\frac{a_0^2}{2} + \sum_{n=1}^{N} (a_n^2 + b_n^2) \right]$$

and since $E_N \underset{N\to+\infty}{\longrightarrow} 0$, it holds that:

$$\int_{\mathbb{T}} f(t)^2 dt = \pi \left[\frac{a_0^2}{2} + \sum_{n=1}^{+\infty} (a_n^2 + b_n^2) \right]$$

This is an identity between an integral and a numerical series, and is particularly useful for determining one of these two objects by calculating the other.

Taking $L^2[a,b]$ and writing $T = b - a$ and $\omega = \frac{2\pi}{T}$, we know that the real Fourier Hilbert basis is:

$$\left\{ \frac{1}{\sqrt{T}}, \sqrt{\frac{2}{T}} \cos(\omega n t), \sqrt{\frac{2}{T}} \sin(\omega n t), \; n = 1, 2, 3, \ldots \right\}$$

With respect to this Hilbert basis, the Fourier series expansion of $f \in L^2([a,b])$, f (T-periodic) is:

$$f(t) \underset{L^2[a,b]}{=} \frac{a_0}{2} + \sum_{n=1}^{+\infty} a_n \cos(\omega n t) + \sum_{n=1}^{+\infty} b_n \sin(\omega n t)$$

with:

$$\frac{a_0}{2} = \frac{1}{T} \int_a^b f(t) dt = \langle f \rangle_{[a,b]} \quad \text{(average of } f\text{)}$$

$$a_n = \frac{2}{T} \int_a^b f(t) \cos(\omega n t) dt, \;\; b_n = \frac{2}{T} \int_a^b f(t) \sin(\omega n t) dt \qquad \forall n = 1, 2, \ldots$$

In this case, the Fourier polynomials are T-periodic functions.

Exercise 5.2

The family $(e_k : [-\pi, \pi] \to \mathbb{C})_{k\in\mathbb{Z}}$ of non-normalized exponentials $\left(e_k(t) := e^{ikt}\right)_{k\in\mathbb{Z}}$ is a Hilbert basis of $L^2[-\pi, \pi]$ if this space is equipped with an inner product defined by $\langle f, g\rangle_0 = \frac{1}{2\pi}\int_{-\pi}^{\pi} f(x)\overline{g(x)}dx$.

1) Write the Fourier series associated with the function $\phi : [-\pi, \pi] \to \mathbb{C}$, $t \mapsto \cos(3t) - \sin(5t)$.

2) Take $\mathbb{N}^* = \mathbb{N}\backslash\{0\}$, and let $(\psi_k : \mathbb{R} \to \mathbb{R})_{k\in\mathbb{N}*}$ be the family defined by $\psi_k(t) = \sin(kt)$.

a) Consider $f \in L^2[0, \pi]$ such that $\int_0^{\pi} f(t)\psi_k(t)dt = 0\ \forall k \in \mathbb{N}^*$ and also

$$g(t) = \begin{cases} f(t) & \text{if } 0 \leqslant t < \pi \\ -f(-t) & \text{if } -\pi < t < 0 \end{cases}$$

Prove that $\int_{-\pi}^{\pi} g(t)e^{-ikt}dt = 0\ \forall k \in \mathbb{N}^*$.

b) Prove that $(\psi_k)_{k\in\mathbb{N}*}$ is a *complete system* in $L^2[0, \pi]$ equipped with the inner product $\langle f, g\rangle = \int_0^{\pi} f(x)\overline{g(x)}dx$, that is a non-orthogonal family of elements in $L^2[0, \pi]$ such that:

$$\overline{\text{span}((\psi_k)_{k\in\mathbb{N}*})} = L^2[0, \pi] \Leftrightarrow (\text{span}((\psi_k)_{k\in\mathbb{N}*}))^{\perp} = \{0_{L^2[0,\pi]}\}$$

$$\Leftrightarrow \langle f, \psi_k\rangle = 0\ \forall f \in L^2[0, \pi], \forall k \in \mathbb{N}^* \Rightarrow f \equiv 0_{L^2[0,\pi]}$$

3) Construct a Hilbert basis of $L^2[0, \pi]$ from the family $(\psi_k)_{k\in\mathbb{N}*}$.

4) Use the result obtained above to determine a sequence of real coefficients $(a_k)_{k\in\mathbb{N}*}$ such that $\sum_{k=1}^{+\infty} a_k\psi_k = 1$ (equality in the sense of $L^2[0, \pi]$).

5) Using Plancherel's identity, prove that the following formula is valid:

$$\sum_{k=0}^{+\infty} \frac{1}{(2k+1)^2} = \frac{\pi^2}{8}$$

Solution to Exercise 5.2

1) We can rewrite ϕ as:

$$\phi(x) = \frac{1}{2}(e^{3it} + e^{-3it}) - \frac{1}{2i}(e^{5it} + e^{-5it})$$

that is, $\phi = \frac{1}{2}(e_3 + e_{-3}) - \frac{1}{2i}(e_5 - e_{-5})$, with the equality in the sense of $L^2[-\pi, \pi]$, is the Fourier series of the function ϕ by the uniqueness of the decomposition.

2) We shall consider these two points separately.

a) By direct calculation:

$$\int_{-\pi}^{\pi} g(t)e^{-ikt}dt = \int_{-\pi}^{0} -f(-t)e^{-ikt}dt + \int_{0}^{\pi} f(t)e^{-ikt}dt$$

if we change the variable in the first integral as follows $s = -t$, $ds = -dt$, we obtain: $\int_{-\pi}^{0} -f(-t)e^{-ikt}dt = \int_{\pi}^{0} f(s)e^{iks}ds = \int_{0}^{\pi} -f(s)e^{iks}ds = \int_{0}^{\pi} -f(t)e^{ikt}dt$ and thus:

$$\int_{-\pi}^{\pi} g(t)e^{-ikt}dt = \int_{0}^{\pi} -f(t)e^{ikt}dt + \int_{0}^{\pi} f(t)e^{-ikt}dt = \int_{0}^{\pi} f(t)(e^{-ikt} - e^{ikt})dt$$

By using Euler's formula for the sine we have:

$$\int_{-\pi}^{\pi} g(t)e^{-ikt}dt = -2i\int_{0}^{\pi} f(t)\sin(kt)dt = -2i\int_{0}^{\pi} f(t)\psi_k(t)dt = 0$$

by definition of the functions ψ_k.

b) The function f defined in 2(a) is, by hypothesis, orthogonal to all the elements $(\psi_k)_{k\in\mathbb{N}*}$, so, to verify that $(\psi_k)_{k\in\mathbb{N}*}$ is a complete system for $L^2[0,\pi]$ we simply have to prove that $f = 0_{L^2[0,\pi]}$. To do that, we use the fact that, by definition, $g|_{[0,\pi]} = f$, thus, if we show that $g = 0_{L^2[\pi,\pi]}$, then, necessarily, $f = 0_{L^2[0,\pi]}$. Thanks to what shown previously, $\langle g, e_k\rangle = \frac{1}{2\pi}\int_{-\pi}^{\pi} g(t)e^{-ikt}dt = 0\ \forall k \in \mathbb{N}^*$, if this holds also for $k = 0$ and $-k$, with $k \in \mathbb{N}^*$, then g is orthogonal to all the elements of the Hilbert basis $(e_k)_{k\in\mathbb{Z}}$ de $L^2[-\pi,\pi]$, which implies $g = 0_{L^2[-\pi,\pi]}$, thanks to theorem 5.11. To resume, the only properties that we have to verify are: $\langle g, e_0\rangle = 0$ and $\langle g, e_{-k}\rangle = 0$ for all $k \in \mathbb{N}^*$:

$$\begin{aligned}\langle g, e_0\rangle = \langle g, e_0\rangle &= \frac{1}{2\pi}\int_{-\pi}^{0}(\cos(-3t) - \sin(-5t))dt \\ &= \frac{1}{2\pi}\int_{-\pi}^{0}(\cos(3t) + \sin(5t))dt \\ &= \frac{1}{2\pi}\left(\frac{\sin(3t)}{3}\bigg|_{-\pi}^{0} - \frac{\cos(5t)}{3}\bigg|_{-\pi}^{0}\right) = 0\end{aligned}$$

$$\langle g, e_{-k}\rangle = \frac{1}{2\pi}\int_{-\pi}^{0}(\cos(3t) + \sin(5t))\overline{e^{-ikt}}dt + \frac{1}{2\pi}\int_{0}^{\pi}(\cos(3t) - \sin(5t))\overline{e^{-ikt}}dt$$

$$= \frac{1}{2\pi}\int_{-\pi}^{0}(\cos(3t) + \sin(5t))e^{ikt}dt + \frac{1}{2\pi}\int_{0}^{\pi}(\cos(3t) - \sin(5t))e^{ikt}dt$$

$$= \frac{1}{2\pi}\int_{\pi}^{0} -(\cos(3s) - \sin(5s))e^{-iks}ds + \frac{1}{2\pi}\int_{0}^{-\pi} -(\cos(3s) + \sin(5s))e^{-iks}ds$$

$$= \frac{1}{2\pi}\int_0^{\pi}(\cos(3t) - \sin(5t))e^{-ikt}dt + \frac{1}{2\pi}\int_{-\pi}^{0}(\cos(3t) + \sin(5t))e^{-ikt}dt$$

$$= \langle g, e_k\rangle = 0 \ \forall k \in \mathbb{N}^*$$

3) The fact that $(\psi_k)_{k\in\mathbb{N}^*}$ is a complete system in $L^2[0,\pi]$ means that we can obtain a Hilbert basis for the same space simply by examining the orthonormal properties of this system. For all $n, m \in \mathbb{N}^*$:

$$\langle \psi_n, \psi_m\rangle = \int_0^{\pi}\sin(nt)\sin(mt)dt \quad (t \mapsto \sin(nt)\sin(mt) \text{ is even})$$

$$= \frac{1}{2}\int_{-\pi}^{\pi}\sin(nt)\sin(mt)dt$$

$$= \frac{1}{2}\int_{-\pi}^{\pi}\frac{e^{int} - e^{-int}}{2i}\frac{e^{imt} - e^{-imt}}{2i}dt$$

$$= -\frac{1}{8}\int_{-\pi}^{\pi}(e^{int} - e^{-int})(e^{imt} - e^{-imt})dt$$

$$= -\frac{1}{8}\int_{-\pi}^{\pi}(e^{int} - e^{-int})(\overline{e^{-imt} - e^{imt}})dt$$

$$= -\frac{2\pi}{8}\langle e_n - e_{-n}, e_{-m} - e_m\rangle_0$$

$$= -\frac{\pi}{4}\left(\langle e_n, e_{-m}\rangle_0 - \langle e_n, e_m\rangle_0 - \langle e_{-n}, e_{-m}\rangle_0 + \langle e_{-n}, e_m\rangle_0\right)$$

$$= \begin{cases} 0 & \text{if } n \neq m \\ -\frac{\pi}{4}(-1-1) = \frac{\pi}{2} & \text{if } n = m, \end{cases}$$

Thus $\|\psi_n\| = \sqrt{\frac{\pi}{2}} \ \forall n \in \mathbb{N}^*$ and so $\left(\sqrt{\frac{2}{\pi}}\psi_n\right)_{n\in\mathbb{N}^*}$ is a Hilbert basis of $L^2[0,\pi]$.

4) Let us interpret 1 as the constant function $1 \in L^2[0,\pi]$, $1(t) = 1 \ \forall t \in [0,\pi]$, which we shall decompose on the Hilbert basis $\left(\sqrt{\frac{2}{\pi}}\psi_n\right)_{n\in\mathbb{N}^*}$ of $L^2[0,\pi]$, determined above:

$$1 = \sum_{k=1}^{+\infty}\langle 1, \sqrt{\frac{2}{\pi}}\psi_k\rangle\sqrt{\frac{2}{\pi}}\psi_k = \sum_{k=1}^{+\infty}\langle 1, \frac{2}{\pi}\psi_k\rangle\psi_k$$

showing us that $1 = \sum_{k=1}^{+\infty} a_k\,\psi_k$, with:

$$a_k = \langle 1, \frac{2}{\pi}\psi_k\rangle = \frac{2}{\pi}\int_0^{\pi}\sin(kt)dt = \frac{2}{\pi k}\left[-\cos(kt)\right]_0^{\pi} = \frac{2}{\pi k}\left[1-(-1)^k\right]$$

that is, the sequence we wanted to find is: $a_k = \begin{cases} 0 & k \text{ even} \\ \frac{4}{\pi k} & k \text{ odd} \end{cases}$.

5) Plancherel's identity for 1 gives us:

$$\|1\|^2 = \sum_{k=1}^{+\infty} \left| \langle 1, \sqrt{\frac{2}{\pi}} \psi_k \rangle \right|^2$$

Moreover, $\|1\|^2 = \int_0^\pi 1dt = \pi$ and $\langle 1, \sqrt{\frac{2}{\pi}} \psi_k \rangle = \sqrt{\frac{\pi}{2}} a_k$, hence:

$$\pi = \sum_{k=0}^{+\infty} \frac{\pi}{2} |a_{2k+1}|^2 = \sum_{k=0}^{+\infty} \frac{\pi}{2} \left(\frac{4}{\pi}\right)^2 \frac{1}{(2k+1)^2} \iff \sum_{k=0}^{+\infty} \frac{1}{(2k+1)^2} = \frac{\pi^2}{8}$$

□

5.6.5. *Pointwise convergence of the real Fourier series: Dirichlet's theorem*

Fourier series were initially met with skepticism by the mathematical community. The idea that series with trigonometric (hence infinitely derivable) functions could be used to approximate non-derivable or, worse, non-continuous functions was considered absurd by many. Furthermore, Fourier did not provide rigorous convergence results for the series that bears his name.

In fact, the theorems that we saw earlier concerning convergence in norm were obtained at a later stage by other mathematicians; furthermore, they are not sufficient to guarantee the pointwise convergence of the series. The first conditions for pointwise convergence of the Fourier series were identified by Dirichlet[6] (b. 1805, Düren; d. 1859; Göttingen) in 1829. Dirichlet's constructive proof is of crucial importance in Fourier analysis; readers who wish to explore the subject further may wish to consult Vretblad (2003).

For the purposes of this book, we shall simply provide a rigorous definition of Dirichlet's theorem, introducing the associated notation and terminology. If t_0 is a point of discontinuity of a real-valued function f of one real variable, then the right and left limits are written as:

$$f(t_0^+) = \lim_{t \to t_0^+} f(t), \quad f(t_0^-) = \lim_{t \to t_0^-} f(t)$$

DEFINITION 5.15 (Dirichlet function).– *Let $f : \mathbb{R} \to \mathbb{R}$. f is a Dirichlet function if it verifies the following conditions:*

6 Remarkably, the "modern" definition of a function, as a univocal correspondence between two sets, was established by Dirichlet as part of his efforts to prove the pointwise convergence of the Fourier series.

1) f is T-periodic, $T \in \mathbb{R}^+$;

2) f is piecewise continuous, that is there is only a finite number of points at which f is not continuous;

3) for all $t_0 \in \mathbb{R}$:

$$f(t_0) = \frac{f(t_0^+) + f(t_0^-)}{2}, \qquad [5.7]$$

that is, at any point $t_0 \in \mathbb{R}$, the value of f in t_0 is the average of the right and left limits of f in t_0.

Condition [5.7] is of course satisfied in any point where f is continuous; however, it is not trivial to requite at any point of discontinuity.

DEFINITION 5.16 (Generalized derivative).– *Let f be a Dirichlet function and take $t_0 \in \mathbb{R}$. f is said to possess a generalized derivative on the right in t_0 if the following (finite) limit exists:*

$$\lim_{h \to 0^+} \frac{f(t_0 + h) - f(t_0^+)}{h}$$

In the same way, f is said to possess a generalized derivative on the left in t_0 if the following (finite) limit exists:

$$\lim_{h \to 0^-} \frac{f(t_0 + h) - f(t_0^-)}{h}$$

These elements are necessary in defining Dirichlet's theorem.

THEOREM 5.17 (Dirichlet's theorem, 1829).– Let f be a Dirichlet function and take $t_0 \in \mathbb{R}$. If the function f possesses generalized derivatives on the right and left at point t_0, then the real Fourier series of f evaluated in t_0 converges to $f(t_0)$.

The conditions of this theorem are known as the Dirichlet conditions; they are sufficient, but not necessary, for the pointwise convergence of the real Fourier series. Conditions which are both necessary and sufficient for the pointwise convergence of the Fourier series have yet to be identified.

Nevertheless – thankfully – the Dirichlet conditions are verified for the vast majority of functions encountered in practical applications.

Note that, if we ignore the requirement [5.7], then the Fourier series converges to $f(t_0) = \frac{f(t_0^+)+f(t_0^-)}{2}$.

One final remark concerning the possible consequences of a lack of continuity in f: In 1923, the great Russian mathematician Kolmogorov (b. 1903, Tambov; d. 1987, Moscow) succeeded in building a function with pathological discontinuities which make its Fourier series diverge at all points.

5.6.6. *The Gibbs phenomenon and Cesàro summation*

Dirichlet's theorem does not imply that the behavior of the Fourier series in the neighborhood of a discontinuity of a function will be "regular"; in fact, as we approach a jump discontinuity, oscillations – known as *Gibbs* oscillations – begin to appear, and remain present even when the number of Fourier coefficients is increased. If a function f is a Dirichlet function, then the oscillations to the left and right of the discontinuity cancel out, and their average coincides with the value of f at the jump.

The difference between the value of the function f and the value of the trigonometric polynomial S_N in an arbitrarily close neighborhood of a jump continuity can be shown to be close to 18 %, even when $N \to +\infty$. The analysis of the Gibbs phenomenon involves mathematical subtleties which lie outside the scope of this book. For a more detailed exploration of the Gibbs phenomenon, readers may wish to consult Vretblad (2003).

Figure 5.2 shows the Gibbs effect for a rectangular pulse function.

Gibbs oscillations can be eliminating by considering a Cesàro (1859, Naples-1906, Torre Annunziata) summation in place of the usual summation; in this case, arithmetic averages of the partial sums are used to "smooth out" oscillations.

5.6.7. *Speed of convergence to 0 of Fourier coefficients*

We begin with a general result.

LEMMA 5.3 (Riemann-Lebesgue lemma).– Taking $f \in L^1[a, b]$, then:

$$\lim_{n \to +\infty} \int_a^b f(t) \cos(nt) dt = \lim_{n \to +\infty} \int_a^b f(t) \sin(nt) dt = \lim_{n \to +\infty} \int_a^b f(t) e^{int} dt = 0$$

The geometric interpretation of the Riemann-Lebesgue lemma is that the function $f(t)\cos(nt)$ or $f(t)\sin(nt)$ oscillates at such a high frequency when $n \to +\infty$ that the values around the average cancel out, and thus the integral converges to 0.

An immediate corollary of this lemma is that *the Fourier coefficients of the Fourier series of a function* $f \in L^1[a, b]$ (and, of course, $(b - a)$-periodic), *decay toward 0* when $n \to +\infty$.

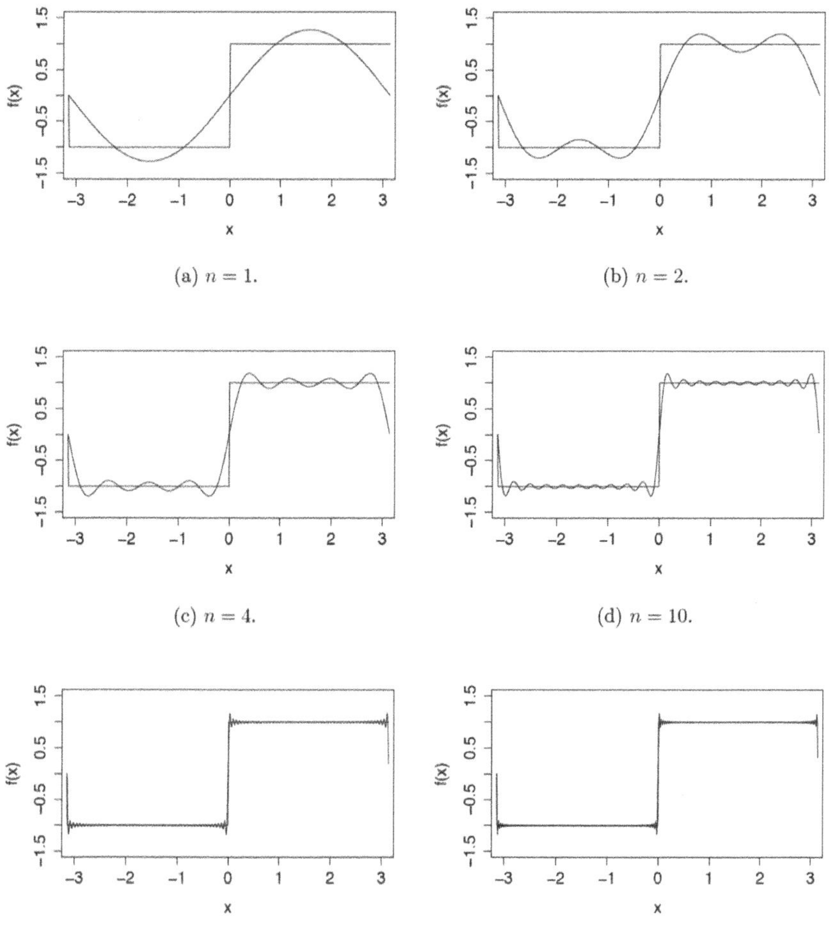

(a) $n = 1$. (b) $n = 2$.

(c) $n = 4$. (d) $n = 10$.

(e) $n = 50$. (f) $n = 80$.

Figure 5.2. *Gibbs phenomenon for the rectangular pulse function (courtesy of Éric Luçon)*

Theorem 5.18 shows that the regularity of f has an important effect on the speed of decay of Fourier coefficients.

THEOREM 5.18.– Let $f : \mathbb{R} \to \mathbb{R}$ be a function that:

– is of class $\mathcal{C}^p([a, b])$, that is f is derivable p times on $[a, b]$ with p continuous derivatives;

– is $(b - a)$-periodic;

– possesses equal generalized derivatives at the extrema of the interval $[a, b]$.

Then, the Fourier coefficients of f, $a_n, b_n, n = 1, 2, \ldots$ verify:

$$a_n, b_n = o\left(\frac{1}{n^p}\right),$$

that is they decay toward 0 faster than $\frac{1}{n^p}$.

This result is very important, as it tells us that if f is "smooth", then it can be approximated in a precise manner even with a small number of Fourier coefficients.

However, if f is not sufficiently smooth, then the convergence to 0 of the Fourier coefficients of f is slow, and a large number of these coefficients is required in order to obtain a good approximation of f.

The inverse is also true under some suitable hypotheses, which space does not permit us to describe here. The most important concept to grasp is that the faster the Fourier coefficients of a function converge to 0, the smoother the function is.

PROOF.–

Let us consider the coefficients a_n; the proof is identical for the coefficients b_n. We can develop our proof, without loosing generality, by considering $b = \pi$, $a = -\pi$, in fact it is always taken back our analysis to these values thanks to the following linear variable change:

$$\boxed{s(t) = \frac{b+a}{2} + \frac{b-a}{2\pi}t}$$

which shows that $s(\pi) = b$ and $s(-\pi) = a$.

Using this convention, the expression of a_n, $n = 0, 1, 2, \ldots$ is integrated by parts, with $u = f(t)$ and $dv = \cos(nt)dt$, hence $du = f'(t)dt$ and $v = \frac{1}{n}\sin(nt)$.

We obtain:

$$a_n = \frac{1}{\pi n}[f(t)\sin(nt)]_{-\pi}^{\pi} - \frac{1}{\pi n}\int_{-\pi}^{\pi} f'(t)\sin(nt)dt$$

$$= \frac{1}{\pi n}\int_{-\pi}^{\pi} f'(t)\cos\left(\frac{\pi}{2} + nt\right)dt$$

since $\sin(n\pi) = \sin(-n\pi) = 0$ and $\cos\left(\frac{\pi}{2} + \alpha\right) = -\sin(\alpha)\ \forall\alpha \in \mathbb{R}$.

After a second integration by parts, we obtain:

$$a_n = \frac{1}{\pi n}\left\{\frac{1}{n}\left[f'(t)\sin\left(\frac{\pi}{2}+nt\right)\right]_{-\pi}^{\pi} - \frac{1}{n}\int_{-\pi}^{\pi} f''(t)\sin\left(\frac{\pi}{2}+nt\right)dt\right\}$$

Since $f'(-\pi) = f'(\pi)$ by hypothesis, the first bracketed term is zero, hence:

$$\left[f'(t)\sin\left(\frac{\pi}{2}+nt\right)\right]_{-\pi}^{\pi} = f'(\pi)\sin\left(\frac{\pi}{2}+n\pi\right) - f'(-\pi)\sin\left(\frac{\pi}{2}-n\pi\right)$$

$$= f'(\pi)\left[\sin\left(\frac{\pi}{2}+n\pi\right) - \sin\left(\frac{\pi}{2}-n\pi\right)\right]$$

$$= f'(\pi)\left[\sin\left(\frac{\pi}{2}+n\pi\right) - \sin\left(\frac{\pi}{2}-n\pi+2n\pi\right)\right]$$

$$= f'(\pi)\left[\sin\left(\frac{\pi}{2}+n\pi\right) - \sin\left(\frac{\pi}{2}+n\pi\right)\right] = 0$$

Furthermore, the second term in brackets can be rewritten as:

$$-\frac{1}{n}\int_{-\pi}^{\pi} f''(t)\sin\left(\frac{\pi}{2}+nt\right)dt = \frac{1}{n}\int_{-\pi}^{\pi} f''(t)\cos\left(\frac{\pi}{2}+\frac{\pi}{2}+nt\right)dt$$

$$= \frac{1}{n}\int_{-\pi}^{\pi} f''(t)\cos\left(\frac{\pi}{2}\cdot 2+nt\right)dt$$

Moreover:

$$a_n = \frac{1}{\pi n^2}\int_{-\pi}^{\pi} f''(t)\cos\left(\frac{\pi}{2}\cdot 2+nt\right)dt$$

In short, integration by parts of a_n gives us the expression:

$$a_n = \frac{1}{\pi n}\int_{-\pi}^{\pi} f'(t)\cos\left(\frac{\pi}{2}+nt\right)dt$$

With two integrations by parts of a_n, we have:

$$a_n = \frac{1}{\pi n^2}\int_{-\pi}^{\pi} f''(t)\cos\left(\frac{\pi}{2}\cdot 2+nt\right)dt$$

With p integrations by parts of a_n, we have:

$$a_n = \frac{1}{\pi n^p}\int_{-\pi}^{\pi} f^{(p)}(t)\cos\left(\frac{\pi}{2}\cdot p+nt\right)dt$$

Similarly, we obtain:

$$b_n = \frac{1}{\pi n^p}\int_{-\pi}^{\pi} f^{(p)}(t)\sin\left(\frac{\pi}{2}\cdot p+nt\right)dt$$

We now see that, by using the trigonometric identities $\cos\left(\frac{\pi}{2}+\alpha\right) = -\sin(\alpha)$ and $\sin\left(\frac{\pi}{2}+\alpha\right) = \cos(\alpha)$, $\forall\alpha \in \mathbb{R}$, the integrals:

$$\varepsilon_n = \frac{1}{\pi}\int_{-\pi}^{\pi} f^{(p)}(t)\cos\left(\frac{\pi}{2}\cdot p + nt\right)dt,\ \ \tilde{\varepsilon}_n = \frac{1}{\pi}\int_{-\pi}^{\pi} f^{(p)}(t)\sin\left(\frac{\pi}{2}\cdot p + nt\right)dt$$

are, by definition, the Fourier coefficients of the function $f^{(p)}$ to within a sign. By hypothesis, $f^{(p)}$ is continuous on $[-\pi,\pi]$ and thus, as the domain $[-\pi,\pi]$ is compact, $f^{(p)} \in L^1[-\pi,\pi]$; hence, by the Riemann-Lebesgue lemma, its Fourier coefficients converge to 0 when $n \to +\infty$. Furthermore, $\varepsilon_n \underset{n\to\infty}{\longrightarrow} 0$ and $\tilde{\varepsilon}_n \underset{n\to\infty}{\longrightarrow} 0$, which means that:

$$a_n = \frac{\varepsilon_n}{n^p} \underset{n\to\infty}{\longrightarrow} 0,\ \ b_n = \frac{\tilde{\varepsilon}_n}{n^p} \underset{n\to\infty}{\longrightarrow} 0$$

that is $a_n, b_n = o\left(\frac{1}{n^p}\right)$. □

This result was used by Krylov (1863–1945) as the foundation of his method for improving the convergence of Fourier series for jump-discontinuous functions.

5.6.8. *Fourier transform in $L^2(\mathbb{T})$ and shift*

Now, let us analyze the relationship between shift and the Fourier transform for a function $f \in L^2(\mathbb{T})$. The result is qualitatively identical to that which we obtained for the DFT in section 2.7.2.

THEOREM 5.19 (Fourier transform and shift).– Taking $f \in L^2(\mathbb{T})$, then:

1) if $g_a(x) = f(x-a)$, $a \in \mathbb{R}$, then: $\hat{g_a}(n) = e^{-ina}\hat{f}(n)$, $\forall n \in \mathbb{Z}$;

2) if $g_k(x) = e^{ikx}f(x)$, $k \in \mathbb{Z}$, then: $\hat{g_k}(n) = \hat{f}(n-k)$, $\forall n \in \mathbb{Z}$.

PROOF.– Only the proof for 1 is shown here, as the proof for 2 is analogous. The proof consists of a direct calculation in which we make use of the shift-invariance of the Lebesgue measure:

$$\begin{aligned}\hat{g_a}(n) &= \langle g_a, u_n\rangle = \int_0^{2\pi} \frac{f(x-a)}{\sqrt{2\pi}} e^{-inx}dx \\ &= \int_{-a}^{2\pi-a} \frac{f(x)}{\sqrt{2\pi}} e^{-in(x+a)}d(x+a) = e^{-ina}\int_0^{2\pi}\frac{f(x)}{\sqrt{2\pi}}e^{-inx}dx \\ &= e^{-ina}\hat{f}(n).\end{aligned}$$

□

DEFINITION 5.17.– *The set* $\left\{|\hat{f}(n)|,\ n \in \mathbb{Z}\right\}$ *is the spectrum (amplitude spectrum) of* $f \in L^2(\mathbb{T})$.

$|\hat{f}(n)|$ represents the weight of importance of the harmonic of frequency n, that is, e^{inx} in reconstructing f, as can be seen in the formula $f = \sum_{n\in\mathbb{Z}} \hat{f}(n)\frac{e^{inx}}{\sqrt{2\pi}}$.

The property which we have just proved shows that the spectrum of f gives us information concerning the presence of certain frequencies in f; however, it tells us nothing about their "position": the shifted signal $g_a(x) = f(x - a)$ has the same spectrum as f, since $|\widehat{g}_a(n)| = |\widehat{f}(n)|$.

Localized information concerning frequency and position can be obtained in the context of wavelet theory.

5.7. Summary

In this chapter, we extended some structural property of finite-dimensional inner product spaces to infinite dimensional Hilbert spaces.

The orthogonal complement to a subset or vector subspace of a Hilbert space plays an important role in this extension.

The theorem of projection onto a closed convex subset of a Hilbert space is essential for extending the geometric structure of finite-dimensional Euclidean spaces to infinite dimensions. The proof of this theorem draws on the parallelogram law, for which a Hilbert norm is required; hence, the theorem is only valid in Hilbert spaces.

When the closed convex subset from the previous theorem is also a vector subspace, then the difference between the original vector and its projection belongs to the orthogonal complement of the subspace, as it does in finite dimensions; this property allows us to extend the orthogonal projection theorem to infinite-dimensional Hilbert spaces.

The orthogonal projection theorem is used to produce an extremely useful characterization of closed vector subspaces in Hilbert spaces, as those which coincide with their biorthogonal complement.

We examined orthonormal systems in separable Hilbert spaces, that is, those which possess at least one countable dense subset. An orthonormal system of a separable Hilbert space is countable. All of the Hilbert spaces discussed here are implicitly considered to be separable unless otherwise stated.

In order for an orthonormal system $(u_n)_{n\in\mathbb{N}}$ to be the generalization of an orthonormal basis to an infinite-dimensional Hilbert space $\mathcal{H}$, we must first guarantee that for all $x \in \mathcal{H}$, the sequence of Fourier coefficients $(\hat{x}(n) = \langle x, u_n \rangle)_{n\in\mathbb{N}}$ decays toward 0; otherwise, the expansion $\sum_{n\in\mathbb{N}} \langle x, u_n \rangle u_n$ would not converge. Bessel's inequality ensures that this is the case, due to the fact that the sequence of Fourier coefficients with respect to any orthonormal system of a Hilbert space belongs to ℓ^2. Bessel's inequality also tells us that Plancherel's identity is not necessarily verified for any orthonormal system, as, in general, it holds that $\sum_{n\in\mathbb{N}} |\langle x, u_n \rangle|^2 u_n \leqslant \|x\|^2$.

The Fischer-Riesz theorem states that Plancherel's identity holds when the series $\sum_{n\in\mathbb{N}} \langle x, u_n \rangle u_n$ converges to x; using a counter-example, we showed that this is not the case for an arbitrary orthonormal system. It turns out that $\sum_{n\in\mathbb{N}} \langle x, u_n \rangle u_n$ is the expansion of x when the orthonormal system $(u_n)_{n\in\mathbb{N}}$ is complete, that is, it is not a proper part of another orthonormal system in $\mathcal{H}$. Complete orthonormal systems are also known as Hilbert bases.

A Hilbert basis $(u_n)_{n\in\mathbb{N}}$ can be characterized using five equivalent conditions: the fact that the zero vector is the only vector which is orthogonal to all elements in a Hilbert basis, the fact that the subspace generated by the Hilbert basis is dense in $\mathcal{H}$, the ability to expand into a generalized Fourier series, Parseval's identity and Plancherel's identity.

An isomorphism of Hilbert spaces is a surjective transformation which preserves the inner product. We saw that the preservation of inner products implies isometry, and thus injectivity; furthermore, the combination of surjectivity and conservation of the inner product implies linearity. All separable Hilbert spaces on the same field are isomorphic to one another; the prototype of an infinite-dimensional, separable Hilbert space on the field $\mathbb{K}$ is $\ell^2(\mathbb{N}, \mathbb{K})$. This result is the extension, to infinite dimensions, of the fact that $\mathbb{K}^n$ is the prototype of all vector spaces of finite dimension n on $\mathbb{K}$.

The classic Fourier series and transform on spaces $L^2([a, b])$ are defined as a special case of the theory developed earlier; their specificity lies in the choice of a Hilbert basis given by complex exponentials, or by a cosine and sine (plus a constant function). This also holds for functions defined on $\mathbb{R}$, as long as they are periodic.

As in the case of sequences in $\ell^2(\mathbb{Z}_N)$, also for the functions of L^2 described earlier, the Fourier spectrum (the set of magnitudes of the Fourier coefficients) is shift-invariant, raising the necessity of an extension of Fourier theory to provide a localized frequency analysis. Wavelet theory responded to this need.

6

Bounded Linear Operators in Hilbert Spaces

A function $A : V \implies W$, with V and W normed vector spaces on the same field $\mathbb{K}$, is known as a *linear operator* between V and W if:

$$\forall \alpha, \beta \in \mathbb{K},\ A(\alpha x + \beta y) = \alpha A(x) + \beta A(y), \qquad \forall x, y \in V$$

To simplify the notation, the parentheses may be omitted in later occurrences, writing Ax in place of $A(x)$. V is the *domain* of A; the set:

$$\text{Im}(A) = A(V) = \{y \in W \ : \ \exists x \in V \ : \ y = Ax\} \subseteq W$$

is the *codomain* or *image* of A, and W is the destination set of A.

Basic examples are shown below.

1) The *identity operator*: $id : V \to V$, $id(x) = x \ \forall x \in V$ and the *null operator*: $0 : V \to V, 0(x) = 0_V \ \forall x \in V$;

2) The *differential operator*: this is defined on a space of differentiable functions which may change according to the particular application we are interested in. As a concrete example, consider the first-order differential operator: $D_1 f(t) = \frac{df}{dt}(t) = f'(t)$. $\text{dom}(D_1) = \{f \in L^2[a,b] \cap C^1[a,b] \ : \ f' \in L^2[a,b]\}$, where $a < b$ are real constants, could be a perfectly valid domain for D_1. Then:

$$\begin{array}{rcl} D_1 : \text{dom}(D_1) \subset L^2[a,b] & \longrightarrow & L^2[a,b] \\ f & \longmapsto & D_1 f \end{array}$$

Similarly, the operator

$$D_n f(t) = \frac{d^n f}{dt^n}(t) = f^{(n)}(t)$$

can be defined on the domain $\mathrm{dom}(D_n) = \{f \in L^2[a,b] \cap C^1[a,b] \;:\; f^{(n)} \in L^2[a,b]\}$, where $a < b$ are real constants, that is:

$$\begin{aligned} D_n : \mathrm{dom}(D_n) \subset L^2[a,b] &\longrightarrow L^2[a,b] \\ f \quad &\longmapsto D_n f \end{aligned}$$

Partial differential operators are defined in a similar way;

3) The *integral operator*: this operator is typically defined by considering a kernel function $k(s,t)$, $k \in L^2([a,b]\times[a,b])$, where $a < b$ are real constants. The integration operator with kernel k is:

$$\begin{aligned} T_k : L^2[a,b] &\longrightarrow L^2[a,b] \\ f \quad &\longmapsto T_k f, \quad \text{where } T_k f(s) = \textstyle\int_a^b k(s,t)f(t)dt \end{aligned}$$

4) *Linear operators in finite dimensions*. Let $A : \mathbb{K}^n \to \mathbb{K}^n$ be a linear operator and let $(u_1, \ldots, u_n)$ be an orthonormal basis in $\mathbb{K}^n$. Any $x \in \mathbb{K}^n$ can be written as $x = \sum_{j=1}^{n} \lambda_j u_j$, with $\lambda_j \in \mathbb{K}\ \forall j$ and, by linearity, $Ax = \sum_{j=1}^{n} \lambda_j A u_j$. Then:

$$\langle Ax, u_j \rangle = \sum_{j=1}^{n} \lambda_j \langle Au_j, u_i \rangle = \sum_{j=1}^{n} \alpha_{ij}\lambda_j, \qquad \forall i = 1, \ldots, n \qquad [6.1]$$

where $\alpha_{ij} = \langle Au_j, u_i \rangle$. This shows that the action of A is entirely determined by the matrix of element $(\alpha_{ij})_{i,j=1,\ldots,n}$ and *vice versa*: for any matrix with elements $(\alpha_{ij})_{i,j=1,\ldots,n}$, formula [6.1] can be used to define a linear operator on $\mathbb{K}^n$.

This last example highlights the well-known relationship between linear operators on $\mathbb{K}^n$ and $n \times n$ matrices with elements in $\mathbb{K}$. Since $\mathbb{K}^n$ is the prototype of all vector spaces V of dimension n on $\mathbb{K}$, we can say that the theory of linear operators on vector spaces in finite dimensions is, in essence, a matrix theory.

As we shall see, the action of bounded linear operators on separable Hilbert spaces can also be expressed using a matrix, but, in this case, the matrix contains a countably infinite number of rows and columns.

The presence of a topology generated by a norm motivates the need to check the continuity of linear operators defined between two normed vector spaces V and W. If V and W have finite dimension, then any linear operator between them is continuous.

However, as we shall see in section 6.2.1, if V is of infinite dimension, then even simple linear operators may not be continuous.

In the following sections, we shall examine the main properties of linear operators starting by showing that a linear operator is continuous if and only if it is bounded.

6.1. Fundamental properties of bounded linear operators between normed vector spaces

We begin by introducing formal definitions for continuous and bounded operators. Let $(V, \| \ \|_V)$ and $(W, \| \ \|_W)$ be two generic normed vector spaces.

DEFINITION 6.1.– *Let* $A : V \Longrightarrow W$ *be a linear operator:*

– A is continuous in $x_0 \in V$ *if:*

$$\forall \varepsilon > 0 \ \exists \delta_\varepsilon > 0 \ : \ \|x - x_0\|_V < \delta_\varepsilon \Longrightarrow \|Ax - Ax_0\|_W = \|A(x - x_0)\|_W < \varepsilon$$

– A is continuous on V *if* A *is continuous in every element of* V*;*

– A is bounded *if* $\exists c \in \mathbb{R}$, $c \geqslant 0$, *such that:*

$$\|Ax\|_W \leqslant c\|x\|_V \qquad \forall x \in V$$

that is, any vector $x \in V$ *is transformed by* A *into a vector* Ax *whose norm in* W *is majorized by a positive multiple of the norm of* x *in* V.

The continuity of a linear operator is equivalent to sequential continuity, just as we saw in the case of functions defined on metric spaces.

THEOREM 6.1.– The linear operator $A : V \Longrightarrow W$ is continuous in $x_0 \in V$ if and only if:

$$\forall (x_n)_{n\in\mathbb{N}} \subset V, \ x_n \underset{n \Longrightarrow +\infty}{\longrightarrow} x_0 \Longrightarrow Ax_n \underset{n \Longrightarrow +\infty}{\longrightarrow} Ax_0$$

that is:

$$\forall (x_n)_{n\in\mathbb{N}} \subset V, \ \|x_n - x_0\|_V \underset{n \Longrightarrow +\infty}{\longrightarrow} 0 \Longrightarrow \|Ax_n - Ax_0\|_W \underset{n \Longrightarrow +\infty}{\longrightarrow} 0$$

Before going into the details concerning the properties of continuous linear operators, we can show that *any continuous linear operator on a Hilbert space can be represented by an infinite matrix*. Let us use the same argument of example 4 previously discussed: let $\mathcal{H}$ be a Hilbert space, $A : \mathcal{H} \to \mathcal{H}$ a continuous linear operator and $(u_n)_{n\in\mathbb{N}}$ a Hilbert basis of $\mathcal{H}$. Then, for all $x \in \mathcal{H}$, $x = \sum_{n\in\mathbb{N}} \langle x, u_n\rangle u_n$ and by the continuity and linearity of A, we have:

$$Ax = A\left(\sum_{n\in\mathbb{N}} \langle x, u_n\rangle u_n\right) \underset{\text{(continuity)}}{=} \sum_{n\in\mathbb{N}} A(\langle x, u_n\rangle u_n) \underset{\text{(linearity)}}{=} \sum_{n\in\mathbb{N}} \langle x, u_n\rangle Au_n$$

Furthermore, by the continuity of the inner product:

$$\langle Ax, u_m\rangle = \langle \sum_{n\in\mathbb{N}} \langle x, u_n\rangle Au_n, u_m\rangle = \sum_{n\in\mathbb{N}} \langle Au_n, u_m\rangle\langle x, u_m\rangle$$

$$= \sum_{n\in\mathbb{N}} \alpha_{nm}\langle x, u_m\rangle, \qquad \forall m \in \mathbb{N}$$

where $\alpha_{nm} = \langle Au_n, u_m \rangle$, thus the infinite matrix with elements $(\alpha_{mn})_{n,m\in\mathbb{N}}$ is the representation of the continuous linear operator A with respect to the Hilbert basis $(u_n)_{n\in\mathbb{N}}$.

Unlike the finite dimensional case, it is not easy to know when an infinite matrix corresponds to a continuous linear operator; this is the reason why infinite matrices are almost never used when studying linear operators in infinite-dimensional Hilbert spaces.

Theorem 6.2 makes it considerably simpler to analyze the continuity of linear operators.

THEOREM 6.2.– Let $A : V \Longrightarrow W$ be a linear operator and $x_0 \in V$ an arbitrary fixed element. Then, A is continuous in x_0 if and only if A is continuous on all V.

This theorem implies that we simply need to prove the continuity of a linear operator at a single, arbitrary point in order to guarantee the continuity over the whole vector space on which it is defined.

PROOF.–

$\boxed{\Longleftarrow}$: trivial, as if A is continuous on V, then, by definition, A is continuous at all points in V.

$\boxed{\Longrightarrow}$: let A be continuous in x_0. To demonstrate that A is continuous in V, we must prove that the continuity of A in x_0 implies its continuity in any arbitrary element $x \in V$. Given any sequence $(x_n)_{n\in\mathbb{N}} \subset V$ such that $x_n \underset{n \Longrightarrow +\infty}{\longrightarrow} x$, we must prove that this implies $\|A(x_n) - A(x)\| \underset{n \Longrightarrow +\infty}{\Longrightarrow} 0$.

We note that the sequence $(x_n - x + x_0)_{n\in\mathbb{N}}$ converges to x_0 since $(x_n)_{n\in\mathbb{N}}$ converges to x. Thus, by the continuity of A in x_0, it holds that $A(x_n - x + x_0) \underset{n \Longrightarrow +\infty}{\longrightarrow} A(x_0)$, that is $\|A(x_n - x + x_0) - A(x_0)\| \underset{n \Longrightarrow +\infty}{\Longrightarrow} 0$; furthermore, by the linearity of A, $\|A(x_n - x + x_0) - A(x_0)\| = \|A(x_n) - A(x) + \cancel{A(x_0)} - \cancel{A(x_0)}\| = \|A(x_n) - A(x)\| \underset{n \Longrightarrow +\infty}{\Longrightarrow} 0$. □

Thus, to verify the continuity[1] (or lack of continuity!) of a linear operator $A : V \Longrightarrow W$, we must simply verify this property for an arbitrary point in V. This point is often chosen to be 0_V, the zero vector in V, as, in many cases, it simplifies the calculations involved.

1 We recall that, for a linear operator, continuity and uniform continuity are equivalent conditions.

This fact is used below to prove a theorem which shows the relationship between continuous and bounded linear operators.

THEOREM 6.3.– A linear operator $A : V \implies W$ is bounded if and only if it is continuous.

PROOF.–

A bounded $\implies$ *A continuous* $\iff$ *A continuous in* 0_V. Take $(x_n)_{n\in\mathbb{N}} \subset V$, $x_n \underset{n \implies +\infty}{\longrightarrow} 0_V$, that is $\|x_n\|_V \underset{n \implies +\infty}{\longrightarrow} 0$; as A is assumed to be bounded, $\exists c \in \mathbb{R}^+$ such that:

$$\|Ax_n - A0_V\|_W \underset{A0_V = 0_W}{=} \|Ax_n\|_W \leqslant c\|x_n\|_V \underset{n \implies +\infty}{\longrightarrow} 0$$

thus, for any sequence $(x_n)_{n\in\mathbb{N}} \subset V$, $x_n \underset{n \implies +\infty}{\longrightarrow} 0_V$, $Ax_n \underset{n \implies +\infty}{\longrightarrow} A(0_V)$, which corresponds to the continuity of A in 0_V, and hence, by the previous theorem, on all V.

A continuous $\iff$ *A continuous in* 0_V $\implies$ *A bounded*. In this case, it is helpful to consider the original definition of continuity, and to express it for $x_0 = 0_V$:

$$\forall \varepsilon > 0 \; \exists \delta_\varepsilon > 0 \; : \; \|x - 0_V\|_V < \delta_\varepsilon \implies \|Ax - A0_V\|_W < \varepsilon$$

that is :

$$\forall \varepsilon > 0 \; \exists \delta_\varepsilon > 0 \; : \; \|x\|_V < \delta_\varepsilon \implies \|Ax\|_W < \varepsilon$$

As the previous expression is valid for all ε, we can consider the case where $\varepsilon = 1$. For simplicity's sake, we shall write $\delta_{\varepsilon=1} \equiv K > 0$. Using these choices, the hypothesis that A is continuous in 0_V gives us the following implication:

$$\|x\|_V < K \implies \|Ax\|_W < 1 \qquad [6.2]$$

Note that we are approaching the definition of a bounded operator. The final step of the proof consists of determining a specific vector x which satisfies [6.2] and that allows us to handle the inequality $\|Ax\|_W < 1$ in order to prove that A is bounded.

To this aim, let us consider a real positive number $0 < \sigma < K$, hence $K - \sigma > 0$, and an *arbitrary* element $y \in V$.

We analyze the norm of the vector $(K - \sigma)\frac{y}{\|y\|_V}$:

$$\left\|(K - \sigma)\frac{y}{\|y\|_V}\right\|_V = \frac{K - \sigma}{\|y\|_V}\|y\|_V = K - \sigma < K$$

that is, $(K - \sigma)\frac{y}{\|y\|_V}$ is a vector in V whose norm is strictly less than K; thus, relationship [6.2] implies:

$$\left\| A\left((K-\sigma)\frac{y}{\|y\|_V}\right)\right\|_W < 1 \Longleftrightarrow \frac{K-\sigma}{\|y\|_V}\|Ay\|_W < 1$$

$$\Longleftrightarrow \|Ay\|_W < \frac{1}{K-\sigma}\|y\|_V$$

Since $y \in V$ is arbitrary and $K - \sigma > 0$, we can take $c \equiv \frac{1}{K-\sigma}$ and obtain the definition of a bounded A:

$$\|Ay\|_W < c\|y\|_W \quad \forall y \in V$$ □

This theorem implies that *the terms "bounded" and "continuous" can be interchanged for linear operators* between normed vector spaces.

So far, we specified the vector space in which the norm in question was considered. From now on, for simplicity's sake, this specification will not be shown and we shall simply write $\| \ \|$.

6.1.1. *Continuity of linear operators defined on a finite-dimensional normed vector space*

The following result shows that all linear operators defined on a finite-dimensional vector space are continuous (and thus bounded).

THEOREM 6.4.– If V is a normed vector space of finite dimension N and W is a normed vector space (of any dimension), then any linear operator $A : V \to W$ is bounded (and thus continuous).

PROOF.– As the space V is of finite dimensions, all norms on V are equivalent by Tychonoff's theorem (Theorem 4.4); thus, we must simply prove that $A : V \to W$ is continuous with respect to one norm, and this proof holds for all other norms.

Let $(u_1, \ldots, u_N)$ be a basis of V, then any $x \in V$ can be written as $x = \sum_{n=1}^{N} x_n u_n$, $x_n \in \mathbb{K}$. Let us consider the following norm on V: $\|x\| = \left\|\sum_{n=1}^{N} x_n u_n\right\| \equiv \sup_{n=1,\ldots,N} |x_n|$. By the linearity of A and the triangular inequality, we have:

$$\begin{aligned}\|Ax\| &= \left\| A\left(\sum_{n=1}^{N} x_n u_n\right)\right\| = \left\|\sum_{n=1}^{N} x_n A u_n\right\| \leqslant \sum_{n=1}^{N} \|x_n A u_n\| = \sum_{n=1}^{N} |x_n|\,\|Au_n\| \\ &\leqslant \sum_{n=1}^{N} \sup_{n=1,\dots,N} |x_n|\,\|Au_n\| \\ &= \left(\sup_{n=1,\dots,N} |x_n|\right)\left(\sum_{n=1}^{N} \|Au_n\|\right) \underset{\text{def. of } \|x\|}{=} \left(\sum_{n=1}^{N} \|Au_n\|\right)\|x\|\end{aligned}$$

this shows us that A is bounded, that is continuous. □

We therefore do not face any continuity problems when considering linear operators defined on finite-dimensional normed vector spaces, *whatever the dimension of the image space*. As we shall see, the situation is much more complicated in the case of infinite-dimensional domains.

6.2. The operator norm, convergence of operator sequences and Banach algebras

DEFINITION 6.2.– *Let* $A : V \to W$, $V \neq \{0_V\}$ *be a bounded linear operator. The* operator norm *of* A *can be defined in four different (equivalent) ways:*

$$\|A\| = \inf\{c \geqslant 0 : \|Ax\| \leqslant c\|x\|,\ \forall x \in V\} = N_1 \qquad [6.3]$$

$$\|A\| = \sup_{\|x\|\leqslant 1} \|Ax\| = N_2 \qquad [6.4]$$

$$\|A\| = \sup_{\|x\|=1} \|Ax\| = N_3 \qquad [6.5]$$

$$\|A\| = \sup_{x\neq 0_V} \frac{\|Ax\|}{\|x\|} = N_4 \qquad [6.6]$$

For a non-bounded operator A, we write $\|A\| = +\infty$; evidently, for the zero operator 0 it holds that $\|0\| = 0$. Theorem 6.5 guarantees that the definition above is well posed.

THEOREM 6.5.– The four definitions given above coincide.

PROOF.– We shall show that $N_1 \leqslant N_4 \leqslant N_3 \leqslant N_2 \leqslant N_1$, working from right to left. In all of these proofs, we shall use the fact that the sup of a set is, by definition, the smallest of the upper bounds of the set itself.

$\boxed{N_2 \leqslant N_1}$: by the definition of N_1 (i.e. equation [6.3]) we can write $\|Ax\| \leqslant N_1\|x\|\ \forall x \in V$, thus, in particular, for vectors x such that $\|x\| \leqslant 1$, it is true that $\|Ax\| \leqslant N_1$, that is, N_1 is an upper bound for the set $\{\|Ax\|,\ x \in V,\ \|x\| \leqslant 1\}$. By definition, the sup is the smallest of the upper bound of a set, hence $N_2 = \sup\limits_{\|x\|\leqslant 1} \|Ax\| \leqslant N_1$.

$\boxed{N_3 \leqslant N_2}$: consider $x \in V$ such that $\|x\| = 1$ and the sequence $x_n = \left(1 - \frac{1}{n}\right) x$, $n \geqslant 1$. On one side: $\|x_n\| = \left(1 - \frac{1}{n}\right)\|x\| = \left(1 - \frac{1}{n}\right) \leqslant 1$ and thus $\|Ax_n\| \leqslant \sup\limits_{\|y\|\leqslant 1} \|Ay\| = N_2$. Passing by the limit, we obtain: $\lim\limits_{n\to+\infty} \|Ax_n\| \leqslant \lim\limits_{n\to+\infty} N_2 = N_2$. On the other side, it is clear that $x_n \underset{n\to+\infty}{\longrightarrow} x$ and thus, by the continuity of A and of the norm, $\lim\limits_{n\to+\infty} \|Ax_n\| = \|A \lim\limits_{n\to+\infty} x_n\| = \|Ax\|$.

Combining this information, we can write $\|Ax\| \leqslant N_2$, that is, N_2 is an upper bound for the set $\{\|Ax\|,\ x \in V,\ \|x\| = 1\}$. The quantity N_3 is defined as the sup of this set, that is the smallest upper bound, hence $N_3 \leqslant N_2$.

$\boxed{N_4 \leqslant N_3}$: let us consider $x \in V$, $x \neq 0_V$, then $\left\|\frac{x}{\|x\|}\right\| = 1$ and $\left\|A\frac{x}{\|x\|}\right\| \leqslant \sup\limits_{\|y\|=1} \|Ay\| = N_3$. Furthermore, $\left\|A\frac{x}{\|x\|}\right\| = \frac{\|Ax\|}{\|x\|}$, hence $\frac{\|Ax\|}{\|x\|} \leqslant N_3$, that is, c is an upper bound for the set $\left\{\frac{\|Ax\|}{\|x\|},\ x \in V,\ x \neq 0_V\right\}$. Since N_4 is the sup of this set, that is the smallest of the upper bounds, then $N_4 \leqslant N_3$.

$\boxed{N_1 \leqslant N_4}$: for all $x \neq 0_V$, $\frac{\|Ax\|}{\|x\|} \leqslant N_4$, and $\|Ax\| \leqslant N_4\|x\|$; moreover, by definition of N_1, it holds that $N_1 \leqslant N_4$. □

REMARK.–

1) The specification $\forall x \in V$ plays an important role in the definition $\|A\| = \inf\{c \geqslant 0 : \|Ax\| \leqslant c\|x\|,\ \forall x \in V\}$. Without this condition, the norm of A would be trivially null for any linear operator, since $A0 = 0$. By considering all of the transformed vectors Ax, $x \in V$, we ensure that the norm of A is $\neq 0$ (except, evidently, in the case where A is the identically null operator).

2) We should also highlight the difference between the expression $\|A\|$, which represents the operator norm of the linear application $A : V \to W$, and the expression $\|x\|$, which represents the norm of a vector $x \in V$. Certain authors use a different symbol for the operator norm, for example $|||A|||$, but we have chosen to retain the same symbol, $\|\ \|$.

We shall now verify that the operator norm is well defined on the set of linear operators from V to W, and that this space is stable with respect to pointwise-defined linear operations, that is $(A+B)x = Ax + Bx$ and $(\alpha A)x = \alpha Ax$, for all $\alpha \in \mathbb{K}$ and for all $x \in V$.

– *Positive definiteness*: evidently, $\|A\| \geqslant 0$ for any bounded operator A by equation [6.3]. Furthermore, by equation [6.6], $\|A\| = \sup_{x \neq 0_V} \frac{\|Ax\|}{\|x\|} = 0$ if and only if $\|Ax\| = 0 \ \forall x \in V, x \neq 0_V$ (if $x = 0_V$ then $Ax = 0$ by linearity). Thus, due to the positive definiteness of the norm of W, $\|A\| = 0 \iff Ax = 0 \ \forall x \in V$, that is, if and only if A is the null operator $\mathbf{0}(x) = 0 \ \forall x \in V$.

– *Homogeneity*: this is a direct consequence of the homogeneity of the norm of W. Using, for example, equation [6.5], we obtain, $\forall \alpha \in \mathbb{K}$:

$$\|\alpha A\| = \sup_{\|x\|=1} \|\alpha Ax\| = \sup_{\|x\|=1} |\alpha|\|Ax\| = |\alpha| \sup_{\|x\|=1} \|Ax\| = |\alpha|\|A\|$$

that is:

$$\boxed{\|\alpha A\| = |\alpha|\|A\|} \qquad \forall \alpha \in \mathbb{K} \qquad [6.7]$$

– *Triangular inequality*: an immediate consequence of equation [6.3] in Definition 6.2 is that we can write:

$$\boxed{\|Ax\| \leqslant \|A\|\|x\|} \quad \forall x \in V \qquad [6.8]$$

Using this alongside the triangular inequality of the norm of W, for any pair of operators $A, B : V \to W$ and for all $x \in V$, we can write:

$$\|(A+B)x\| = \|Ax + Bx\| \leqslant \|Ax\| + \|Bx\| \leqslant \|A\|\|x\| + \|B\|\|x\| = (\|A\| + \|B\|)\|x\|$$

By equation [6.3], this implies:

$$\boxed{\|A+B\| \leqslant \|A\| + \|B\|} \qquad [6.9]$$

The inequality [6.9] and the property of homogeneity [6.7] show that the set of bounded linear operators is invariant with respect to linear combinations, and is thus itself a vector space; this space becomes normed by the operator norm.

DEFINITION 6.3.– *The normed vector space of bounded linear operators from V to W endowed with the operator norm is noted $\mathcal{B}(V, W)$. If $V = W$, we simply write $\mathcal{B}(V)$.*

In the literature, the letter $\mathcal{B}$ is used to denote *bounded*. The notation $\mathcal{L}(V, W)$ is also used in this sense.

Definition 6.4 is an immediate consequence of the fact that $\mathcal{B}(V, W)$ is a normed vector space.

DEFINITION 6.4 (Convergence in $\mathcal{B}(V, W)$).– *A sequence of bounded operators* $(A_n)_{n\in\mathbb{N}} \subset \mathcal{B}(V, W)$ *converges to the bounded operator* $A \in \mathcal{B}(V, W)$ *if:*

$$\|A_n - A\| \underset{n\to+\infty}{\longrightarrow} 0$$

where $\|A_n - A\|$ *is the operator norm of the difference between* A_n *and* A.

Exercise 6.1

Using Definition 6.4, prove that a necessary condition for the convergence of a sequence of operators from $(A_n)_{n\in\mathbb{N}} \subset \mathcal{B}(V, W)$ to $A \in \mathcal{B}(V, W)$ is:

$$\lim_{n\to+\infty} \|(A_n - A)x\| = 0 \quad \forall x \in \overline{B(0,1)} \tag{6.10}$$

Solution to Exercise 6.1

We start by noting that, since the sup is a majorant of a set, it holds that:

$$\boxed{\|A\| \geqslant \|Ax\| \quad \forall x \in \overline{B(0,1)}}, \qquad \overline{B(0,1)} := \{x \in V \ : \ \|x\| \leqslant 1\} \tag{6.11}$$

Inequality [6.11] implies $\|A_n - A\| \geqslant \|(A_n - A)x\|$ $\forall x \in \overline{B(0,1)}$, thus, if there exists at least one $x \in \overline{B(0,1)}$ such that $\lim_{n\to+\infty} \|(A_n - A)x\| > 0$, then $\lim_{n\to+\infty} \|A_n - A\| > 0$ which prevents the convergence of the sequence $(A_n)_{n\in\mathbb{N}}$ to A. Property [6.10] is thus necessary for $(A_n)_{n\in\mathbb{N}} \subset \mathcal{B}(V, W)$ to converge to $A \in \mathcal{B}(V, W)$. □

In the case where $V = W$, we can add a third operation on $\mathcal{B}(V)$, the *product*:

$$(AB)(x) := (A \circ B)(x) = A(B(x)) \quad \forall x \in V$$

that is *the product in* $\mathcal{B}(V)$ *corresponds to the operation of functional composition between linear operators*. We observe that:

$$\|(AB)x\| = \|A(Bx)\| \underset{(Bx \text{ is a vector of } V)}{\leqslant} \|A\|\|Bx\| \leqslant \|A\|\|B\|\|x\| \quad \forall x \in V$$

and thus, by Definition 6.3:

$$\boxed{\|AB\| \leqslant \|A\|\|B\|} \tag{6.12}$$

Hence, taking $A = B$, $\|A^2\| \leqslant \|A\|^2$, by iterating these considerations we obtain the formula:

$$\boxed{\|A^n\| \leqslant \|A\|^n} \qquad \forall n \in \mathbb{N}.$$

Thus $\mathcal{B}(V)$ is invariant with respect to the product operation defined above, and, consequently, $\mathcal{B}(V)$ *is a normed associative unital algebra*, where the unit is the identity operator.

We recall that an *algebra* $\mathfrak{A}$ on the field $\mathbb{K}$ is a vector space on $\mathbb{K}$ equipped with a binary operation $\cdot : \mathfrak{A} \times \mathfrak{A} \to \mathfrak{A}$, commonly called the *product*, which is compatible with linear operations; this is equivalent to requiring that $\cdot$ is bilinear, that is, for all $a, b, c \in \mathfrak{A}$ and $k \in \mathbb{K}$ it holds that:

– $(a + b) \cdot c = a \cdot c + b \cdot c$ and $a \cdot (b + c) = a \cdot b + a \cdot c$;

– $(ka) \cdot b = k(a \cdot b)$, $a \cdot (kb) = k(a \cdot b)$.

THEOREM 6.6.– Let $(V, \| \ \|)$ be an arbitrary normed vector space on the field $\mathbb{K}$. The sum, product by a scalar of $\mathbb{K}$ and product in the algebra $\mathcal{B}(V)$ are continuous with respect to the operator norm.

PROOF.– Theorem 4.2 also applies in the case of the algebra $\mathcal{B}(V)$, so the sum and product by a scalar are continuous and only the continuity of the product must be proven. If $(A_n)_{n\in\mathbb{N}}$ and $(B_n)_{n\in\mathbb{N}}$ are two sequences of operators of $\mathcal{B}(V)$ which converge to $A \in \mathcal{B}(V)$ and $B \in \mathcal{B}(V)$, respectively, that is $\|A_n - A\| \underset{n\to+\infty}{\longrightarrow} 0$, $\|B_n - B\| \underset{n\to+\infty}{\longrightarrow} 0$, then we must show that $A_nB_n \underset{n\to+\infty}{\longrightarrow} AB$, that is, $\|A_nB_n - AB\| \underset{n\to+\infty}{\longrightarrow} 0$:

$$\|A_nB_n - AB\| = \|A_n(B_n - B) + (A_n - A)B\| \underset{[6.9],[6.12]}{\leqslant} \|A_n\|\|B_n - B\|$$

$$+\|A_n - A\|\|B\| \underset{n\to+\infty}{\longrightarrow} 0 \qquad \square$$

The presence of a norm on $\mathcal{B}(V, W)$ generates a topology, and this naturally leads us to examine the conditions under which this space is complete. The following result provides a sufficient condition for $\mathcal{B}(V, W)$ to be complete.

THEOREM 6.7.– Let V, W be two normed vector spaces. If W is complete, then $\mathcal{B}(V, W)$ is complete.

Before proving this theorem, we wish to highlight the fact that *the theorem holds for $\mathcal{B}(\mathcal{H})$ or $\mathcal{B}(\mathcal{H}_1, \mathcal{H}_2)$*, if $\mathcal{H}, \mathcal{H}_1, \mathcal{H}_2$ are Hilbert spaces.

PROOF.– Let $(A_n)_{n\in\mathbb{N}}$ be a Cauchy sequence of operators in $\mathcal{B}(V, W)$, that is:

$$\forall \varepsilon > 0 \quad \exists N_\varepsilon > 0 \ : \ \forall m, n \geqslant N_\varepsilon \ : \ \|A_n - A_m\| < \varepsilon$$

To prove the theorem, we must show that $(A_n)_{n\in\mathbb{N}}$ converges in $\mathcal{B}(V, W)$ using the hypothesis of completeness of W.

We begin by noting that for all fixed $x \in V$, it holds that:

$$\forall m, n \geqslant N_\varepsilon \ : \ \|A_n x - A_m x\| \leqslant \|A_n - A_m\| \|x\| < \varepsilon \|x\| \qquad [6.13]$$

and thus, by the arbitrary nature of ε, $(A_n x)_{n\in\mathbb{N}}$ is a Cauchy sequence in W.

By hypothesis, W is complete, thus there exists $\lim\limits_{n \Longrightarrow \infty} A_n x \in W$; this means that we can define the limit operator A associated with the sequence $(A_n)_{n\in\mathbb{N}}$:

$$\begin{aligned} A : V &\longrightarrow W \\ x &\longmapsto A(x) = \lim_{n \Longrightarrow +\infty} A_n x \end{aligned}$$

We shall show that $(A_n)_{n\in\mathbb{N}}$ converges in operator norm to A, and that $A \in \mathcal{B}(V, W)$, completing our proof.

We begin by noting that $\forall n \geqslant N_\varepsilon$, it holds that:

$$\begin{aligned} \|A_n(x) - A(x)\| &= \|A_n(x) - \lim_{m \Longrightarrow +\infty} A_m(x)\| \\ &\underset{(\text{continuity of } \| \ \|)}{=} \lim_{m \Longrightarrow +\infty} \|A_n(x) - A_m(x)\| \underset{[6.13]}{<} \varepsilon \|x\| \end{aligned} \qquad [6.14]$$

The final equality draws on the fact that m tends toward $+\infty$, so we know that $m \geqslant N_\varepsilon$.

Hence,

$$\forall \varepsilon > 0 \ \exists N_\varepsilon > 0 \ : \ n \geqslant N_\varepsilon \implies \|A_n - A\| = \sup_{\|x\|=1} \|(A_n - A)x\|$$

$$= \sup_{\|x\|=1} \|A_n(x) - A(x)\| < \varepsilon$$

that is, $(A_n)_{n\in\mathbb{N}}$ converges in operator norm to A.

Finally, we must verify that $A \in \mathcal{B}(V, W)$. Taking an arbitrary $x \in V$, then, since inequality [6.14] holds for all $n \geqslant N_\varepsilon$, we can write:

$$\|Ax\| = \|Ax - A_{N_\varepsilon} x + A_{N_\varepsilon} x\| \leqslant \|Ax - A_{N_\varepsilon} x\| + \|A_{N_\varepsilon} x\| \underset{[6.14]}{<} \varepsilon \|x\| + \|A_{N_\varepsilon}\| \|x\|$$

that is, $\|Ax\| < (\varepsilon + \|A_{N_\varepsilon}\|)\|x\| \ \forall x \in V$, thus A is bounded. □

DEFINITION 6.5 (Banach algebra).– *An algebra $\mathfrak{A}$ on the field $\mathbb{K}$ is a Banach algebra if the following properties are verified $\forall a, b, c \in \mathfrak{A}$:*

– $\mathfrak{A}$ *is an associative algebra, that is* $a \cdot (b \cdot c) = (a \cdot b) \cdot c$;

– $\mathfrak{A}$, *as a vector space, admits a norm* $\| \ \|$ *with respect to which it is a Banach space;*

– $\|a \cdot b\| \leqslant \|a\| \, \|b\|$.

From what we have already seen, we know that if V is a Banach space, $\mathcal{B}(V)$ is a complete, associative unital algebra with respect to the operator norm ; hence, $\mathcal{B}(V)$ is a unital Banach algebra. Evidently, for any Hilbert space $\mathcal{H}$, $\mathcal{B}(\mathcal{H})$ *is a unital Banach algebra.*

A particularly important property of the kernel of the operators of $\mathcal{B}(V, W)$ is shown below.

THEOREM 6.8.– Let V, W be two normed vector spaces and take $A \in \mathcal{B}(V, W)$, then $\ker(A)$ *is a closed vector subspace* of V.

PROOF.– Let $(v_n)_{n\in\mathbb{N}} \subset \ker(A)$ be an arbitrary convergent sequence. We must prove that its limit, $\bar{v} = \lim_{n\to+\infty} v_n$, remains within $\ker(A)$. A is bounded, and thus continuous, so $\lim_{n\to+\infty} Av_n = A(\lim_{n\to+\infty} v_n) = A\bar{v}$. Furthermore, $Av_n = 0 \ \forall n \in \mathbb{N}$ since $v_n \in \ker(A)$, hence $A\bar{v} = \lim_{n\to+\infty} 0 = 0$, which implies $\bar{v} \in \ker(A)$. □

The usefulness of this theorem is shown in the following exercise, which highlights the fact that *the theorem of projection onto a closed proper vector subspace is not valid without the completeness hypothesis.*

Exercise 6.2

Let T be the linear operator (actually, a linear functional) defined by:

$$\begin{array}{rrcl} T: & \ell^2(\mathbb{N}, \mathbb{C}) & \longrightarrow & \mathbb{C} \\ & x = (x_n)_{n\in\mathbb{N}} & \longmapsto & T(x) = \sum\limits_{n\in\mathbb{N}} \frac{x_n}{n+1} \end{array}$$

1) Show that T is continuous.

2) Taking $F = \left\{ (x_n)_{n\in\mathbb{N}} \in \ell^2(\mathbb{N}, \mathbb{C}) \ : \ \sum\limits_{n\in\mathbb{N}} \frac{x_n}{n+1} = 0 \right\}$, show that F is a closed proper vector subspace of $\ell^2(\mathbb{N}, \mathbb{C})$.

3) Prove the existence of $u \in \ell^2(\mathbb{N}, \mathbb{C})$ such that $F = \{u\}^\perp$ and use your result to deduce the explicit expression of $F^\perp$.

4) We know (see the definition corresponding to [4.26]) that $\ell^0(\mathbb{N}, \mathbb{C})$ is the vector subspace of $\ell^2(\mathbb{N}, \mathbb{C})$ made up of sequences $(x_n)_{n\in\mathbb{N}}$ which are zero after a

certain index, which we equip with the topology induced by $\ell^2(\mathbb{N}, \mathbb{C})$. Take $G = F \cap \ell^0(\mathbb{N}, \mathbb{C})$.

a) Show that G is a closed proper vector subspace of $\ell^0(\mathbb{N}, \mathbb{C})$.

b) Using formula [5.4], show that the orthogonal complement of G in $\ell^0(\mathbb{N}, \mathbb{C})$, that is $G^{\perp_{\ell^0}} := G^{\perp} \cap \ell^0$, is reduced to the zero vector: $G^{\perp_{\ell^0}} = \{0_{\ell^2(\mathbb{N},\mathbb{C})}\}$.

c) Use your findings to deduce that $\ell^0(\mathbb{N}, \mathbb{C})$ *is not complete in the topology inherited from* $\ell^2(\mathbb{N}, \mathbb{C})$.

Solution to Exercise 6.2

1) Let $u = (u_n)_{n\in\mathbb{N}}$ denote the sequence defined by $u_n = \frac{1}{n+1}$ $\forall n \in \mathbb{N}$, which obviously belongs to $\ell^2(\mathbb{N}, \mathbb{C})$. For all $x \in \ell^2(\mathbb{N}, \mathbb{C})$, it holds that:

$$|T(x)| = \left|\sum_{n\in\mathbb{N}} x_n \frac{1}{n+1}\right| = \left|\sum_{n\in\mathbb{N}} x_n u_n\right| = |\langle x, u\rangle| \underset{\text{Cauchy-Schwarz}}{\leqslant} \|x\| \|u\|$$

thus T is bounded, with $\|T\| \leqslant \|u\|$, that is continuous.

2) By definition, $F = \ker(T)$ and thus it forms a closed vector subspace in $\ell^2(\mathbb{N}, \mathbb{C})$, since we have just proved that T is continuous. One example of an element in $\ell^2(\mathbb{N}, \mathbb{C})$ that does not belong to F is the first vector in the canonical basis of $\ell^2(\mathbb{N}, \mathbb{C})$, that is, $e_1 = (1, 0, 0, \dots)$, since $\sum\limits_{n\in\mathbb{N}} \frac{e_1(n)}{n+1} = \frac{1}{2} \neq 0$. Thus, F is a closed proper vector subspace of $\ell^2(\mathbb{N}, \mathbb{C})$.

3) Taking u in the same way as in question 1, we know that:

$$F = \{x \in \ell^2(\mathbb{N}, \mathbb{C}) \ : \ \langle x, u\rangle = 0\} \equiv \{u\}^{\perp}$$

so $F^{\perp} = \{u\}^{\perp\perp} \underset{[5.5]}{=} \overline{\text{span}\{u\}}$; moreover, $\text{span}\{u\} = \left\{\frac{\lambda}{n+1}, \ \lambda \in \mathbb{C}\right\}$ is a one-dimensional vector subspace, and thus it is closed. Hence $\overline{\text{span}\{u\}} = \text{span}\{u\}$ and $F^{\perp} = \left\{\frac{\lambda}{n+1}, \ \lambda \in \mathbb{C}\right\}$.

4) a) We can rewrite G in an explicit form as:

$$G = F \cap \ell^0(\mathbb{N}, \mathbb{C}) = \{(x_n)_{n\in\mathbb{N}}, \ \exists\, N \in \mathbb{N} \ : \ x_n = 0 \ \forall n > N \text{ and } \sum_{n=0}^{N} \frac{x_n}{n+1} = 0\}$$

showing that $G = \ker T|_{\ell^0(\mathbb{N},\mathbb{C})}$. As the restriction of a continuous linear operator is itself continuous, G must be a closed vector subspace of $\ell^0(\mathbb{N}, \mathbb{C})$. To prove that G is proper, we consider $e_1 : e_1 \in \ell^0(\mathbb{N}, \mathbb{C})$, and $\sum\limits_{n=0}^{N} \frac{e_1(n)}{n+1} = \frac{1}{2} \neq 0$.

b) We have:

$$\begin{aligned} G^{\perp_{\ell^0}} &= G^{\perp} \cap \ell^0(\mathbb{N},\mathbb{C}) = (F \cap \ell^0(\mathbb{N},\mathbb{C}))^{\perp} \cap \ell^0(\mathbb{N},\mathbb{C}) \\ &\underset{[5.4]}{=} \overline{\operatorname{span}(F^{\perp} \cup \ell^0(\mathbb{N},\mathbb{C}))^{\perp}} \cap \ell^0(\mathbb{N},\mathbb{C}) \end{aligned}$$

Knowing (from Theorem 4.21) that $\ell^0(\mathbb{N},\mathbb{C})$ is dense in $\ell^2(\mathbb{N},\mathbb{C})$, we have $\ell^0(\mathbb{N},\mathbb{C}))^{\perp} = \{0_{\ell^2(\mathbb{N},\mathbb{C})}\}$, which is already included in $F^{\perp}$ as a vector subspace of $\ell^2(\mathbb{N},\mathbb{C})$. Furthermore:

$$\overline{\operatorname{span}(F^{\perp} \cup \ell^0(\mathbb{N},\mathbb{C})^{\perp})} = \overline{\operatorname{span}(F^{\perp})} = \overline{F^{\perp}} = F^{\perp} = \left\{\frac{\lambda}{n+1},\ \lambda \in \mathbb{C}\right\}$$

since $F^{\perp}$ is closed, from the answer to question 3. Then:

$$G^{\perp_{\ell^0}} = \left\{\frac{\lambda}{n+1},\ \lambda \in \mathbb{C}\right\} \cap \ell^0(\mathbb{N},\mathbb{C}) = \{0_{\ell^2(\mathbb{N},\mathbb{C})}\}$$

since it is clear that the sequence $\frac{\lambda}{n+1} \notin \ell^0(\mathbb{N},\mathbb{C})$.

c) G is a closed, proper vector subspace in $\ell^0(\mathbb{N},\mathbb{C})$ equipped with the topology inherited by $\ell^2(\mathbb{N},\mathbb{C})$; nevertheless, we have just shown that $G^{\perp_{\ell^0}}$, the orthogonal complement of G in $\ell^0(\mathbb{N},\mathbb{C})$, consists of the zero vector alone. This contradicts the result of Theorem 5.4 (a corollary of the theorem of projection onto a closed, convex proper part of a Hilbert space) which states that the orthogonal complement of a closed, proper vector subspace does not solely consist of the zero vector. Clearly, the only hypothesis which is not respected here is the completeness of $\ell^0(\mathbb{N},\mathbb{C})$ with respect to the inherited topology of $\ell^2(\mathbb{N},\mathbb{C})$. □

Our next step is to consider the way in which a continuous linear operator between two normed vector spaces interacts with Cauchy sequences.

THEOREM 6.9.– Let V and W be two arbitrary normed vector spaces, $A \in \mathcal{B}(V,W)$, and let $(x_n)_{n\in\mathbb{N}} \subset V$ be a Cauchy sequence; then $(Ax_n)_{n\in\mathbb{N}}$ is a Cauchy sequence in W.

PROOF.– By hypothesis: $\forall\varepsilon > 0\ \exists N_\varepsilon > 0 : \forall n,m \geqslant N_\varepsilon : \|x_n - x_m\| < \varepsilon$. Now, let us consider $(Ax_n)_{n\in\mathbb{N}}$ and analyze $\|Ax_n - Ax_m\| = \|A(x_n - x_m)\| \leqslant \|A\|\|x_n - x_m\| < \|A\|\varepsilon, \forall n,m \geqslant N_\varepsilon$. By the arbitrary nature of ε, $(Ax_n)_{n\in\mathbb{N}}$ is a Cauchy sequence of elements in W. □

This result can *help to prove the completeness of a normed vector space*, as we shall see in Exercise 6.3, which may be seen as a continuation of Exercise 4.2.

Exercise 6.3

Given a fixed sequence $a = (a_n)_{n\in\mathbb{N}}$ of strictly positive real numbers, we write:

$$\ell_a^2(\mathbb{N},\mathbb{C}) := \{u \in \mathbb{C}^{\mathbb{N}} : \sum_{n\in\mathbb{N}} a_n|u_n|^2 < +\infty \iff \sqrt{a}u \in \ell^2(\mathbb{N},\mathbb{C})\}$$

In Exercise 4.2, we verified that:

$$\langle u, v\rangle_{\ell_a^2} = \sum_{n\in\mathbb{N}} a_n u_n \overline{v_n} \quad \text{and} \quad \|u\|_{\ell_a^2}^2 = \sum_{n\in\mathbb{N}} a_n|u_n|^2$$

are an inner product and a norm on $\ell_a^2(\mathbb{N},\mathbb{C})$, respectively.

1) Show that the operator

$$\begin{aligned} \imath_a : \ell_a^2(\mathbb{N},\mathbb{C}) &\hookrightarrow \ell^2(\mathbb{N},\mathbb{C}) \\ u &\mapsto \imath_a(u) := \sqrt{a}u \equiv (\sqrt{a_n}u_n)_{n\in\mathbb{N}} \end{aligned}$$

is linear, continuous, has unit norm and is bijective. Give the explicit expression of the inverse operator of $\imath_a$; verify that this is continuous and has a norm of 1.

2) Using your findings, deduce that $\ell_a^2(\mathbb{N},\mathbb{C})$ is a Hilbert space.

3) Let a and b be two sequences of strictly positive real numbers such that $a_n \underset{n\to+\infty}{=} O(b_n)$. Show that $\ell_b^2(\mathbb{N},\mathbb{C}) \subset \ell_a^2(\mathbb{N},\mathbb{C})$, and that the canonical injection is continuous.

Solution to Exercise 6.3

1) Linearity can be shown by simple rewriting: if $u, v \in \ell_a^2(\mathbb{N},\mathbb{C})$ and $\lambda \in C$, then $\imath(u+\lambda v) = \sqrt{a}(u+\lambda v) = \sqrt{a}u + \lambda\sqrt{a}v = \imath(u) + \lambda\imath(v)$. Concerning continuity, for all $u \in \ell_a^2(\mathbb{N},\mathbb{C})$, $\imath(u) \in \ell^2(\mathbb{N},\mathbb{C})$ and the norm is:

$$\|\imath_a(u)\|_{\ell^2}^2 = \sum_{n\in\mathbb{N}} |\sqrt{a_n}u_n|^2 = \sum_{n\in\mathbb{N}} a_n|u_n|^2 = \|u\|_{\ell_a^2}^2 \iff \|\imath_a(u)\|_{\ell^2} = \|u\|_{\ell_a^2}^2$$

This shows that $\imath_a$ is continuous, and that its norm is 1. The final condition to prove is bijectivity. We note that the operator:

$$\begin{aligned} \jmath_{1/a} : \ell^2(\mathbb{N},\mathbb{C}) &\Longrightarrow \ell_a^2(\mathbb{N},\mathbb{C}) \\ v &\mapsto \imath_{1/a}(v) := \tfrac{1}{\sqrt{a}}v \equiv (\tfrac{1}{\sqrt{a_n}}v_n)_{n\in\mathbb{N}} \end{aligned}$$

is well defined, since $a > 0$ and $v/\sqrt{a} \in \ell_a^2(\mathbb{N},\mathbb{C}) \iff \sqrt{a}v/\sqrt{a} = v \in \ell^2(\mathbb{N},\mathbb{C})$. Furthermore, it is such that $\jmath_{1/a} \circ \imath_a : \ell_a^2(\mathbb{N},\mathbb{C}) \to \ell_a^2(\mathbb{N},\mathbb{C})$, $\jmath_{1/a} \circ \imath_a(u) = \frac{\sqrt{a}}{\sqrt{a}}u = u$ $\forall u \in \ell_a^2(\mathbb{N},\mathbb{C})$; *vice versa*, for all $v \in \ell^2(\mathbb{N},\mathbb{C})$, $\imath_a \circ \jmath_{1/a} : \ell^2(\mathbb{N},\mathbb{C}) \to \ell^2(\mathbb{N},\mathbb{C})$, $\imath_a \circ \jmath_{1/a}(v) = \frac{\sqrt{a}}{\sqrt{a}}v = v$, that is, $\jmath_{1/a} \circ \imath_a = id_{\ell_a^2(\mathbb{N},\mathbb{C})}$ and $\imath_a \circ \jmath_{1/a} = id_{\ell^2(\mathbb{N},\mathbb{C})}$. Thus,

$\imath_a$ is bijective with inverse $\jmath_{1/a}$. The inverse is also clearly continuous and possesses a unit norm, since:

$$\|\jmath_{1/a}(v)\|^2_{\ell^2_a} = \sum_{n\in\mathbb{N}} \frac{a_n}{a_n}|v_n|^2 = \|v\|^2_{\ell^2} \iff \|\jmath_{1/a}(v)\|_{\ell^2_a} = \|v\|^2_{\ell^2} \qquad \forall v \in \ell^2(\mathbb{N},\mathbb{C})$$

[6.15]

2) By the continuity of $\imath_a$ and by Theorem 6.9, $\imath_a$ transforms the Cauchy sequences in $\ell^2_a(\mathbb{N},\mathbb{C})$ into Cauchy sequences in $\ell^2(\mathbb{N},\mathbb{C})$. Now, let $(u_m)_{m\in\mathbb{N}}$ be an arbitrary Cauchy sequence of elements in $\ell^2_a(\mathbb{N},\mathbb{C})$; $\imath_a((u_m)_{m\in\mathbb{N}})$ is a Cauchy sequence in $\ell^2(\mathbb{N},\mathbb{C})$, which we know to be complete, thus $\exists\, L \in \ell^2(\mathbb{N},\mathbb{C})$ such that $\imath_a((u_m)_{m\in\mathbb{N}}) \underset{m\to+\infty}{\longrightarrow} L$, that is:

$$\begin{aligned} 0 &= \lim_{m\to+\infty} \|\imath_a((u_m)_{m\in\mathbb{N}}) - L\|_{\ell^2} \underset{[6.15]}{=} \lim_{m\to+\infty} \|\jmath_{1/a}(\imath_a((u_m)_{m\in\mathbb{N}}) - L)\|_{\ell^2_a} \\ &\underset{\jmath_{1/a}\text{ linear}}{=} \lim_{m\to+\infty} \|\jmath_{1/a}\circ\imath_a((u_m)_{m\in\mathbb{N}}) - \jmath_{1/a}(L)\|_{\ell^2_a} \\ &= \lim_{m\to+\infty} \|(u_m)_{m\in\mathbb{N}} - \jmath_{1/a}(L)\|_{\ell^2_a} \end{aligned}$$

that is, $(u_m)_{m\in\mathbb{N}}$ converges in $\ell^2_a(\mathbb{N},\mathbb{C})$ to $\jmath_{1/a}(L)$, hence $\ell^2_a(\mathbb{N},\mathbb{C})$ is a Hilbert space.

3) We must show that if $a_n \underset{n\to+\infty}{=} O(b_n)$, then $u \in \ell^2_b(\mathbb{N},\mathbb{C}) \implies \ell^2_a(\mathbb{N},\mathbb{C})$, that is:

$$\sum_{n\in\mathbb{N}} b_n|u_n|^2 < +\infty \implies \sum_{n\in\mathbb{N}} a_n|u_n|^2 < +\infty$$

for all $u \in \ell^2_b(\mathbb{N},\mathbb{C})$. By definition, $a_n \underset{n\to+\infty}{=} O(b_n)$ if and only if there exist $C_1 > 0$ and $N \in \mathbb{N}$ such that, for all $n \geqslant N$, it holds that $a_n \leqslant C_1 b_n$.

For the purposes of this demonstration, we must multiply both sides of the previous inequality by $|u_n|^2$, giving us $a_n|u_n|^2 \leqslant C_1 b_n|u_n|^2$ for all $n \geqslant N$, that is, $\sum_{n=N}^{+\infty} a_n|u_n|^2 \leqslant C_1 \sum_{n=N}^{+\infty} b_n|u_n|^2$. The summation of the first N terms, from $n = 0$ to $n = N-1$, is finite, so there must be a constant $C_2 > 0$ which is sufficiently large to result in $\sum_{n=0}^{N-1} a_n|u_n|^2 \leqslant C_2 \sum_{n=0}^{N-1} b_n|u_n|^2$; we therefore take $C := \max(C_1, C_2) > 0$, giving us $\sum_{n\in\mathbb{N}} a_n|u_n|^2 \leqslant C \sum_{n\in\mathbb{N}} b_n|u_n|^2$. This tells us that if $u \in \ell^2_b(\mathbb{N},\mathbb{C})$, then $u \in \ell^2_a(\mathbb{N},\mathbb{C})$. Furthermore, the previous inequality can be rewritten as $\|u\|^2_{\ell^2_a} \leqslant C\|u\|^2_{\ell^2_b}$, thus the canonical injection $\iota : \ell^2_b(\mathbb{N},\mathbb{C}) \hookrightarrow \ell^2_a(\mathbb{N},\mathbb{C})$ verifies $\|\iota(u)\|_{\ell^2_a} \leqslant \sqrt{C}\|u\|_{\ell^2_b}$ for all $u \in \ell^2_b(\mathbb{N},\mathbb{C})$, meaning that it is bounded and thus continuous. □

We shall conclude this section by presenting an *extremely useful result* which can be used to characterize the equality between continuous operators on an inner product space of arbitrary dimensions, *via* the equality of their action on vectors within an inner product.

THEOREM 6.10.– Let $A, B : V \to W$ be two linear operators defined on an inner product space of arbitrary dimension. Then:

$$A = B \iff \langle x, Ay\rangle = \langle x, By\rangle \quad \forall x, y \in V$$

PROOF.– By linearity of A, it holds that $\langle x, Ay\rangle = \langle x, By\rangle \; \forall x, y \in V \iff \langle x, (A - B)y\rangle = 0 \; \forall x, y \in V$. Let us take an arbitrary but fixed element $y \in V$ and write $u = (A - B)y \in V$, then $\langle x, u\rangle = 0 \; \forall x \in V$ holds true if and only if $u = 0$, that is $(A - B)y = 0 \; \forall y \in V$, that is $A - B = 0$, implying $A = B$. □

6.2.1. *A classical example of a non-bounded linear operator on a vector space of infinite dimension*

Although we have chosen to focus only on bounded linear operators on Hilbert spaces, it is important to show at least one example of a non-bounded linear operator. Actually, we are going to prove that one of the simplest operations – the derivation – on the simplest Hilbert basis – the Fourier basis – does not produce a bounded operator.

Let $u_n(x) = \frac{1}{\sqrt{2\pi}} e^{inx}$, $(n \in \mathbb{Z})$, be the Fourier basis of $L^2[0, 2\pi]$. Let us consider the first derivation operator on the infinite-dimensional vector space generated by the Fourier basis:

$$\begin{aligned} D : \text{span}((u_n)_{n\in\mathbb{Z}}) &\longrightarrow L^2[0, 2\pi] \\ u_n &\longmapsto Du_n = \frac{d}{dx} u_n \end{aligned}$$

where $\frac{d}{dx} u_n(x) = \frac{in}{\sqrt{2\pi}} e^{inx}$, which is square integrable on $[0, 2\pi]$. Of course, the previous definition of D is extended by linearity on the whole span.

We can show that the norm of D is not finite. To calculate it, we may use equation [6.5] in Definition 6.2 of an operator norm, taken v in the domain of D we have :

$$\|D\| = \sup_{\|v\|=1} \|Dv\| \geqslant \sup_{\|u_n\|=1} \|Du_n\|$$

where the inequality is motivated by the fact that the sup on the right hand side is computed over a subset of the domain of D.

However, the condition $\|u_n\| = 1$ does not determine any constraints, as any element u_n in the Fourier Hilbert basis of $L^2[0, 2\pi]$ has a unit norm, thus $\|D\|$ is simply the sup of the set of values $\|Du_n\|$ with respect to the integer index n, that is:

$$\|D\| \geqslant \sup_{n\in\mathbb{Z}} \|Du_n\| = \sup_{n\in\mathbb{Z}} \left(\int_0^{2\pi} \left| \frac{in}{\sqrt{2\pi}} e^{inx} \right|^2 dx \right)^{1/2}$$

$$= \sup_{n\in\mathbb{Z}} \left(|in|^2 \int_0^{2\pi} \left| \frac{1}{\sqrt{2\pi}} e^{inx} \right|^2 dx \right)^{1/2}$$

that is:

$$\|D\| \geqslant \sup_{n\in\mathbb{Z}} |in| = \sup_{n\in\mathbb{Z}} |n| = +\infty$$

which implies that the derivation operator defined above is not bounded, and is therefore not continuous.

6.3. Invertibility of linear operators

Exercise 6.3 highlighted the importance of analyzing the inverse of a linear operator. This subject will be examined in greater detail in this section.

DEFINITION 6.6.– *Let V, W be two normed vector spaces on the same field $\mathbb{K}$ and let $A : V \to \mathrm{Im}(A) \subseteq W$ be a linear operator. The inverse operator of A is A^{-1} : $\mathrm{Im}(A) \subseteq W \to V$ such that $\forall x \in V$:*

$$\begin{aligned} A^{-1} : \mathrm{Im}(A) \subseteq W &\longrightarrow V \\ Ax &\longmapsto A^{-1}(Ax) = x \end{aligned}$$

If there exists A^{-1}, then A is invertible.

For all $x \in V$, it holds that $A^{-1}(Ax) = x$ and $A(A^{-1}(Ax)) = A(x)$, thus the invertibility of A can be defined in an equivalent manner with the conditions:

$$A^{-1} \circ A = id_V \text{ and } A \circ A^{-1} = id_{\mathrm{Im}(A)}$$

In the specific case where $W = V$ and $\mathrm{Im}(A) = V$, the invertibility of A is equivalent to the existence of an operator $A^{-1} : V \to V$ such that:

$$A \circ A^{-1} = A^{-1} \circ A = id_V$$

If $A : V \to V$, the symbol $\mathrm{GL}(V)$ is used to designate *the set of continuous bijective linear operators with a continuous inverse*, known as the *set of regular elements in $\mathcal{B}(V)$*.

Theorem 6.11 summarizes the elementary properties of the inverse (the proofs of these properties are identical to those performed in finite dimension).

THEOREM 6.11.– Let V, W be two normed vector spaces and let $A : V \to Im(A) \subseteq W$ be linear:

1) If A^{-1} exists, then it is unique;

2) If A^{-1} exists, then it is a linear operator;

3) A^{-1} exists if and only if $\ker(A) = \{0_V\}$, that is a necessary and sufficient condition for A to be invertible on its image is that its kernel is reduced to the zero vector of V.

PROOF.–

1) Let $B_1, B_2 : \text{Im}(A) \subseteq W \to V$ be two inverse operators of A, then: $B_1 = B_1 \circ id_{\text{Im}(A)} = B_1 \circ (A \circ B_2) = (B_1 \circ A) \circ B_2 = id_{\text{Im}(A)} \circ B_2 = B_2$.

2) For all $w_1, w_2 \in \text{Im}(A)$ and $k \in \mathbb{K}$, we have:

$$\begin{aligned} A^{-1}(w_1 + kw_2) &= A^{-1}(AA^{-1}(w_1) + kAA^{-1}(w_2)) \\ &\quad \text{(linearity of } A\text{)} \\ &= A^{-1}A(A^{-1}(w_1) + kA^{-1}(w_2)) \\ &= A^{-1}(w_1) + kA^{-1}(w_2) \end{aligned}$$

3) We know that the inverse of A can be defined on its image if and only if A is injective. Let us verify that this is equivalent to $\ker(A) = 0_V$. On one side, if $Ax = 0_W$, then $x = A^{-1}Ax = A^{-1}0_W = 0_V$ by linearity of A^{-1}, so if there exists A^{-1}, the kernel of A is reduced to the zero vector of V. On the other side, taking $\ker(A) = \{0_V\}$ and $x_1, x_2 \in V$ such that $Ax_1 = Ax_2$, then $Ax_1 - Ax_2 = 0_W$, that is by linearity of A, $A(x_1 - x_2) = 0_W$, but if $\ker(A) = \{0_V\}$ then $x_1 - x_2 = 0_V$, that is, $x_1 = x_2$, proving the injectivity of A. □

The condition $\ker(A) = \{0_V\}$ is necessary and sufficient for the invertibility of a linear operator on its image space $\text{Im}(A)$ in finite and infinite dimensions. In finite dimensions, the inverse of a linear operator, if it exists, is always bounded.

In infinite dimensions, on the other hand, the condition $\ker(A) = \{0_V\}$ does not imply any relationship between the continuity of A and that of A^{-1}: *A may be bounded and have a non-bounded inverse* or, conversely, *A may be non-bounded and have a bounded inverse*. One classic example of this situation is given by the derivation and integral operators. An easier example is provided by the linear operator $A : \ell^2(\mathbb{N}, \mathbb{K}) \to \ell^2(\mathbb{N}, \mathbb{K})$ defined by $A(x_1, x_2, x_3, \ldots, x_n, \ldots) = (x_1, x_2/2, x_3/3, \ldots, x_n/n, \ldots)$, that is $A((x_n)_{n \in \mathbb{N}*}) = (x_n/n)_{n \in \mathbb{N}*}$. A is bounded and $\|A\| \leqslant 1$. For all $x = (x_n)_{n \in \mathbb{N}} \in \ell^2(\mathbb{N}, \mathbb{K})$:

$$\|Ax\|_{\ell^2}^2 = \sum_{n \in \mathbb{N}} \frac{|x_n|^2}{n^2} \leqslant \sum_{n \in \mathbb{N}} |x_n|^2 = \|x\|_{\ell^2}^2$$

The operator $A^{-1} : \ell^2(\mathbb{N}, \mathbb{K}) \to \ell^2(\mathbb{N}, \mathbb{K})$, $A^{-1}((y_n)_{n\in\mathbb{N}*}) = (ny_n)_{n\in\mathbb{N}*}$ is evidently the inverse of A. Nevertheless, A^{-1} is not bounded: we can verify this by considering the general element of the canonical basis of $\ell^2(\mathbb{N}, \mathbb{K})$, that is $e_n = (0, 0, \ldots, 1, 0, 0, \ldots)$, where 1 is in the position n. We see that, on one side, $\|e_n\|_{\ell^2} = 1 \ \forall n \in \mathbb{N}$, and, on the other side, $\|A^{-1}e_n\|_{\ell^2} = n$, hence $\|A^{-1}\| = \sup_{n\in\mathbb{N}} \|A^{-1}e_n\|_{\ell^2} = +\infty$.

A very useful characterization exists for the bounded invertibility of linear operators. It is important to note that this characterization holds independently of the continuity of the operator, making it particularly helpful in practical applications.

THEOREM 6.12 (Bounded invertibility of a linear operator).– If V and W are two normed vector spaces and $A : V \to W$ is a linear operator (not necessarily bounded), then $\exists A^{-1} \in \mathcal{B}(\mathit{Im}(A), V)$ if and only if $\exists \mu > 0$ such that $\|Ax\| \geqslant \mu\|x\| \ \forall x \in V$.

PROOF.–

$\boxed{\Longrightarrow}$: suppose that $\exists A^{-1} \in \mathcal{B}(\mathrm{Im}(A), W)$, then, by definition, $\exists m > 0$ such that $\forall y \in \mathrm{Im}(A)$: $\|A^{-1}y\| \leqslant m\|y\|$. Since A is invertible and $y \in \mathrm{Im}(A)$, $\exists x \in V$ such that we can write $y = Ax$, then $\underbrace{\|A^{-1}Ax\|}_{\|x\|} \leqslant m\|Ax\|$, that is $\|Ax\| \geqslant \underbrace{\frac{1}{m}}_{=\mu>0} \|x\|$ and, since y is an arbitrary element in $\mathrm{Im}(A)$, the inequality holds for all $x \in V$.

$\boxed{\Leftarrow}$: suppose that $\|Ax\| \geqslant \underset{(>0)}{\mu} \|x\| \ \forall x \in V$, then, in particular, if we consider $x \in \ker(A)$:

$$\|Ax\| = \|0\| = 0 \geqslant \mu\|x\| \underset{(\mu>0)}{\iff} \|x\| = 0 \iff x = 0_V \implies \ker(A) = \{0_V\}$$

that is $\exists A^{-1} : \mathrm{Im}(A) \implies V$. We must therefore prove that A^{-1} is bounded. For all $y \in \mathrm{Im}(A)$ such that $x = A^{-1}y$, we have: $\|Ax\| \geqslant \mu\|x\| \iff \|AA^{-1}y\| \geqslant \mu\|A^{-1}y\| \iff \|y\| \geqslant \mu\|A^{-1}y\| \iff \|A^{-1}y\| \leqslant \frac{1}{\mu}\|y\|$, $\forall y \in \mathrm{Im}(A)$, that is A^{-1} is bounded. □

The condition of the theorem is interpreted as follows. First, the fact that $\|Ax\| \geqslant \mu\|x\|$ guarantees that the kernel of A consists solely of the zero vector. Furthermore, the inequality $\|Ax\| \geqslant \mu\|x\|$ is inverted with respect to the inequality which defines a bounded operator, that it is well suited to guarantee that the inverse operator of A is bounded.

One immediate consequence of the theorem shown above is that a linear operator $A : V \to W$ is bounded and has a bounded inverse if and only if it satisfies the following condition:

$$\exists a, b > 0, \ a \leqslant b : \boxed{a\|x\| \leqslant \|Ax\| \leqslant b\|x\|} \quad \forall x \in V$$

that is, the norm of all of the vectors of V, transformed by the action of A, is bounded by the norm of the vector itself multiplied by two positive constants. This consideration has an important consequence for the images of bounded linear operators defined on Banach spaces, as stated in the next theorem.

THEOREM 6.13.– Let V be a Banach space and W an arbitrary normed vector space. Take $A \in \mathcal{B}(V, W)$. If A is invertible with a bounded inverse, then $Im(A)$ is a closed vector subspace of W.

PROOF.– From Theorem 6.12, we know that the condition $\exists A^{-1} \in \mathcal{B}(\mathrm{Im}(A), V)$ is equivalent to:

$$\exists a > 0 \ : \ \|x\| \leqslant a\|Ax\| \quad \forall x \in V$$

We must prove that this condition implies that $\mathrm{Im}(A)$ is closed, that is, if $(y_n)_{n\in\mathbb{N}} \subset \mathrm{Im}(A)$ is such that $y_n \underset{n\to+\infty}{\longrightarrow} y$, then $y \in \mathrm{Im}(A)$. Since $y_n \in \mathrm{Im}(A)$, then there exists $(x_n)_{n\in\mathbb{N}} \subset V$ such that $y_n = Ax_n \; \forall n \in \mathbb{N}$, hence:

$$\|x_n - x_m\| \leqslant a\,\|A(x_n - x_m)\| = \|Ax_n - Ax_m\| = \|y_n - y_m\| \underset{n,m\to+\infty}{\longrightarrow} 0$$

because $(y_n)_{n\in\mathbb{N}}$ is a convergent, and thus Cauchy sequence. The sequence $(x_n)_{n\in\mathbb{N}}$ must therefore also be Cauchy and, since V is a Banach space, there exists $x \in V$ such that $x_n \underset{n\to+\infty}{\longrightarrow} x$. By the continuity of A, we obtain:

$$Ax = A \lim_{n\to+\infty} x_n = \lim_{n\to+\infty} Ax_n = \lim_{n\to+\infty} y_n = y$$

that is, $y \in \mathrm{Im}(A)$. □

There is a second condition which is sufficient to ensure the continuity of the inverse of a linear operator. The presentation of this condition relies on an intermediary result, which is, itself, one of the most important theorems in functional analysis (the proof of this theorem is beyond the scope of this book, we simply note that it is a consequence of Baire's category theorem).

THEOREM 6.14 (Open mapping theorem – Banach-Schauder).– Let V and W be two Banach spaces. If $A \in \mathcal{B}(V, W)$ is surjective, then A is an open mapping, that is A transforms open subsets of V into open subsets of W.

THEOREM 6.15 (Continuous inverse operator theorem in Banach spaces).– Let V and W be two Banach spaces. If $A \in \mathcal{B}(V, W)$ is bijective, that is $\ker(A) = \{0_V\}$, and A is surjective, then $A^{-1} \in \mathcal{B}(W, V)$, that is A^{-1} is continuous.

PROOF.– Recall the topological characterization of continuity: a function between two topological spaces is continuous if and only if the counterimage of any open subset is open. By definition, the counterimages of A^{-1} are the images of A, hence A^{-1} is continuous if and only if any image of open *via* A is open; this property is guaranteed by the open mapping theorem. □

The continuous inverse theorem can be used to characterize operators belonging to the set $\mathrm{GL}(V)$ for any given Banach space V.

THEOREM 6.16 (Characterization of $\mathrm{GL}(V)$).– Let V be a Banach space and $\mathrm{GL}(V)$ the set of regular elements of the Banach algebra $\mathcal{B}(V)$ (linear bijections with continuous inverse). For an operator $A \in \mathcal{B}(V)$, the following two conditions are equivalent:

1) $A \in GL(V)$;

2) $\exists$ a linear operator B defined on all V such that $BA = id_V$ and $AB = id_V$.

If one of the two conditions is satisfied, then B is unique and $B = A^{-1}$.

PROOF.–

1) $\Longrightarrow$ 2) If $A \in \mathrm{GL}(V)$, then we must simply consider $B = A^{-1}$ to prove the implication.

2) $\Longrightarrow$ 1) The hypothesis $BA = id_V$ implies that $\ker(A) = \{0\}$, that is, A is injective. Reasoning by the absurd, if $x \neq 0$ and $Ax = 0$, then we would have $BAx = 0$, which contradicts the fact that $BAx = id_V(x) = x \neq 0$. Furthermore, the hypothesis $AB = id_V$ implies that $\mathrm{Im}(A) = V$; for all $x \in V$, it holds that $A(Bx) = AB(x) = id_V(x) = x$, so any $x \in V$ can be seen as the image *via* A of an element in V, that is, Bx, meaning that A is surjective. Thus, the existence of B such that the hypotheses $BA = id_V$ and $AB = id_V$ are valid implies that A is a linear bijection, and that $\forall x \in V$, $B(Ax) = x$, that is, $B = A^{-1}$. Hence A is bounded by hypothesis, invertible and surjective; by the continuous inverse theorem, $B = A^{-1}$, and therefore $A \in \mathrm{GL}(V)$.

The final step is to prove uniqueness. Let B and B' be two operators which verify 2; then $A^{-1} = A^{-1}AB$ and $A^{-1} = A^{-1}AB'$, hence $A^{-1} = B = B'$. □

Clearly, if $A \in \mathrm{GL}(V)$, then we also have $A^{-1} \in \mathrm{GL}(V)$ and if $A, B \in \mathrm{GL}(V)$, then $AB \in \mathrm{GL}(V)$ since $(AB)^{-1} = B^{-1}A^{-1}$ given that $ABB^{-1}A^{-1} = id_V$ and $B^{-1}A^{-1}AB = id_V$. $\mathrm{GL}(V)$ is therefore stable with respect to the product and inversion, and its unit element is id_V, that is $\mathrm{GL}(V)$ is a group.

DEFINITION 6.7.– *The group $GL(V)$ is called the general linear group of V.*

6.4. The dual of a Hilbert space and the Riesz representation theorem

Again, let us consider $\mathcal{B}(V, W)$, where V, W are two normed vector spaces. We know that $\mathcal{B}(V, W)$ is a Banach space with respect to the operator norm if W is a Banach space. Consider the specific case in which W is the field $\mathbb{K}$ on which V is defined as a vector space.

As $\mathbb{K} = \mathbb{R}$ or $\mathbb{C}$ is complete, $\mathcal{B}(V, \mathbb{K})$ is a Banach space, known as the *dual* of V and noted V^* (the notation V' is sometimes used in the literature to denote a dual space). The elements of V^* are known as the *bounded linear functionals* on V.

We could ask ourselves how the "dualization" process of V can be iterated. For Hilbert spaces, the answer to this question is quite surprising: *the dualization of any Hilbert space $\mathcal{H}$ is an involution*, that is, $\mathcal{H}^{**} \simeq \mathcal{H}$, where $\simeq$ is an isomorphism between Hilbert spaces. $\mathcal{H}^{**}$ is called the *bidual* of $\mathcal{H}$.

This is not true, in general, for Banach spaces; those which are isomorphic to their bidual are known as *reflexive Banach spaces*. The Banach spaces $L^p(X, \mathcal{A}, \mu)$ are reflexive for $1 < p < \infty$, but $L^1(X, \mathcal{A}, \mu)$ and $L^\infty(X, \mathcal{A}, \mu)$ are not.

Each functional $\varphi \in V^*$ transforms an element of V into a scalar of $\mathbb{K}$. This transformation is represented using the following notation:

$$\begin{aligned} \varphi : V &\longrightarrow \mathbb{K} \\ x &\longmapsto \varphi(x) = \langle \varphi, x \rangle \end{aligned}$$

The notation $\langle \varphi, x \rangle$ comes from the fact that if V is a Hilbert space, then any continuous linear functional $\varphi \in V^*$ acts as an inner product on the vectors of V. This statement forms the basis for a famous result first identified by Riesz, which will be shown and proved below.

THEOREM 6.17 (Riesz representation theorem).– Let $\mathcal{H}$ be a Hilbert space on $\mathbb{K} = \mathbb{R}$ or $\mathbb{C}$, and let $\mathcal{H}^*$ be the dual of $\mathcal{H}$. Then:

$$\begin{aligned} T : \mathcal{H} &\longrightarrow \mathcal{H}^* \\ x &\longmapsto T_x \end{aligned}$$

where:

$$\begin{aligned} T_x : \mathcal{H} &\longrightarrow \mathbb{K} \\ y &\longmapsto T_x(y) = \langle y, x \rangle \end{aligned}$$

is an isomorphism between $\mathcal{H}$ and $\mathcal{H}^*$ interpreted as Banach spaces, that is, T is bijective, preserves the norms and:

– if $\mathbb{K} = \mathbb{R}$, then T is linear;

– if $\mathbb{K} = \mathbb{C}$, then T is antilinear.

The functional T_x is called the *Riesz representative* of x in $\mathcal{H}^*$.

Before presenting the proof, it is important to understand the reason for the antilinearity in the case $\mathbb{K} = \mathbb{C}$. We shall begin by analyzing the summation operation:

$$\begin{aligned} T: \quad \mathcal{H} &\longrightarrow \mathcal{H}^* \\ x_1 + x_2 &\longmapsto T_{x_1+x_2} \end{aligned}$$

$$\begin{aligned} T_{x_1+x_2} : \mathcal{H} &\longrightarrow \mathbb{C} \\ y &\longmapsto T_{x_1+x_2}(y) \\ &\quad = \langle y, x_1 + x_2 \rangle = \langle y, x_1 \rangle + \langle y, x_2 \rangle = T_{x_1}(y) + T_{x_2}(y) \end{aligned}$$

thus $\boxed{T_{x_1+x_2} = T_{x_1} + T_{x_2}}$.

Now, consider the multiplication by a scalar using $k \in \mathbb{C}$:

$$\begin{aligned} T: \mathcal{H} &\longrightarrow \mathcal{H}^* \\ kx &\longmapsto T_{kx} \end{aligned}$$

$$\begin{aligned} T_{kx} : \mathcal{H} &\longrightarrow \mathbb{C} \\ y &\longmapsto T_{kx}(y) = \langle y, kx \rangle = \bar{k}\langle y, x \rangle = \bar{k}T_x(y) \end{aligned}$$

thus $\boxed{T_{kx} = \bar{k}T_x}$.

Therefore:

$$\begin{aligned} T: \quad \mathcal{H} &\longrightarrow \mathcal{H}^* \\ x_1 + x_2 &\longmapsto T_{x_1} + T_{x_2}, \end{aligned} \qquad \begin{aligned} T: \quad \mathcal{H} &\longrightarrow \mathcal{H}^* \\ kx &\longmapsto \bar{k}T_x \end{aligned}$$

which explains why T is antilinear if $\mathbb{K} = \mathbb{C}$. Evidently, if $\mathbb{K} = \mathbb{R}$, this distinction has no place and T is linear.

The Riesz representation theorem owes its name to the fact that it allows all continuous linear functions on a Hilbert space to be represented *via* inner products; notably, for any continuous linear function φ on $\mathcal{H} = L^2(X, \mathcal{A}, \mu)$ there exists a single element $f \in L^2(X, \mathcal{A}, \mu)$ such that $\varphi = T_f$ with:

$$\begin{aligned} T_f : L^2(X, \mathcal{A}, \mu) &\longrightarrow \mathbb{K} \\ g \quad &\longmapsto T_f(g) = \langle g, f \rangle = \textstyle\int_X g\bar{f}d\mu \end{aligned}$$

More generally, we know that all separable, infinite dimensional Hilbert spaces are isomorphic to $\ell^2(\mathbb{N}, \mathbb{K})$, for which the inner product is defined by a series.

These observations are the reason why *continuous linear functionals are very often represented by finite sums, series or integrals in applications of functional analysis.*

One final aspect to note before moving on to the proof is that if we consider the inner product in the way it is used in physics, that is, as antilinear with respect to the first entry and linear with respect to the second entry, then the definition of T_x becomes $T_x(y) = \langle x, y\rangle$.

PROOF.– Since the linear or antilinear character of T has already been examined, we shall start by verifying that T is well defined, that is, T_x is a bounded linear functional on $\mathcal{H}$. Taking $\alpha, \beta \in \mathbb{K}$, $y, y_1, y_2 \in \mathcal{H}$:

– T_x is linear[2]:

$$T_x(\alpha y_1 + \beta y_2) = \langle \alpha y_1 + \beta y_2, x\rangle = \alpha\langle y_1, x\rangle + \beta\langle y_2, x\rangle = \alpha T_x(y_1) + \beta T_x(y_2)$$

– T_x is bounded: We begin by observing that $\|T_x(y)\| = |T_x(y)|$ since $T_x(y) \in \mathbb{K}$. Thus:

$$\|T_x(y)\| = |T_x(y)| = |\langle y, x\rangle| \underset{\text{(Cauchy-Schwarz)}}{\leqslant} \|x\|\|y\| \qquad [6.16]$$

The fact that T_x is a bounded linear operator between the Hilbert spaces $\mathcal{H}$ and $\mathbb{K}$ allows us to calculate the operator norm of T_x. With respect to this norm, T *is an isometry*, that is, $\|T_x\|_{\mathcal{B}(\mathcal{H},\mathbb{K})} = \|x\|_{\mathcal{H}}$ $\forall x \in \mathcal{H}$. The case of the zero vector is straightforward: if $x = 0_{\mathcal{H}}$ then $T_{0_{\mathcal{H}}}$ is the zero functional since $T_{0_{\mathcal{H}}}(y) = \langle y, 0_{\mathcal{H}}\rangle = 0$ $\forall y \in \mathcal{H}$, thus: $\|0_{\mathcal{H}}\| = 0 = \|T_{0_{\mathcal{H}}}\|$.

Taking $x \in \mathcal{H}$, $x \neq 0_{\mathcal{H}}$, let us prove that $\|T_x\| \leqslant \|x\|$ and that $\|x\| \leqslant \|T_x\|$, in that order:

– $\|T_x\| \leqslant \|x\|$: by [6.16] we can write $\|T_x(y)\| \leqslant \|x\|\|y\|$ $\forall y \in \mathcal{H}$, hence:

$$\|T_x\| = \sup_{\|y\|=1} |T_x(y)| \leqslant \sup_{\|y\|=1} \|x\|\|y\| = \|x\|$$

– $\|x\| \leqslant \|T_x\|$: in this case, we can write:

$$\|x\|^2 = \langle x, x\rangle \underset{\text{(def. of } T_x)}{=} T_x(x) \underset{T_x(x)=\|x\|^2\geqslant 0\,!}{=} |T_x(x)| = \|T_x(x)\| \underset{(T_x \text{ bounded})}{\leqslant} \|T_x\|\|x\|$$

and since $\|x\| \neq 0$, the first and last members of the expression above can be divided by $\|x\|$, giving us $\|x\| \leqslant \|T_x\|$.

In summary, $\|T_x\| = \|x\|$ $\forall x \in \mathcal{H}$, hence T is an isometry and consequently T is injective.

2 If we had defined $T_x(y) = \langle x, y\rangle$, then we would have $T_x(\alpha y_1 + \beta y_2) = \bar{\alpha}\langle x, y_1\rangle + \bar{\beta}\langle x, y_2\rangle = \bar{\alpha}T_x(y_1) + \bar{\beta}T_x(y_2)$, that is, T_x would be an antilinear functional. It is thus impossible to avoid antilinearity either in T or T_x.

The final step in the proof is to demonstrate that T *is surjective*, that is, for all $\varphi \in \mathcal{H}^*$ there exists $x \in \mathcal{H}$ such that $\varphi = T_x$. The argument which Riesz used to demonstrate the surjectivity of T is particularly elegant.

First, if φ is the identically zero functional $\mathbf{0}$, then $\varphi = T_{0_\mathcal{H}}$.

Now, let φ be a non-identically zero function, and *consider its kernel*:

– $0_\mathcal{H} \in \ker(\varphi)$ by linearity of φ, thus $\ker(\varphi) \neq \varnothing$;

– since $\varphi \neq 0$, there exists at least one vector in $\mathcal{H}$ that is not nullified by φ, that is, $\ker(\varphi) \neq \mathcal{H}$;

– as we saw in Theorem 6.8, $\ker(\varphi)$ is always closed.

Thus, $\ker(\varphi)$ *is a closed proper subspace of* $\mathcal{H}$; based on this observation, Theorem 5.4 can be used to guarantee that $\ker(\varphi)^\perp \neq \{0_\mathcal{H}\}$, that is, there exists at least one $u \neq 0_\mathcal{H}$, $u \in \ker(\varphi)^\perp$.

Now, we note that since $\ker(\varphi) \cap \ker(\varphi)^\perp = \{0_\mathcal{H}\}$ and since $u \neq 0_\mathcal{H}$, $u \notin \ker(\varphi)$, for all $y \in \mathcal{H}$, the vector $z = y - \frac{\varphi(y)}{\varphi(u)} u$ is well defined.

$z \in \ker(\varphi)$, and by linearity, $\varphi(z) = \varphi(y - \frac{\varphi(y)}{\varphi(u)}u) = \varphi(y) - \frac{\varphi(y)}{\cancel{\varphi(u)}}\cancel{\varphi(u)} = 0$; in short:

$$\begin{cases} u \in \ker(\varphi)^\perp \\ z = y - \frac{\varphi(y)}{\varphi(u)} u \in \ker(\varphi) \end{cases}$$

hence:

$$0 = \langle z, u\rangle = \langle y - \frac{\varphi(y)}{\varphi(u)} u, u\rangle = \langle y, u\rangle - \langle \frac{\varphi(y)}{\varphi(u)} u, u\rangle = \langle y, u\rangle - \frac{\varphi(y)}{\varphi(u)} \|u\|^2$$

that is:

$$\varphi(y) = \frac{\varphi(u)}{\|u\|^2}\langle y, u\rangle = \langle y, \frac{\overline{\varphi(u)}}{\|u\|^2} u\rangle \qquad \forall y \in \mathcal{H}$$

Hence, for any vector $u \in \ker(\varphi)^\perp$, $u \neq 0_\mathcal{H}$, the vector $x = \frac{\overline{\varphi(u)}}{\|u\|^2} u$ is such that:

$$\varphi(y) = \langle y, x\rangle = T_x(y), \qquad \forall y \in \mathcal{H}$$

that is, $\varphi = T_x$. This proves that T is surjective and concludes the proof. □

The final step of the proof above actually demonstrates an even finer result: *the orthogonal complement of the kernel of a bounded linear function on a Hilbert space* $\mathcal{H}$ *is a straight line in* $\mathcal{H}$.

COROLLARY 6.1.– Let $\mathcal{H}$ be a Hilbert space and take $\varphi \in \mathcal{H}^*$, $\varphi \not\equiv \mathbf{0}$. Then $\ker(\varphi)^\perp$ is a one-dimensional vector subspace of $\mathcal{H}$, that is, $dim(\ker(\varphi)^\perp) = 1$. One generator of this space is the residual vector $x - P_{\ker \varphi} x$, where $x \in \mathcal{H}$ is such that $\varphi = T_x$ *via* the Riesz isomorphism.

PROOF.– In the final part of the proof of the Riesz representation theorem, we showed that if φ is not identically null functional, then for any given $u \in \ker(\varphi)^\perp$, $u \neq 0_\mathcal{H}$, $x = \frac{\overline{\varphi(u)}}{\|u\|^2} u$ is the vector in $\mathcal{H}$, which is identified with φ *via* the formula $\varphi = T_x$.

Reasoning by the absurd, if $\ker(\varphi)^\perp$ has a dimension greater than 1, then there exists at least one other generator, which we shall note $u' \neq u$, $u' \neq 0_\mathcal{H}$, $u' \in \ker(\varphi)^\perp$, where u and u' are linearly independent. Since $\ker(\varphi)^\perp$ is a vector space, the Gram-Schmidt algorithm can be applied to orthonormalize the pair (u, u') and obtain the pair $(\tilde{u}, \tilde{u}') \in \ker(\varphi)^\perp \times \ker(\varphi)^\perp$, $\|\tilde{u}\| = \|\tilde{u}'\| = 1$ and $\tilde{u} \perp \tilde{u}'$. We define the vectors:

$$x = \frac{\overline{\varphi(\tilde{u})}}{\|\tilde{u}\|^2} \tilde{u} = \overline{\varphi(\tilde{u})} \tilde{u}, \quad x' = \frac{\overline{\varphi(\tilde{u}')}}{\|\tilde{u}'\|^2} \tilde{u}' = \overline{\varphi(\tilde{u}')} \tilde{u}'$$

which are themselves orthogonal, so Pythagoras' theorem can be used to estimate the squared norm of their difference:

$$\|x - x'\|^2 = \|x + (-x')\|^2 = \|x\|^2 + \|x'\|^2 = |\varphi(\tilde{u})|^2 \|\tilde{u}\|^2 + |\varphi(\tilde{u}')|^2 \|\tilde{u}'\|^2 > 0$$

since $\varphi(\tilde{u})$, $\varphi(\tilde{u}')$ and the norms of $\tilde{u}$ and $\tilde{u}'$ are $\neq 0$. Consequently, $x \neq x'$, so we would have two different vectors in $\mathcal{H}$, x and x', associated with the same functional $\varphi \in \mathcal{H}^*$. This is incompatible with the injectivity of the Riesz map.

Furthermore, since $x = \frac{\overline{\varphi(u)}}{\|u\|^2} u$ and $u \in \ker(\varphi)^\perp$, $x \notin \ker \varphi$ and so Theorem 5.4 tells us that the residual vector of the orthogonal projection of x onto $\ker \varphi$, that is, $x - P_{\ker \varphi} x$, belongs to $\ker(\varphi)^\perp$. □

REMARK.– In light of this discussion, the inverse of the Riesz map can be expressed as:

$$\begin{aligned} T^{-1} : \mathcal{H}^* &\longrightarrow \mathcal{H} \\ \varphi &\longmapsto T^{-1}(\varphi) = x = \frac{\overline{\varphi(u)}}{\|u\|^2} u \end{aligned}$$

where $u \neq 0_\mathcal{H}$ is an arbitrary vector in $\ker(\varphi)^\perp$. Since $\dim(\ker(\varphi)^\perp) = 1$, in order to verify that this definition is well established, we must simply verify that if $k \in \mathbb{K}$, $k \neq 0$, then the vector x associated with φ *via* $u' = ku$ (as an arbitrary element of the one-dimensional subspace $\ker(\varphi)^\perp$) is the same:

$$x' = \frac{\overline{\varphi(u')}}{\|u'\|^2} u' = \frac{\bar{k}\, \overline{\varphi(u)}}{|k|^2 \|u\|^2} k u == \frac{\cancel{\bar{k} k}\, \overline{\varphi(u)}}{\cancel{|k|^2} \|u\|^2} u = \frac{\overline{\varphi(u)}}{\|u\|^2} u = x$$

hence the definition of T^{-1} does not depend on the choice of the vector $u \neq 0_\mathcal{H} \in \ker(\varphi)^\perp$.

6.4.1. *The scalar product induced on the dual of a Hilbert space*

In the context of the Riesz representation theorem, we saw that a Hilbert space $\mathcal{H}$ and its dual $\mathcal{H}^*$ can be identified as Banach spaces, since the isometry of the transformation T draws only on the norm of $\mathcal{H}$ and $\mathcal{H}^*$. It is possible to go even further, and identify these as Hilbert spaces.

The first step is to introduce an inner product on $\mathcal{H}^*$. This can be done using the Riesz isomorphism $T : \mathcal{H} \to \mathcal{H}^*$: any bounded linear functional of $\mathcal{H}^*$ is the image of a vector in $\mathcal{H}$ and, as we know the inner product of $\mathcal{H}$, there is no risk of ambiguity if we define the inner product on $\mathcal{H}^*$ as:

$$\langle \varphi, \psi \rangle_{\mathcal{H}^*} := \langle T^{-1}\varphi, T^{-1}\psi \rangle_{\mathcal{H}}, \qquad \forall \varphi, \psi \in \mathcal{H}^*$$

The fact that T preserves the norm guarantees that this definition of inner product will be compatible with the pre-existing Banach space structure on $\mathcal{H}^*$. If $\varphi = T_x$, that is, φ is the functional which can be identified with the image of the vector $x \in \mathcal{H}$ *via* T, then:

$$\|\varphi\|^2 = \langle T^{-1}(T_x), T^{-1}(T_x) \rangle = \langle x, x \rangle = \|x\|^2 = \|T_x\|^2$$

where the final equality is a consequence of the Riesz representation theorem.

The compatibility between the co-existing structures of inner product space and complete normed space implies that $\mathcal{H}^*$, equipped with the inner product induced by the Riesz isomorphism T, is itself a Hilbert space; thus, T becomes an (antilinear) isomorphism between the Hilbert spaces $\mathcal{H}$ and $\mathcal{H}^*$.

The Riesz representation theorem is one of the most important results of functional analysis. In the following two sections, we discuss an extension of this result (called the Lax-Milgram theorem) and an extremely significant consequence of Riesz's theorem: each operator in $\mathcal{B}(\mathcal{H})$ can be unambiguously associated with another operator, called its adjoint, which plays a fundamental role in the analysis of projection and unitary operators, among other things.

6.5. Bilinear forms, sesquilinear forms and associated quadratic forms

The concept of a quadratic form associated with a bilinear or sesquilinear form could have been introduced in Chapter 1. However, we have decided to discuss this subject here because the connection between bounded linear operators in Hilbert spaces and quadratic forms leads directly to the definition of the adjoint operator, which will be presented in section 6.6.

DEFINITION 6.8 (quadratic form).– *Let $\phi : V \times V \to \mathbb{R}$ (resp. $\phi : V \times V \to \mathbb{C}$) be a bilinear (resp. sesquilinear) form on the real (resp. complex) vector space V. The* *function* $\Phi : V \to \mathbb{R}$, *resp.* $\Phi : V \to \mathbb{C}$, *defined by restriction of* ϕ *on the diagonal of* $V \times V$, *that is:*

$$\Phi(x) := \phi(x, x),$$

is called the quadratic form associated with ϕ.

With the addition of positive-definiteness and symmetry (resp. conjugate symmetry) requirements, ϕ becomes an inner product $\langle\ ,\ \rangle$ and, in this case, $\Phi(x) = \langle x, x \rangle = \|x\|^2$ for all $v \in V$, that is, Φ is the square of the norm canonically associated with ϕ. This observation is the reason why Φ is known as the quadratic form.

Now, let us consider the concept of bounded forms.

DEFINITION 6.9.– *If* $(V, \|\ \|)$ *is a normed vector space, then the form* $\phi : V \times V \to \mathbb{K}$, *taken to be bilinear if* $\mathbb{K} = \mathbb{R}$ *and sesquilinear if* $\mathbb{K} = \mathbb{C}$, *is said to be bounded if there exists a constant* $m > 0$ *such that:*

$$|\phi(x, y)| \leqslant m\|x\|\|y\|, \qquad \forall x, y \in V$$

Where applicable, the norm of ϕ *is defined by the formula:*

$$\|\phi\| := \inf\{m > 0 \ : \ |\phi(x, y)| \leqslant m\|x\|\|y\|, \ \forall x, y \in V\}$$

As in the case of operators in $\mathcal{B}(\mathcal{H})$, the norm of ϕ can be rewritten in an equivalent, and highly useful, form:

$$\|\phi\| = \sup_{\|x\|=\|y\|=1} \phi(x, y)$$

giving us:

$$|\phi(x, y)| \leqslant \|\phi\| \, \|x\| \, \|y\| \, , \qquad \forall x, y \in \mathcal{H}$$

DEFINITION 6.10 (bounded quadratic forms and their norm).– *If* $(V, \|\ \|)$ *is a normed vector space, then the quadratic form* Φ *is said to be bounded if there exists a constant* $k > 0$ *such that:*

$$|\Phi(x)| \leqslant k\|x\|^2, \qquad \forall x \in V$$

The norm of a bounded quadratic form is defined by:

$$\|\Phi\| := \inf\{k > 0 \ : \ |\Phi(x)| \leqslant k\|x\|^2, \ \forall x \in V\}$$

As we saw with the norm of ϕ, the norm of Φ can be rewritten as:

$$\|\Phi\| := \sup_{\|x\|=1} |\Phi(x)|$$

giving us:

$$\|\Phi(x)\| \leqslant \|\Phi\|\|x\|^2, \qquad \forall x \in V \tag{6.17}$$

As in the case of inner products and their norms, the polarization formula can be used to completely describe a bilinear (sesquilinear) form *via* its associated quadratic form.

THEOREM 6.18.– Let ϕ be a bilinear (resp. sesquilinear) form on V and let Φ be its associated quadratic form. Then, for all $x, y \in V$:

$$4\phi(x, y) = \Phi(x + y) - \Phi(x - y)$$

respectively:

$$4\phi(x, y) = \Phi(x + y) - \Phi(x - y) + i\Phi(x + iy) - i\Phi(x - iy)$$

The proof is identical to that presented in section 1.2.1, where we saw that the bilinearity or sesquilinearity of the form ϕ is the only aspect required to prove the polarization formula.

The following result is an immediate corollary of the polarization formula, and gives a condition which is equivalent to that set out in Theorem 6.10 for bilinear or sesquilinear forms.

COROLLARY 6.2.– Let ϕ_1 and ϕ_2 be two bilinear or sesquilinear forms on V. Then:

$$\phi_1 = \phi_2 \iff \Phi_1 = \Phi_2, \text{ that is } \phi_1(x, y) = \phi_2(x, y)$$

$$\forall x, y \in V \iff \phi_1(x, x) = \phi_2(x, x) \ \forall x \in V$$

that is, the equality of the quadratic forms is necessary and sufficient to characterize the equality of the forms with which they are associated.

Now, let us consider an important consequence of this corollary.

THEOREM 6.19.– A sesquilinear form $\phi : V \times V \to \mathbb{C}$ is Hermitian if and only if its associated quadratic form Φ is real, that is if $\Phi(x) \in \mathbb{R} \ \forall x \in V$.

PROOF.– Let us prove these two implications.

$\boxed{\Longrightarrow}$:Let ϕ be Hermitian, that is, $\phi(x,y) = \overline{\phi(y,x)}$ $\forall x, y \in V$. Then:

$$\Phi(x) = \phi(x,x) = \overline{\phi(x,x)} = \overline{\Phi(x)}, \qquad \forall x \in V$$

that is, Φ is real.

$\boxed{\Longleftarrow}$: Now, taking $\Phi(x) = \overline{\Phi(x)}$, let us define a sesquilinear form $\psi : V \times V \to \mathbb{C}$ as follows: $\psi(x,y) = \overline{\phi(y,x)}$. If we can show that $\psi = \phi$, this will prove that ϕ is sesquilinear. To do this, we examine the quadratic form Ψ associated with ψ:

$$\Psi(x) = \overline{\phi(x,x)} = \overline{\Phi(x)} = \Phi(x), \qquad \forall x \in V$$

and, by Corollary 6.2, $\Psi = \Phi$ implies $\psi = \phi$. □

As a special case of the theorem just proven, if a sesquilinear form ϕ is positive, and thus real, it must necessarily be Hermitian. This consideration provides additional justification for the definition of complex inner product given in Chapter 1 as a sesquilinear positive-definite Hermitian form.

Theorem 6.20 relates to the relationship between the boundedness of a bilinear or sesquilinear form ϕ and that of its associated quadratic form.

THEOREM 6.20.– A bilinear or sesquilinear form ϕ on a normed vector space $(V, \| \ \|)$ is bounded if and only if the associated quadratic form Φ is bounded. Furthermore:

– if ϕ is real, then: $\|\phi\| = \|\Phi\|$;

– if ϕ is complex, then its norm is contained in the interval between the norm of Φ and its double: $\|\Phi\| \leqslant \|\phi\| \leqslant 2\|\Phi\|$.

PROOF.– We shall prove the first inequality by considering a real bilinear or complex sesquilinear form.

$\boxed{\|\Phi\| \leqslant \|\phi\|,\ \phi \text{ real or complex}}$: by definition we have:

$$\|\Phi\| = \sup_{\|x\|=1} |\Phi(x)| = \sup_{\|x\|=1} |\phi(x,x)| \underset{(*)}{\leqslant} \sup_{\|x\|=\|y\|=1} |\phi(x,y)| = \|\phi\|$$

where $(*)$ is due to the fact that the upper bound is calculated on a larger set of values. If ψ is bounded, then Φ is also bounded, and the first inequality is valid.

$\boxed{\|\phi\| \leqslant \|\Phi\|,\ \phi \text{ real bilinear}}$: now, taking Φ to be bounded, then, by the polarization formula, we have:

$$\begin{aligned}|\phi(x,y)| &= \tfrac{1}{4}|\Phi(x+y) - \Phi(x-y)| \underset{[6.17]}{\leqslant} \tfrac{1}{4}\|\phi\|(\|x+y\|^2 + \|x-y\|^2)\\ &\leqslant \tfrac{1}{4}\|\phi\|2(\|x\|^2 + \|y\|^2) = \tfrac{1}{2}\|\phi\|(\|x\|^2 + \|y\|^2)\end{aligned}$$

by applying the parallelogram formula $\|x+y\|^2 + \|x-y\|^2 = 2(\|x\|^2 + \|y\|^2)$. Hence:

$$\|\phi\| = \sup_{\|x\|=\|y\|=1} \frac{1}{2}|\phi(x,y)| \leqslant \sup_{\|x\|=\|y\|=1} \|\Phi\|(\|x\|^2 + \|y\|^2) = \|\Phi\|$$

Hence, a bounded Φ implies a bounded ϕ and it holds that $\|\varphi\| \leqslant \|\Phi\|$.

$\boxed{\|\phi\| \leqslant 2\|\Phi\|,\ \phi \text{ complex sesquilinear}}$: taking Φ to be bounded, using the polarization formula, we have:

$$\begin{aligned}|\phi(x,y)| &= \tfrac{1}{4}|\Phi(x+y) - \Phi(x-y) + i\Phi(x+iy) - i\Phi(x-iy)| \\ &\underset{[6.17]}{\leqslant} \tfrac{1}{4}\|\phi\|(\|x+y\|^2 + \|x-y\|^2 + \|x+iy\|^2 + \|x-iy\|^2)\end{aligned}$$

In this case, the parallelogram formula gives us: $\|x+iy\|^2 + \|x-iy\|^2 = 2(\|x\|^2 + \|iy\|^2) = 2(\|x\|^2 + |i|^2\|y\|^2) = 2(\|x\|^2 + \|y\|^2)$, thus $\|x+y\|^2 + \|x-y\|^2 + \|x+iy\|^2 + \|x-iy\|^2 = 4(\|x\|^2 + \|y\|^2)$ and so:

$$|\phi(x,y)| \leqslant \|\Phi\|(\|x\|^2 + \|y\|^2)$$

which implies:

$$\|\phi\| = \sup_{\|x\|=\|y\|=1} |\phi(x,y)| \leqslant \sup_{\|x\|=\|y\|=1} \|\Phi\|(\|x\|^2 + \|y\|^2) = 2\|\Phi\|$$

Thus, a bounded Φ implies that ϕ is bounded, and it holds that $\|\varphi\| \leqslant 2\|\Phi\|$. □

If a (complex) sesquilinear form ϕ is also Hermitian, then we know that its associated quadratic form Φ is real. The theorem proved above guarantees the equality of the norms of ϕ and Φ when ϕ is a real bilinear form (and thus Φ is also real). These considerations naturally lead to the idea that a Hermitian (complex) sesquilinear form might have a norm which coincides with that of its (real) quadratic form. The following result confirms that this is the case.

THEOREM 6.21.– If a sesquilinear form $\phi : V \times V \to \mathbb{C}$, where $(V, \| \ \|)$ is a normed vector space, is bounded and Hermitian, then $\|\phi\| = \|\Phi\|$.

PROOF.– We have seen that the inequality $\|\Phi\| \leqslant \|\phi\|$ is always valid, so we must simply show that the opposite inequality is valid when ϕ is Hermitian.

Once again, consider the polarization formula:

$$\phi(x,y) = \frac{1}{4}(\Phi(x+y) - \Phi(x-y) + i\Phi(x+iy) - i\Phi(x-iy))$$

Since Φ is real, the real part of both sides is:

$$\Re(\phi(x,y)) = \frac{1}{4}(\Phi(x+y) - \Phi(x-y))$$

Using equation [6.17] and the parallelogram formula, we can write the following inequality:

$$|\Re(\phi(x,y))| \leqslant \frac{1}{4}\|\Phi\|(\|x+y\|^2 + \|x-y\|^2) = \frac{1}{2}\|\Phi\|(\|x\|^2 + \|y\|^2) \qquad [6.18]$$

If $\theta \in [0, 2\pi)$ is such that $\phi(x,y) = |\phi(x,y)|e^{i\theta}$, then, by linearity on the first entry of ϕ:

$$0 \leqslant |\phi(x,y)| = e^{-i\theta}\phi(x,y) = \phi(e^{-i\theta}x, y)$$

that is, $\phi(e^{-i\theta}x, y)$ is a real positive quantity, and thus it coincides with its real part and also with its magnitude, hence $|\phi(x,y)| = |\Re(\phi(e^{-i\theta}x, y))|$. Using equation [6.18], we obtain:

$$\begin{aligned} \|\phi\| &= \sup_{\|x\|=\|y\|=1} |\phi(x,y)| = \sup_{\|x\|=\|y\|=1} |\Re(\phi(e^{-i\theta}x, y))| \\ &\leqslant \sup_{\|x\|=\|y\|=1} \frac{1}{2}\|\Phi\|(\|x\|^2 + \|y\|^2) = \|\Phi\| \end{aligned}$$

□

Now, let us consider the important relationship between bounded bilinear or sesquilinear forms defined on a Hilbert space $\mathcal{H}$ and the operators of $\mathcal{B}(\mathcal{H})$. The two results presented below are essential for defining the adjoint of a bounded operator.

THEOREM 6.22.– For all fixed $A \in \mathcal{B}(\mathcal{H})$, the bilinear form (if $\mathcal{H}$ is real) or sesquilinear form (if $\mathcal{H}$ is complex) ϕ_A on $\mathcal{H}$ defined by:

$$\phi_A(x,y) = \langle Ax, y\rangle \quad \text{ou} \quad \phi_A(x,y) = \langle x, Ay\rangle$$

is bounded, and it holds that $\|\phi_A\| = \|A\|$.

PROOF.– Consider the definition $\phi_A(x,y) = \langle Ax, y\rangle$: the proof for the other definition is similar. We observe that:

$$|\phi_A(x,y)| = |\langle Ax, y\rangle| \underset{\text{(Cauchy-Schwarz)}}{\leqslant} \|Ax\|\|y\| \underset{[6.11]}{\leqslant} \|A\|\|x\|\|y\|, \quad \forall x, y \in \mathcal{H}$$

hence ϕ_A is bounded and:

$$\|\phi_A\| = \sup_{\|x\|=\|y\|=1} |\phi_A(x,y)| \leqslant \sup_{\|x\|=\|y\|=1} \|A\|\|x\|\|y\| = \|A\|$$

thus $\|\phi_A\| \leqslant \|A\|$. Now, we shall prove the equality of the norms by demonstrating that $\|A\| \leqslant \|\phi_A\|$. First, we note that $\phi_A(x, Ax) = \langle Ax, Ax\rangle = \|Ax\|^2 \geqslant 0$, so it holds that $\|Ax\|^2 = |\phi_A(x, Ax)|$. Then, given that ϕ_A is bounded:

$$\|Ax\|^2 = |\phi_A(x, Ax)| \leqslant \|\phi_A\|\|x\|\|Ax\|$$

If $Ax \neq 0$, then both sides of the previous inequality can be divided by $\|Ax\|$, giving us $\|Ax\| \leqslant \|\phi_A\|\|x\|$. If $Ax = 0$, then the inequality $\|Ax\| \leqslant \|\phi_A\|\|x\|$ is written as $0 \leqslant \|\phi_A\|\|x\|$, which is trivially true. Thus, the inequality $\|Ax\| \leqslant \|\phi_A\|\|x\|$ holds with no constraints, and we can write:

$$\|A\| = \sup_{\|x\|=1} \|Ax\| \leqslant \sup_{\|x\|=1} \|\phi_A\|\|x\| = \|\phi_A\|$$

that is, $\|A\| \leqslant \|\phi_A\|$. □

If we write $\text{Bil}_b(\mathcal{H})$, resp. $\text{Sesq}_b(\mathcal{H})$, to denote the vector space (with respect to the pointwise defined linear operations) of the bounded bilinear, or sesquilinear, forms on $\mathcal{H}$, then the mapping:

$$\begin{array}{ccc} \mathcal{B}(\mathcal{H}) & \longrightarrow & \text{Bil}_b(h) \\ A & \longmapsto & \phi_A \end{array}, \quad \text{or :} \quad \begin{array}{ccc} \mathcal{B}(\mathcal{H}) & \longrightarrow & \text{Sesq}_b(h) \\ A & \longmapsto & \phi_A \end{array}$$

is an isometric inclusion.

The mapping defined by $\mathcal{B}(\mathcal{H}) \ni A \mapsto \phi_A \in \text{Bil}_b(\mathcal{H})$ is linear. The mapping given by $\mathcal{B}(\mathcal{H}) \ni A \mapsto \phi_A \in \text{Sesq}_b(\mathcal{H})$ is also linear if we define $\phi_A(x, y) = \langle Ax, y\rangle$, but it is antilinear if we define $\phi_A(x, y) = \langle x, Ay\rangle$.

By isometry, we can add a further characterization of the norm of an operator $A \in \mathcal{B}(\mathcal{H})$.

COROLLARY 6.3 (Fifth characterization of the norm of an operator in $\mathcal{B}(\mathcal{H})$).– For all $A \in \mathcal{B}(\mathcal{H})$ it holds that:

$$\boxed{\|A\| = \sup_{\|x\|=\|y\|=1} |\langle Ax, y\rangle|} \qquad [6.19]$$

The following result tells us that the application which associates a bounded operator with a bounded bilinear or sesquilinear form is not only an isometric inclusion, but is also surjective, that is any bounded bilinear or sesquilinear form on a Hilbert space $\mathcal{H}$ is defined by one, and only one, operator in $\mathcal{B}(\mathcal{H})$. In short, *the correspondence bounded operator* $\Longleftrightarrow$ *bounded bilinear or sesquilinear form is an isometric isomorphism.*

THEOREM 6.23.– Let $\mathcal{H}$ be a Hilbert space on $\mathbb{K} = \mathbb{R}, \mathbb{C}$. For any bounded bilinear form $\phi : \mathcal{H} \times \mathcal{H} \to \mathbb{K}$ if $\mathbb{K} = \mathbb{R}$, or any bounded sesquilinear form $\mathbb{K} = \mathbb{C}$, there exists a unique operator $B \in \mathcal{B}(\mathcal{H})$ such that $\phi = \phi_B$, that is:

$$\phi(x, y) = \langle Bx, y\rangle \qquad \text{or:} \qquad \phi(x, y) = \langle x, By\rangle, \qquad \forall x, y \in \mathcal{H}$$

PROOF.– For the purposes of our proof, let us consider the definition $\phi_B(x, y) = \langle x, By\rangle$; the proof for the other one is analogous.

Injectivity: Theorem 6.23 guarantees that, for all $B \in \mathcal{B}(\mathcal{H})$, $\phi_B(x, y) = \langle x, By\rangle$ is a bounded bilinear or sesquilinear form. Now, take $B_1, B_2 \in \mathcal{B}(\mathcal{H})$ such that $\phi = \phi_{B_1} = \phi_{B_2}$, that is $\phi(x, y) = \langle x, B_1 y\rangle = \langle x, B_2 y\rangle$ $\forall x, y \in \mathcal{H}$, then, by Theorem 6.10, $B_1 = B_2$.

Surjectivity: Taking an arbitrary fixed bilinear or sesquilinear form $\phi : \mathcal{H}\times\mathcal{H} \to \mathbb{K}$ and an arbitrary element $y \in \mathcal{H}$, the application :

$$\begin{aligned} \phi_y : \mathcal{H} &\longrightarrow \mathbb{K} \\ x &\longmapsto \phi_y(x) := \phi(x, y) \end{aligned}$$

is clearly a bounded linear functional on $\mathcal{H}$, that is, $\phi_y \in \mathcal{H}^*$. By the Riesz representation theorem, there exists one single element $\xi_y \in \mathcal{H}$ such that $\phi_y = T_{\xi_y} = T(\xi_y)$, where T is the Riesz isomorphism and $T_{\xi_y} \in \mathcal{H}^*$ is the Riesz representative of $\xi_y \in \mathcal{H}$, which has an action on any $x \in \mathcal{H}$ defined by $T_{\xi_y}(x) = \langle x, \xi_y\rangle$.

In short, $\forall x, y \in \mathcal{H}$, we know that $\phi(x, y) = \phi_y(x) = \langle x, \xi_y\rangle$, and thus the property of surjectivity will be proven if we can show that the application:

$$\begin{aligned} B : \mathcal{H} &\longrightarrow \mathcal{H} \\ y &\longmapsto By := \xi_y \end{aligned}$$

is a bounded linear operator on $\mathcal{H}$, since in this case it holds that $\phi(x, y) = \langle x, By\rangle$ $\forall x, y \in \mathcal{H}$.

Taking arbitrary $x, y_1, y_2 \in \mathcal{H}$ and $\alpha_1, \alpha_2 \in \mathbb{K}$, we have:

$$\begin{aligned} \langle x, \xi_{\alpha_1 y_1 + \alpha_2 y_2}\rangle &= \phi(x, \alpha_1 y_1 + \alpha_2 y_2) = \overline{\alpha_1}\phi(x, y_1) + \overline{\alpha_2}\phi(x, y_2) \\ &= \overline{\alpha_1}\langle x, \xi_{y_1}\rangle + \overline{\alpha_2}\langle x, \xi_{y_2}\rangle\langle x, \alpha_1\xi_{y_1}\rangle + \langle x, \alpha_2\xi_{y_2}\rangle \\ &- \langle x, \alpha_1\zeta_{y_1} + \alpha_2\zeta_{y_2}\rangle \end{aligned}$$

which shows the linearity of the correspondence $\mathcal{H} \ni y \mapsto \xi_y = By \in \mathcal{H}$.

To show that B is bounded, we observe that, since ϕ is bounded, there exists $k > 0$ such that :

$$|\langle x, By\rangle| = |\phi(x, y)| < k\|x\|\|y\| \qquad \forall x, y \in \mathcal{H}$$

Due to the arbitrary nature of x, we know that the inequality also holds when $x = Ay$, that is:

$$\|By\|^2 = |\langle By, By\rangle| < k\|By\|\|y\| \qquad \forall y \in \mathcal{H}$$

hence $\|By\| < k\|y\|$ $\forall y \in \mathcal{H}$ such that $By \neq 0$, and when $By = 0$ the inequality $\|By\| < k\|y\|$ is trivially true, so it holds that $\|By\| < k\|y\|$ $\forall y \in \mathcal{H}$, that is, B is bounded. □

6.5.1. *The Lax-Milgram theorem and its consequences*

In 1954, Peter Lax and Arthur Milgram presented a simple and elegant proof of a remarkable consequence of Theorem 6.23, generalizing the Riesz representation theorem to bilinear or sesquilinear forms.

One of the hypotheses required to obtain this result is defined below.

DEFINITION 6.11 (coercive or V-elliptical forms).– *Let $(V, \| \, \|)$ be a normed vector space. A bilinear or sesquilinear form $\phi : V \times V \to \mathbb{K}$, $\mathbb{K} = \mathbb{R}$ or $\mathbb{C}$ is said to be coercive or V-elliptical if there exists a constant $K > 0$ such that:*

$$\Phi(x) \geqslant K\|x\|^2, \qquad \forall x \in V$$

It is evident that *an inner product on V is a coercive form*, as, in this case, $\Phi(x) = \langle x, x\rangle = \|x\|^2 \geqslant K\|x\|^2$ $\forall x \in V$ with $0 < K \leqslant 1$.

The following example is less trivial. If $z \in \mathcal{C}([0,1], \mathbb{R})$ is such that $\min_{t\in[0,1]} z(t) > 0$, then the bilinear form:

$$\begin{aligned} \phi_z : L^2[0,1] \times L^2[0,1] &\longrightarrow \mathbb{R} \\ (x, y) &\longmapsto \phi_z(x,y) := \textstyle\int_0^1 x(t)\overline{y}(t)z(t)dt \end{aligned}$$

is coercive since:

$$\Phi_z(x) = \int_0^1 |x(t)|^2 z(t)dt \geqslant \int_0^1 |x(t)|^2 \min_{t\in[0,1]} z(t)dt$$

$$= \min_{t\in[0,1]} z(t) \int_0^1 |x(t)|^2 dt = K\|x\|^2$$

where $K = \min_{t\in[0,1]} z(t)$.

THEOREM 6.24 (Lax-Milgram theorem).– Let $\mathcal{H}$ be a Hilbert space on $\mathbb{K} = \mathbb{R}$ or $\mathbb{C}$ and let $\phi : \mathcal{H} \times \mathcal{H} \to \mathbb{K}$ be a bounded and coercive bilinear form if $\mathbb{K} = \mathbb{R}$, or a bounded and coercive sesquilinear form if $\mathbb{K} = \mathbb{C}$. Then, for any bounded functional $\varphi \in \mathcal{H}^*$, there exists a single element $u_\varphi \in \mathcal{H}$ such that:

$$\varphi(x) = \phi(x, u_\varphi), \qquad \forall x \in \mathcal{H}$$

PROOF.– We know from Theorem 6.23 that there exists an operator $A \in \mathcal{B}(\mathcal{H})$ such that:

$$\phi(x, y) = \langle x, Ay\rangle, \qquad \forall x, y \in \mathcal{H} \tag{6.20}$$

On the other side, the Riesz representation theorem guarantees that, for any bounded linear functional $\varphi \in \mathcal{H}^*$, there exists a single element $T^{-1}(\varphi) \in \mathcal{H}$, where T is the Riesz isomorphism, such that:

$$\varphi(x) = \langle x, T^{-1}(\varphi)\rangle, \qquad \forall x \in \mathcal{H} \tag{6.21}$$

The main idea behind the proof of this theorem is to compare equations [6.20] and [6.21]. If the operator $A : \mathcal{H} \to \mathcal{H}$ is an isomorphism, then there exists a unique element in $\mathcal{H}$, written as $u_\varphi \in \mathcal{H}$ since it depends on φ, which satisfies $Au_\varphi = T^{-1}(\varphi)$; then:

$$\varphi(x) \underset{[6.21]}{=} \langle x, T^{-1}(\varphi)\rangle \underset{(T^{-1}(\varphi)=Au_\varphi)}{=} \langle x, Au_\varphi\rangle \underset{[6.20]}{=} \phi(x, u_\varphi), \qquad \forall x \in \mathcal{H}$$

that is, the thesis of the Lax-Milgram theorem.

Now, let us show that A is an isomorphism. Injectivity is a simple consequence of coercivity:

$$0 \leqslant K\|x\|^2 \leqslant \Phi(x) = \phi(x, x) = \langle x, Ax\rangle \underset{\langle x, Ax\rangle \geqslant 0}{=} |\langle x, Ax\rangle| \underset{\text{Cauchy-Schwarz}}{\leqslant} \|x\|\,\|Ax\|$$

hence $\|x\| \leqslant \frac{1}{K}\|Ax\|$ for all $x \neq 0$, and for $x = 0$ the inequality is trivial, so it holds for all $x \in \mathcal{H}$. This implies the injectivity of A: given arbitrary $x_1, x_2 \in \mathcal{H}$, by linearity, the condition $Ax_1 = Ax_2$ implies that $A(x_1 - x_2) = 0$; then $\|x_1 - x_2\| \leqslant \frac{1}{K}\|A(x_1 - x_2)\| = 0$, that is $x_1 = x_2$.

The surjectivity of A, that is, the fact that $\mathrm{Im}(A) = \mathcal{H}$, is slightly harder to prove. The first argument used here reposes on the inequality proven above. More precisely, let $(x_n)_{n\in\mathbb{N}} \subset \mathcal{H}$ be an arbitrary sequence of elements in $\mathcal{H}$, then $(Ax_n)_{n\in\mathbb{N}} \subset \mathrm{Im}(A)$ is an arbitrary sequence of elements in the image of A. Now, let us suppose that this sequence is convergent in $\mathcal{H}$, that is there exists $y \in \mathcal{H}$ such that $\|Ax_n - y\| \underset{n\to+\infty}{\to} 0$. Notably, as a convergent sequence, $(Ax_n)_{n\in\mathbb{N}}$ is Cauchy, that is, for all $\varepsilon > 0$ $\exists N_\varepsilon \in \mathbb{N}$ such that $n, m \geqslant N_\varepsilon$ implies $\|Ax_n - Ax_m\| < \varepsilon$. It therefore also holds that $\|x_n - x_m\| \leqslant \frac{1}{K}\|Ax_n - Ax_m\| < \varepsilon$ for all $n, m \geqslant N_\varepsilon$, that is, if $(Ax_n)_{n\in\mathbb{N}}$ converges in $\mathcal{H}$, then $(x_n)_{n\in\mathbb{N}}$ is a Cauchy sequence in $\mathcal{H}$. Since $\mathcal{H}$ is complete, $(x_n)_{n\in\mathbb{N}}$ itself converges in $\mathcal{H}$, that is, there exists $x \in \mathcal{H}$ such that $\lim_{n\to+\infty} \|x_n - x\| = 0$. A is bounded and therefore continuous, so:

$$A\left(\lim_{n\to+\infty} \|x_n - x\|\right) = \lim_{n\to+\infty} \|Ax_n - Ax\| = 0$$

Furthermore, by the uniqueness of the limit in a metric space, we obtain $y = Ax \in \operatorname{Im}(A)$, that is $\operatorname{Im}(A)$ is a closed vector subspace of $\mathcal{H}$ as it contains the limits of all of its sequences.

The closure of $\operatorname{Im}(A)$ means that we can use Theorem 5.4. Reasoning by the absurd, if $\operatorname{Im}(A)$ is a proper vector subspace of $\mathcal{H}$, then there exists a non-zero vector $\xi \in \mathcal{H}\setminus \operatorname{Im}(A)$ that is orthogonal to $\operatorname{Im}(A)$, that is, $\langle \xi, Ay\rangle = 0\ \forall y \in \mathcal{H}$. Taking $y = \xi$, we obtain:

$$0 = \langle \xi, A\xi\rangle = \Phi(\xi) \underset{\text{coercivity}}{\geqslant} K\|\xi\|^2 > 0$$

since $\xi \neq 0$ and $K > 0$, which is absurd. □

The Lax-Milgram theorem is widely used in solving partial differential equations (PDE) expressed in variational form. Roughly speaking, this approach involves rewriting a PDE as the problem of minimization of a functional expressed by an integral, and looking for the so-called weak solution of the PDE, which takes the form of a minimizer of the functional.

In this type of approach, one almost immediate corollary of the Lax-Milgram theorem (often cited as an integral part of the theorem) proves extremely useful.

COROLLARY 6.4 (Lax-Milgram: symmetric case).– Take:

– $\mathcal{H}$ a real Hilbert space;

– $\varphi \in \mathcal{H}^*$;

– $\phi : \mathcal{H} \times \mathcal{H} \to \mathbb{R}$ a bounded, coercive and symmetrical bilinear form;

– $\Phi : \mathcal{H} \to \mathbb{R}$ the quadratic form associated with ϕ.

Then the vector $u_\varphi \in \mathcal{H}$ such that $\varphi(x) = \phi(x, u_\varphi)\ \forall x \in \mathcal{H}$ is the only element in $\mathcal{H}$ which minimizes the linear functional:

$$\begin{aligned} J_\varphi : \mathcal{H} &\longrightarrow \mathbb{R} \\ x &\longmapsto J_\varphi(x) := \tfrac{1}{2}\Phi(x) - \varphi(x) \end{aligned}$$

that is, $\exists!\ u_\varphi \in \mathcal{H}$ such that:

$$J_\varphi(u_\varphi) = \min_{x\in\mathcal{H}} J_\varphi(x) \quad\Longleftrightarrow\quad u_\varphi = \arg\min_{x\in\mathcal{H}} J_\varphi(x)$$

PROOF.– We perform a shift in a neighborhood of u_φ with an arbitrary vector $w \in \mathcal{H}$ and compute J_φ:

$$
\begin{aligned}
J_\varphi(u_\varphi + w) &= \tfrac{1}{2}\phi(u_\varphi + w, u_\varphi + w) - \varphi(u_\varphi + w) \\
&= \frac{1}{2}[\phi(u_\varphi, u_\varphi) + \phi(u_\varphi, w) + \phi(w, u_\varphi) + \phi(w, w)] - \varphi(u_\varphi) - \varphi(w) \\
&\underset{(\phi \text{ symmetric})}{=} \frac{1}{2}[\phi(u_\varphi, u_\varphi) + 2\phi(u_\varphi, w) + \phi(w, w)] - \varphi(u_\varphi) - \varphi(w)
\end{aligned}
$$

$$
= \frac{1}{2}\phi(u_\varphi, u_\varphi) - \varphi(u_\varphi) + \frac{1}{2}\phi(w, w) + \phi(w, u_\varphi) - \varphi(w)
$$

$$
\left(\frac{1}{2}\phi(u_\varphi, u_\varphi) - \varphi(u_\varphi) = J(u_\varphi) \text{ and } u_\varphi \text{ satisfies } \varphi(w) = \phi(w, u_\varphi)\right)
$$

$$
= J(u_\varphi) + \frac{1}{2}\Phi(w)
$$

$$
\underset{(\phi \text{ coercive})}{\geqslant} J(u_\varphi) + \frac{K}{2}\|w\|^2 \underset{(\frac{K}{2}\|w\|^2 \geqslant 0)}{\geqslant} J(u_\varphi)
$$

that is, $J(u_\varphi) \leqslant J_\varphi(u_\varphi + w)\ \forall w \in \mathcal{H}$, thus u_φ is the only minimizer of J. □

Since a real inner product is a bounded, coercive and symmetrical bilinear form, and since its associated quadratic form is the square of the norm (typically expressed in integral form), this result guarantees that, for any real functional of the form:

$$
J_\varphi(x) = \frac{1}{2}\|x\|^2 - \varphi(x)
$$

where $\varphi \in \mathcal{H}^*$, there exists a single minimizer $u_\varphi \in \mathcal{H}$.

The Lax-Milgram theorem and its symmetric variant form the basis for *finite element methods*, which are based around the following idea: If φ does not have a simple expression, then looking directly for the minimizer (weak solution of a PDE) u_φ in the whole Hilbert space $\mathcal{H}$ may be very complicated and time-consuming. In this case, the answer can be approximated by looking for a sequence $u_\varphi^{(n)}$ in $\mathcal{H}_n$, a finite-dimensional subspace of $\mathcal{H}$ (hence the term "finite elements").

In the case where ϕ is symmetrical and definite-positive, $u_\varphi^{(n)}$ is the orthogonal projection of u on $\mathcal{H}_n$ in the sense of the inner product defined by ϕ.

Once we have defined a basis $(h_i)_{i=1}^n$ (which is typically orthonormal) on $\mathcal{H}_n$, the problem consists of solving the linear system $Au_\varphi^{(n)} = b$, where$A_{ij} = \phi(h_j, h_i)$ and $b_i = \varphi(h_i)$.

Finally, note that the Lax-Milgram theorem presented here may be obtained as a corollary of a theorem proven by Lions and Stampacchia in 1967 in the context of variational inequalities.

6.6. The adjoint operator: presentation and properties

In this section, we shall examine a particularly important consequence of the Riesz representation theorem and of the results presented in section 6.5.1: the possibility of associating A with another operator, called "adjoint", which is of fundamental importance in functional analysis and its applications.

Consider an operator $A \in \mathcal{B}(\mathcal{H})$. By Theorem 6.22, the bilinear or sesquilinear form defined by $\phi(x, y) = \langle x, Ay \rangle$ is bounded. By Theorem 6.23, there exists a single bounded operator B such that, for all $x, y \in \mathcal{H}$, it holds that $\phi(x, y) = \langle Bx, y \rangle$, hence: $\langle x, Ay \rangle = \phi(x, y) = \langle Bx, y \rangle$. By the same arguments, if we select the alternative options in theorems 6.22 and 6.23, we obtain the equation: $\langle Ax, y \rangle = \phi(x, y) = \langle x, By \rangle \; \forall x, y \in \mathcal{H}$.

The operator B has a specific name and symbol.

DEFINITION 6.12.– *Take $A \in \mathcal{B}(\mathcal{H})$. The* adjoint operator *of A, noted*[3] $A^\dagger$, *is* $A^\dagger \in \mathcal{B}(\mathcal{H})$ *such that:*

$$\boxed{\langle A^\dagger x, y \rangle = \langle x, Ay \rangle} \quad \textit{and} \quad \boxed{\langle Ax, y \rangle = \langle x, A^\dagger y \rangle} \qquad \forall x, y \in \mathcal{H}$$

The application $^\dagger : \mathcal{B}(\mathcal{H}) \to \mathcal{B}(\mathcal{H})$, $A \mapsto A^\dagger$ *is known as* adjunction.

THEOREM 6.25.– The adjunction is an antilinear automorphism of $\mathcal{B}(\mathcal{H})$ and it verifies the following properties: for all $A, B \in \mathcal{B}(\mathcal{H})$ and $k \in \mathbb{K}$:

1) $(A + B)^\dagger = A^\dagger + B^\dagger$;

2) $(kA)^\dagger = \bar{k} A^\dagger$;

3) $(AB)^\dagger = B^\dagger A^\dagger$;

4) $(A)^{\dagger\dagger} = A$;

5) $\|A^\dagger A\| = \|A\|^2$, $\|AA^\dagger\| = \|A^\dagger\|^2$;

6) $\|A^\dagger\| = \|A\|$.

---

3 The origin of the symbol †, the dagger, reflects the close relationship between the adjoint operator $A^\dagger$ and the transposed or dual operator A^t. For more information, see Appendix 2. The symbol A^* is also widely used.

PROOF.–

1) and 2) are immediate consequences of the sesquilinearity of the complex inner product (if the Hilbert space is real, then evidently, $\bar{k} = k$, as a consequence of bilinearity).

3) $\langle (AB)^\dagger x, y\rangle = \langle x, ABy\rangle = \langle A^\dagger x, By\rangle = \langle B^\dagger A^\dagger x, y\rangle$ $\forall x, y \in \mathcal{H}$, hence property 3.

4) Since $A^{\dagger\dagger} = (A^\dagger)^\dagger$, $\langle A^{\dagger\dagger}x, y\rangle = \langle (A^\dagger)^\dagger x, y\rangle = \langle x, A^\dagger y\rangle = \overline{\langle A^\dagger y, x\rangle} = \overline{\langle y, Ax\rangle} = \langle Ax, y\rangle$ $\forall x, y \in \mathcal{H}$, hence property 4.

5) Let us begin by showing that $\|A\|^2 \leqslant \|A^\dagger A\|$: taking $x \in \mathcal{H}$, $\|x\| = 1$, we have:

$$\begin{aligned}\|Ax\|^2 &= \langle Ax, Ax\rangle = |\langle Ax, Ax\rangle| = |\langle A^\dagger Ax, x\rangle| \\ &\underset{\text{Cauchy-Schwarz}}{\leqslant} \|A^\dagger Ax\|\|x\| = \|A^\dagger Ax\| \\ &\leqslant \|A^\dagger A\|\|x\| = \|A^\dagger A\|\end{aligned}$$

thus, since $\|A\|^2 = \sup\limits_{\|x\|=1} \|Ax\|^2$, $\|A\|^2 \leqslant \|A^\dagger A\|$.

Now, let us show that $\|A^\dagger A\| \leqslant \|A\|^2$. We begin by noting that, for all $x, y \in \mathcal{H}$, $\|x\| = \|y\| = 1$, it holds that:

$$\begin{aligned}\Re\langle Ax, Ay\rangle &\leqslant \sqrt{(\Re\langle Ax, Ay\rangle)^2 + (\Im\mathrm{m}(\langle Ax, Ay\rangle)^2} = |\langle Ax, Ay\rangle| \\ &\underset{\text{Cauchy-Schwarz}}{\leqslant} \|Ax\|\|Ay\| \leqslant \|A\|\|x\|\|A\|\|y\| = \|A\|^2\end{aligned} \qquad [6.22]$$

If $\langle A^\dagger Ax, y\rangle = |\langle A^\dagger Ax, y\rangle| e^{i\vartheta}$, with ϑ the phase of $\langle A^\dagger Ax, y\rangle$, then :

$$\mathbb{R} \ni |\langle A^\dagger Ax, y\rangle| = e^{-i\vartheta}\langle A^\dagger Ax, y\rangle = \langle A^\dagger Ax, e^{i\vartheta} y\rangle$$

that is, $\langle A^\dagger Ax, e^{i\vartheta}y\rangle \in \mathbb{R}$ and thus $\langle A^\dagger Ax, e^{i\vartheta}y\rangle = \Re\langle A^\dagger Ax, e^{i\vartheta}y\rangle \underset{[6.22]}{\leqslant} \|A\|^2$, since $\|e^{i\vartheta}y\| = 1$. Using the fact that $\langle A^\dagger Ax, y\rangle = \langle A^\dagger Ax, e^{i\vartheta}y\rangle$, we can write:

$$|\langle A^\dagger Ax, y\rangle| \leqslant \|A\|^2, \quad \forall x, y \in \mathcal{H}, \ \|x\| = \|y\| = 1 \qquad [6.23]$$

Now, let us take an arbitrary $\xi \in \mathcal{H}$ and use this last inequality to estimate the norm of $A^\dagger A\xi$:

$$\begin{aligned}\|A^\dagger A\xi\| &= \frac{1}{\|A^\dagger A\xi\|}\frac{1}{\|\xi\|}\|A^\dagger A\xi\|^2\|\xi\| = \frac{1}{\|A^\dagger A\xi\|}\frac{1}{\|\xi\|}\langle A^\dagger A\xi, A^\dagger A\xi\rangle\|\xi\| \\ &= \frac{1}{\|A^\dagger A\xi\|}\frac{1}{\|\xi\|}|\langle A^\dagger A\xi, A^\dagger A\xi\rangle|\|\xi\| = \left|\left\langle A^\dagger A\frac{\xi}{\|\xi\|}, \frac{A^\dagger A\xi}{\|A^\dagger A\xi\|}\right\rangle\right| \|\xi\|\end{aligned}$$

Writing $x = \frac{\xi}{\|\xi\|}$ and $y = \frac{A^\dagger A\xi}{\|A^\dagger A\xi\|}$ and observing that these two vectors are unitary, we can use inequality [6.23] to write $\|A^\dagger A\xi\| \leqslant \|A\|^2\|\xi\|$, for all $\xi \in \mathcal{H}$, which implies that $\|A^\dagger A\| = \sup_{\|\xi\|=1}\|A^\dagger A\xi\| \leqslant \|A\|^2$. Hence $\|A^\dagger A\| = \|A\|^2$ $\forall A \in \mathcal{B}(\mathcal{H})$. If we write $B = A^\dagger$, then $B \in \mathcal{B}(\mathcal{H})$ and $\|B^\dagger B\| = \|B\|^2$, that is, $\|A^{\dagger\dagger}A^\dagger\| = \|A^\dagger\|^2$; moreover, $A^{\dagger\dagger} = A$, thus $\|AA^\dagger\| = \|A^\dagger\|^2$ for all $A \in \mathcal{B}(\mathcal{H})$.

6) On one side, we have:

$$\|A\|^2 = \|A^\dagger A\| \underset{[6.12]}{\leqslant} \|A^\dagger\|\|A\| \implies \frac{\|A\|^2}{\|A\|} \leqslant \frac{\|A^\dagger\|\|A\|}{\|A\|} \iff \|A\| \leqslant \|A^\dagger\|$$

and on the other side we have:

$$\|A^\dagger\|^2 = \|AA^\dagger\| \underset{[6.12]}{\leqslant} \|A\|\|A^\dagger\| \implies \frac{\|A^\dagger\|^2}{\|A^\dagger\|} \leqslant \frac{\|A\|\|A^\dagger\|}{\|A^\dagger\|} \iff \|A^\dagger\| \leqslant \|A\|$$

□

An immediate corollary of properties 1 and 6 is that *the adjunction* $^\dagger : \mathcal{B}(\mathcal{H}) \to \mathcal{B}(\mathcal{H})$ *is a continuous function*, in fact, if $(A_n)_{n\in\mathbb{N}} \subset \mathcal{B}(\mathcal{H})$ is a sequence in $\mathcal{B}(\mathcal{H})$ which converges toward $A \in \mathcal{B}(\mathcal{H})$, that is, $\|A_n - A\| \underset{n\to+\infty}{\longrightarrow} 0$, then:

$$\|A_n{}^\dagger - A^\dagger\| \underset{(1)}{=} \|(A_n - A)^\dagger\| \underset{(6)}{=} \|A_n - A\| \underset{n\to+\infty}{\longrightarrow} 0$$

The Banach algebra $\mathcal{B}(\mathcal{H})$ equipped with the adjunction operation becomes a $\mathbb{C}^*$-algebra, as formalized below.

DEFINITION 6.13 ($\mathbb{C}^*$-algebra).– *A Banach algebra* $\mathfrak{A}$ *is called a* $\mathbb{C}^*$*-algebra if it is possible to equip it with a map* $j : \mathfrak{A} \to \mathfrak{A}$ *such that,* $\forall a, b \in \mathfrak{A}$ *and* $\forall k \in \mathbb{C}$*:*

1) $j(a + b) = j(a) + j(b)$*;*

2) $j(ka) = \bar{k}j(a)$*;*

3) $j(ab) = j(b)j(a)$*;*

4) $j(j(a)) = a$*.*

$\mathbb{C}^*$-algebra theory is extremely important in functional analysis and its applications, in particular in quantum mechanics; however, a thorough discussion $\mathbb{C}^*$-algebras lies outside of the scope of this work.

Let us now consider the class of operators that are invariant with respect to adjunction.

DEFINITION 6.14 (self-adjoint or Hermitian operators).– $A \in \mathcal{B}(\mathcal{H})$ *is a self-adjoint (s.a.) or Hermitian operator if* $A^\dagger = A$*, that is, if:*

$$\langle Ax, y\rangle = \langle x, Ay\rangle, \qquad \forall x, y \in \mathcal{H}$$

To understand the importance of self-adjoint operators, we just quote the fact that *the physical observables in quantum mechanics are represented by self-adjoint operators on a Hilbert space.*

Two particularly remarkable self-adjoint operators are $A^\dagger A$ and $AA^\dagger$.

THEOREM 6.26.– Taking $A \in \mathcal{B}(\mathcal{H})$, then $A^\dagger A$ and $AA^\dagger$ are self-adjoint.

PROOF.– We simply apply the properties $(AB)^\dagger = B^\dagger A^\dagger$ and $A^{\dagger\dagger} = A$:

$$(A^\dagger A)^\dagger = A^\dagger A^{\dagger\dagger} = A^\dagger A$$

and:

$$(AA^\dagger)^\dagger = A^{\dagger\dagger} A^\dagger = AA^\dagger$$ □

The following theorem establishes the conditions under which the self-adjoint property is stable with respect to the operations of the algebra $\mathcal{B}(\mathcal{H})$. The following notation will be used: $\forall A, B \in \mathcal{B}(\mathcal{H})$, we define the operator $[A, B] := AB - BA$, called the *commutator between* A *and* B. A and B are said to commute if $[A, B] = 0$, the null operator; in this case, $AB = BA$.

THEOREM 6.27.– If $A, B \in \mathcal{B}(\mathcal{H})$, A, B are self-adjoint, then:

– $\alpha A + \beta B$ is self-adjoint if and only if $\alpha, \beta \in \mathbb{R}$;

– AB is self-adjoint if and only if $[A, B] = 0$.

PROOF.– The first property is a straightforward consequence of property 2 concerning the adjunction and sesquilinearity of the inner product. The second property is proven below.

$\boxed{\Longrightarrow}$: AB s.a., that is $AB = (AB)^\dagger$, then $(AB)^\dagger = B^\dagger A^\dagger \underset{A,B \text{ s.a.}}{=} BA$, thus $AB = BA$.

$\boxed{\Longleftarrow}$: $\forall x, y \in \mathcal{H}$ it holds that: $\langle ABx, y\rangle = \langle Bx, A^\dagger y\rangle = \langle x, B^\dagger A^\dagger y\rangle \underset{A,B \text{ s.a.}}{=} \langle x, BAy\rangle \underset{[A,B]=0}{=} \langle x, ABy\rangle$, hence $AB = (AB)^\dagger$. □

The following exercise makes use of many of the results presented above.

Exercise 6.4

Let $(u_n)_{n\in\mathbb{N}}$ be an orthonormal system in the Hilbert space $\mathcal{H}$, $(\lambda_n)_{n\in\mathbb{N}} \subset \mathbb{C}$ and $A : \mathcal{H} \to \mathcal{H}$:

$$Ax = \sum_{n\in\mathbb{N}} \lambda_n \langle x, u_n\rangle u_n, \qquad \forall x \in \mathcal{H}$$

1) Show that, if the sequence $(\lambda_n)_{n\in\mathbb{N}}$ is bounded, then $A \in \mathcal{B}(\mathcal{H})$.

2) Calculate the adjoint $A^\dagger$ of A. Using your result, deduce a necessary and sufficient condition for operator A to be anti-self-adjoint, that is, $A + A^\dagger = 0$.

3) For all $n \in \mathbb{N}$, consider the operator A_n defined by:

$$A_n x = \sum_{k=0}^{n} \lambda_k \langle x, u_k \rangle u_k$$

a) Calculate $A_n u_{n+1} - A u_{n+1}$ for all $n \in \mathbb{N}$. Using your result, deduce a necessary condition to have $A_n \underset{n\to+\infty}{\longrightarrow} A$ in $\mathcal{B}(\mathcal{H})$.

b) Supposing that $\lim_{n\to+\infty} \lambda_n = 0$, prove that $A_n \underset{n\to+\infty}{\longrightarrow} A$ in $\mathcal{B}(\mathcal{H})$.

Solution to Exercise 6.4

1) Since $(u_n)_{n\in\mathbb{N}}$ is an orthonormal system of a Hilbert space, the Fischer-Riesz theorem guarantees that Ax is well defined, that is, the convergence (in $\mathcal{H}$) of the series $\sum_{n\in\mathbb{N}} \lambda_n \langle x, u_n \rangle u_n$ is equivalent to the convergence (in $\mathbb{C}$) of the series $\sum_{n\in\mathbb{N}} |\lambda_n \langle x, u_n \rangle|^2 = \sum_{n\in\mathbb{N}} |\lambda_n|^2 |\langle x, u_n \rangle|^2$. If $(\lambda_n)_{n\in\mathbb{N}}$ is a bounded sequence, that is, $\sup_{n\in\mathbb{N}} |\lambda_n| = M < +\infty$, then $\sum_{n\in\mathbb{N}} |\lambda_n \langle x, u_n \rangle|^2 \leqslant M^2 \|x\|^2 < +\infty$, by Bessel's inequality [5.5].

Now, let us analyze the conditions under which A is bounded:

$$\|Ax\|^2 = \left\| \sum_{n\in\mathbb{N}} \lambda_n \langle x, u_n \rangle u_n \right\|^2 \underset{\text{(Pythagorean th.)}}{=} \sum_{n\in\mathbb{N}} \|\lambda_n \langle x, u_n \rangle u_n\|^2 = \sum_{n\in\mathbb{N}} |\lambda_n|^2 |\langle x, u_n \rangle|^2$$

The fact that $(\lambda_n)_{n\in\mathbb{N}}$ is bounded and Bessel's inequality can also be used to write $\|Ax\| \leqslant M\|x\|$, $\forall x \in \mathcal{H}$; furthermore, $\|A\| = \sup_{\|x\|=1} \|Ax\| \leqslant M$, showing that $A \in \mathcal{B}(\mathcal{H})$.

2) Taking $x, y \in \mathcal{H}$, we have:

$$\langle Ax, y \rangle = \sum_{n\in\mathbb{N}} \lambda_n \langle x, u_n \rangle \langle u_n, y \rangle = \left\langle x, \sum_{n\in\mathbb{N}} \overline{\lambda_n}\, \overline{\langle u_n, y \rangle} u_n \right\rangle$$

$$= \left\langle x, \sum_{n\in\mathbb{N}} \overline{\lambda_n} \langle y, u_n \rangle u_n \right\rangle$$

Hence $A^\dagger x = \sum_{n\in\mathbb{N}} \overline{\lambda_n} \langle x, u_n \rangle u_n$, $\forall x \in \mathcal{H}$.

By continuity, we can write $(A + A^\dagger)x = \sum_{n\in\mathbb{N}} (\lambda_n + \overline{\lambda_n}) \langle x, u_n \rangle u_n$, thus $A + A^\dagger$ is the zero operator if and only if $\lambda_n + \overline{\lambda_n} = 0$ $\forall n \in \mathbb{N}$. Writing $\lambda_n = a_n + ib_n$,

$a_n, b_n \in \mathbb{R}\ \forall n \in \mathbb{N}$, we see that the condition $\lambda_n + \overline{\lambda_n} = 0$ is equivalent to $a_n = -a_n$ $\forall n \in \mathbb{N}$, that is $a_n = 0\ \forall n \in \mathbb{N}$, whereas there are no constraints on b_n. Thus, A is anti-self-adjoint if and only if $\lambda_n \in i\mathbb{R}$ for all $n \in \mathbb{N}$, that is, λ_n is a pure imaginary sequence.

3) a) Using the following facts:

$$\|u_n\| = 1, \quad \langle u_k, u_{n+1}\rangle = 0 \qquad \forall n \in \mathbb{N},\ k \neq n+1$$

we deduce that $A_n u_{n+1} = 0$, $Au_{n+1} = \lambda_{n+1}u_{n+1}\ \forall n \in \mathbb{N}$. Then, by [6.10], we can write:

$$\|A_n - A\|_{\mathcal{B}(\mathcal{H})} \geqslant \|(A_n - A)u_{n+1}\|_{\mathcal{H}} = |\lambda_{n+1}| \qquad \forall n \in \mathbb{N}$$

that is, $\lim_{n\to+\infty} \lambda_n = 0$ is a necessary condition for $\lim_{n\to+\infty} \|A_n - A\|_{\mathcal{B}(\mathcal{H})} = 0$.

b) For all $n \in \mathbb{N}$ and $x \in \mathcal{H}$, we calculate:

$$\|A_n x - Ax\|^2 = \left\| \sum_{k=n+1}^{\infty} \lambda_n \langle x, u_k\rangle u_k \right\|^2 = \sum_{k=n+1}^{\infty} |\lambda_k|^2\, |\langle x, u_k\rangle|^2$$
$$\leqslant \left(\sup_{k\geqslant n+1} |\lambda_k| \right)^2 \|x\|^2$$

by Bessel's inequality. Thus, $\|A_n - A\|_{\mathcal{B}(\mathcal{H})} \leqslant \sup_{k\geqslant n+1} |\lambda_k|$, $\forall n \in \mathbb{N}$. Using the fact that $\lim_{n\to+\infty} \lambda_n = 0$, we obtain the required result:

$$\lim_{n\to+\infty} \|A_n - A\|_{\mathcal{B}(\mathcal{H})} = 0$$

□

Now, let us consider the norm of self-adjoint operators.

THEOREM 6.28.– Let $A \in \mathcal{B}(\mathcal{H})$ be a self-adjoint operator, then:

$$\|A\| = \sup_{\|x\|=1} |\langle Ax, x\rangle|$$

PROOF.– For simplicity's sake, we write:

$$s_A = \sup_{\|x\|=1} |\langle Ax, x\rangle|$$

$\boxed{s_A \leqslant \|A\|}$: by the Cauchy-Schwartz inequality, we have:

$$|\langle Ax, x\rangle| \leqslant \|x\|\|Ax\| \leqslant \|A\|\|x\|^2$$

thus:

$$s_A = \sup_{\|x\|=1} |\langle Ax, x\rangle| \leqslant \sup_{\|x\|=1} \|A\|\|x\|^2 = \|A\|$$

and so $s_A \leqslant \|A\|$.

$\boxed{\|A\| \leqslant s_A}$: using the fact that $\forall z \in \mathbb{C}$, $z + \bar{z} = 2\Re(z)$, we can write $\forall x, y \in \mathcal{H}$:

$$4\Re(\langle Ax, y\rangle) = 4\frac{1}{2}[\langle Ax, y\rangle + \overline{\langle Ax, y\rangle}] = 2[\langle Ax, y\rangle + \langle y, Ax\rangle]$$

By direct calculation, we can verify that the following equality holds true:

$$2[\langle Ax, y\rangle + \langle y, Ax\rangle] = \langle A(x+y), x+y\rangle - \langle A(x-y), x-y\rangle$$

thus:

$$4\Re(\langle Ax, y\rangle) = \langle A(x+y), x+y\rangle - \langle A(x-y), x-y\rangle$$

$$= \|x+y\|^2\langle A\frac{x+y}{\|x+y\|}, \frac{x+y}{\|x+y\|}\rangle - \|x-y\|^2\langle A\frac{x-y}{\|x-y\|}, \frac{x-y}{\|x-y\|}\rangle$$

$$\leqslant \|x+y\|^2 s_A + \|x-y\|^2 s_A = s_A(\|x+y\|^2 + \|x-y\|^2) \underset{[1.6]}{=} s_A 2(\|x\|^2 + \|y\|^2)$$

that is, $\Re(\langle Ax, y\rangle) \leqslant \frac{1}{2}s_A(\|x\|^2 + \|y\|^2)$ $\forall x, y \in \mathcal{H}$. Since the inequality is valid for any pair of vectors in $\mathcal{H}$, let us consider the pair x, z, where $z = e^{i\vartheta}y$ with arbitrary $\vartheta \in \mathbb{R}$. Given that $\|z\| = \|y\|$, the previous inequality becomes:

$$\Re(\langle Ax, e^{i\vartheta}y\rangle) \leqslant \frac{1}{2}s_A(\|x\|^2 + \|y\|^2) \qquad [6.24]$$

We can now use a similar argument to that used to prove property 5 in the case of adjunction: we write $\langle Ax, y\rangle = |\langle Ax, y\rangle|e^{i\vartheta}$, where ϑ is the phase of $\langle Ax, y\rangle$, then:

$$\mathbb{R} \ni |\langle Ax, y\rangle| = e^{-i\vartheta}\langle Ax, y\rangle = \langle Ax, e^{i\vartheta}y\rangle \underset{\text{(being real)}}{=} \Re(\langle Ax, e^{i\vartheta}y\rangle)$$

thus $|\langle Ax, y\rangle| = \Re(\langle Ax, e^{i\vartheta}y\rangle)$, and so inequality [6.24] may be rewritten as:

$$|\langle Ax, y\rangle| \leqslant \frac{1}{2}s_A(\|x\|^2 + \|y\|^2)$$

Now, let us introduce the vector $y = \frac{\|x\|}{\|Ax\|}Ax$ into this inequality. On the left side, we obtain:

$$|\langle Ax, y\rangle| = |\langle Ax, \frac{\|x\|}{\|Ax\|}Ax\rangle| = \frac{\|x\|}{\|Ax\|}|\langle Ax, Ax\rangle| = \frac{\|x\|}{\|Ax\|}\|Ax\|^2 = \|x\|\|Ax\|$$

while on the right side, we have:

$$\frac{1}{2}s_A(\|x\|^2 + \|y\|^2) = \frac{1}{2}s_A(\|x\|^2 + \frac{\|x\|^2}{\|Ax\|^2}\|Ax\|^2) = \frac{1}{2}s_A(\|x\|^2 + \|x\|^2) = s_A\|x\|^2$$

thus, $\forall x \in \mathcal{H}$, it holds that $\|x\|\|Ax\| \leqslant s_A\|x\|^2$, and if $x \neq 0_{\mathcal{H}}$, then $\|Ax\| \leqslant s_A\|x\|$, hence:

$$\|A\| = \sup_{x\neq 0_{\mathcal{H}}} \frac{\|Ax\|}{\|x\|} \leqslant s_A \sup_{x\neq 0_{\mathcal{H}}} \frac{\|x\|}{\|x\|} = s_A$$

and finally $\|A\| \leqslant s_A$. □

Theorem 6.29 points out a property of the adjoint operator which is *of fundamental importance in optimization.*

THEOREM 6.29.– Taking $A \in \mathcal{B}(\mathcal{H})$, then:

$$\ker(A) = (Im(A^\dagger))^\perp \quad and \quad \overline{Im(A^\dagger)} = (\ker(A))^\perp$$

thus:

$$\mathcal{H} = \ker(A) \oplus \overline{Im(A^\dagger)} \quad and \quad \mathcal{H} = \ker(A^\dagger) \oplus \overline{Im(A)}$$

PROOF.–

$\boxed{\ker(A) \subseteq (\mathrm{Im}(A^\dagger))^\perp}$: taking any $x \in \mathcal{H}$ and $y \in \ker(A)$, then $Ay = 0_{\mathcal{H}}$ and so we can write:

$$0 = \langle x, 0_{\mathcal{H}}\rangle = \langle x, Ay\rangle = \langle A^\dagger x, y\rangle$$

that is, $y \perp A^\dagger x \ \forall x \in \mathcal{H}$. Since $\mathrm{Im}(A^\dagger) = \{A^\dagger x, \ x \in \mathcal{H}\}$, this implies that $y \in (\mathrm{Im}(A^\dagger))^\perp$.

$\boxed{(\mathrm{Im}(A^\dagger))^\perp \subseteq \ker(A)}$: taking $y \in (\mathrm{Im}(A^\dagger))^\perp$, then $\langle A^\dagger x, y\rangle = 0 \ \forall x \in \mathcal{H}$, and since $\langle A^\dagger x, y\rangle = \langle x, Ay\rangle$, then $\langle x, Ay\rangle = 0 \ \forall x \in \mathcal{H}$, that is, $Ay = 0_{\mathcal{H}}$, therefore $y \in \ker(A)$.

Therefore: $\ker(A) = (\mathrm{Im}(A^\dagger))^\perp = (\overline{\mathrm{Im}(A^\dagger)})^\perp$. Taking the orthogonal complement again: $\ker(A)^\perp = (\overline{\mathrm{Im}(A^\dagger)})^{\perp\perp} = \overline{\mathrm{Im}(A^\dagger)}$. We see that it is essential to consider the closure of $\mathrm{Im}(A^\dagger)$, since $\ker(A)^\perp$ is a closed subspace in $\mathcal{H}$ and, in general, $\mathrm{Im}(A^\dagger)$ is not.

The orthogonal decompositions of $\mathcal{H}$ into a direct sum of subspaces are an immediate consequence of the orthogonal projection theorem. □

Finally, let us analyze the relationship between inversion and adjunction.

Recall that, as we saw in section 6.3, if V is a Banach space, then $\mathrm{GL}(V)$ is its general linear group, that is, the group of continuous bijective linear operators with continuous inverses.

THEOREM 6.30.– Let $\mathcal{H}$ be a Hilbert space and let $A \in \mathrm{GL}(\mathcal{H})$. Then $A^\dagger$ is invertible and:

1) it holds that:

$$\boxed{(A^\dagger)^{-1} = (A^{-1})^\dagger}$$

that is, for the operators in $\mathrm{GL}(\mathcal{H})$, inversion and adjunction commute: the inverse of the adjoint is the adjoint of the inverse;

2) if $A \in \mathrm{GL}(\mathcal{H})$ is self-adjoint, then A^{-1} is also self-adjoint.

PROOF.–

1) We need to prove that, for all $x \in \mathcal{H}$, $(A^{-1})^\dagger A^\dagger x = A^\dagger (A^{-1})^\dagger x = x$. To do this, let us consider, $\forall x, y \in \mathcal{H}$:

$$\langle y, (A^{-1})^\dagger A^\dagger x\rangle = \langle A^{-1}y, A^\dagger x\rangle = \langle AA^{-1}y, x\rangle = \langle y, x\rangle$$
$$\langle y, A^\dagger (A^{-1})^\dagger x\rangle = \langle Ay, (A^{-1})^\dagger x\rangle = \langle A^{-1}Ay, x\rangle = \langle y, x\rangle$$

hence, by [6.10] $(A^{-1})^\dagger A^\dagger x = A^\dagger (A^{-1})^\dagger x = x$.

2) An immediate consequence of property 1 is that if $A = A^\dagger$, then $A^{-1} = (A^{-1})^\dagger$. □

6.7. Orthogonal projection operators in a Hilbert space

We have already examined the concept of orthogonal projection in a Hilbert space $\mathcal{H}$. Here we wish to characterize orthogonal projections from an operator point of view. We will see that the adjoint operator will play a crucial role.

A clear, simple way of understanding projection operators (orthogonal or otherwise) is to imagine that we are in a finite-dimensional Euclidean space, for example $\mathbb{R}^2$, and to project a vector in the direction of another vector.

Now, imagine that we want to repeat the process, that is, we want to "project the projection"; clearly, this operation has no effect on the projected vector. This property is used to define the concept of projection itself[4].

4 Many authors refer to this as oblique projection to distinguish it from the more restrictive concept of orthogonal projection.

DEFINITION 6.15.– *An operator $A \in \mathcal{B}(\mathcal{H})$ is called a projector, or a projection operator, if it is idempotent, i.e. $A^2 = A$.*

The presence of an inner product in $\mathcal{H}$ allows us to target a specific projection: the orthogonal projection. The results presented in Chapter 5 showed that the completeness of $\mathcal{H}$ with respect to the topology generated by the inner product allows us to give two equivalent definitions of orthogonal projection.

DEFINITION 6.16.– *Let $\mathcal{H}$ be a Hilbert space and S a closed proper subspace of $\mathcal{H}$. The function:*

$$\begin{aligned} P_S : \mathcal{H} &\longrightarrow S \\ x &\longmapsto P_S(x) \end{aligned}$$

is the orthogonal projector on the subspace S if $\|x - P_S(x)\| = \inf_{y \in S} \|x - y\|$, that is, $P_S(x)$ is the element in S which minimizes the distance from $x \in \mathcal{H}$ with respect to the norm induced by the inner product of $\mathcal{H}$.

In an equivalent manner, if we consider the decomposition of $\mathcal{H}$: $\mathcal{H} = S \oplus S^{\perp}$ and note $x = x' + x''$, with $x \in \mathcal{H}$, $x' \in S$, $x'' \in S^{\perp}$, then the orthogonal projection operator P_S is defined via the formula $P_S(x) = x'$.

Let us consider an example of a projector. Take $\mathcal{H} = L^2[-a, a]$, with $a \in \mathbb{R}$ equipped with the Borel σ-algebra and the Lebesgue measure. The odd and even functions can be easily verified to be orthogonal for the inner product of $L^2[-a, a]$. We then have the following decomposition:

$$f(x) = \underbrace{\frac{f(x) + f(-x)}{2}}_{\text{even part}} + \underbrace{\frac{f(x) - f(-x)}{2}}_{\text{odd part}}, \qquad \forall x \in [-a, a],$$

thus the projector of $f \in L^2[-a, a]$ on the subspace $\mathcal{P} \subset L^2[-a, a]$ of even functions is defined by $P_{\mathcal{P}} f(x) = \frac{f(x)+f(-x)}{2}$, and the projector on the subspace $\mathcal{I} \subset L^2[-a, a]$ of odd functions is defined by $P_{\mathcal{I}} f(x) = \frac{f(x)-f(-x)}{2}$, $\forall x \in [-a, a]$.

Now, let us examine the properties of the operator P_S.

1) $\boxed{P_S|_S = id_S}$. This is trivial: the element $P_S(x) \in S$ which minimizes the distance to $x \in S$ is itself. In other words, if $x = x' \in S$, then $P_S(x) = P_S(x') = x'$.

2) $\boxed{P_S^2 = P_S}$ (*idempotence*). $\forall x \in \mathcal{H}$, we have $P_S^2(x) = P_S(\underbrace{P_S(x)}_{\in S,\text{ by definition}}) = P_S(x)$. Thus P_S is indeed a projector.

3) *P_S is a continuous linear operator*. Let $x_1 = x_1' + x_1'' \in \mathcal{H}$, with $x_1' \in S$ and $x_1'' \in S^\perp$, and let $x_2 = x_2' + x_2'' \in \mathcal{H}$, with $x_2' \in S$ and $x_2'' \in S^\perp$. For all $\alpha, \beta \in \mathbb{K}$ we have:

$$\alpha x_1 + \beta x_2 = \underbrace{\alpha x_1' + \beta x_2'}_{\in S} + \underbrace{\alpha x_1'' + \beta x_2''}_{\in S^\perp}$$

and thus:

$$\begin{aligned} P_S(\alpha x_1 + \beta x_2) &= \alpha x_1' + \beta x_2' \\ &= \alpha P_S(x_1) + \beta P_S(x_2) \end{aligned}$$

P_S is thus a linear operator. Its continuity can be proven by showing that it is bounded: taking any $x = x' + x'' \in \mathcal{H}$, with $x' \in S$, $x'' \in S^\perp$, then, by the Pythagorean theorem, $\|x\|^2 = \|x'\|^2 + \|x''\|^2$ and $\|P_S x\|^2 = \|x'\|^2 \leqslant \|x'\|^2 + \|x''\|^2 = \|x\|^2$, i.e. $\|P_S x\| \leqslant \|x\|\ \forall x \in \mathcal{H}$;

4) $\boxed{P_S = P_S^\dagger}$ (*self-adjoint*). To prove this, we use the projection theorem twice, on $x \in \mathcal{H}$ and on $y \in \mathcal{H}$: $x = x' + x''$, $y = y' + y''$, $x', y' \in S$, $x'', y'' \in S^\perp$:

$$\begin{aligned} &\langle P_S x, y\rangle = \langle x', y' + y''\rangle = \langle x', y'\rangle + \cancelto{0}{\langle x', y''\rangle} = \langle x - x'', y'\rangle \\ &= \langle x, y'\rangle - \cancelto{0}{\langle x'', y'\rangle} = \langle x, y'\rangle \end{aligned}$$

and since $y' = P_S y$, then: $\langle P_S x, y\rangle = \langle x, P_S y\rangle\ \forall x, y \in \mathcal{H}$;

5) *P_S is a 1-Lipschitz function*, that is:

$$\|P_S(x) - P_S(y)\| \leqslant \|x - y\|, \quad \forall x, y \in \mathcal{H}$$

We simply note that $\forall x, y \in \mathcal{H}$, the projection of $x - y$, i.e. $P_S(x - y)$, and the residual vector $(x - y) - P_S(x - y)$ are orthogonal, since one belongs to S and the other to $S^\perp$. Thus, we can apply the Pythagorean theorem and write:

$$\begin{aligned} \|x - y\|^2 &= \|(x - y) - P_S(x - y) + P_S(x - y)\|^2 \\ &= \|(x - y) - P_S(x - y)\|^2 + \|P_S(x - y)\|^2 \\ &\geqslant \|P_S(x - y)\|^2 \\ &= \|P_S(x) - P_S(y)\|^2 \end{aligned}$$

So, we obtain: $\|P_S(x) - P_S(y)\| \leqslant \|x - y\|, \forall x, y \in \mathcal{H}$.

6) The non-trivial orthogonal projectors have a unitary norm:

$$\boxed{\|P_S\| = \begin{cases} 1 \text{ if } S \neq \{0_{\mathcal{H}}\} \\ 0 \text{ if } S = \{0_{\mathcal{H}}\} \end{cases}}$$

If $S = \{0_{\mathcal{H}}\}$ then $P_S \equiv 0$, thus its norm is 0. Otherwise, by setting $y = 0$ in the 1-Lipschitz property of P_S we have that $\|P_S x\| \leqslant \|x\|\ \forall x \in \mathcal{H}$, i.e. $\frac{\|P_S x\|}{\|x\|} \leqslant 1$

$\forall x \in \mathcal{H} \backslash \{0_{\mathcal{H}}\}$. Furthermore, if $S \neq \{0_{\mathcal{H}}\}$, then there exists $\bar{x} \in S$, $\bar{x} \neq 0_{\mathcal{H}}$ and $P_S \bar{x} = \bar{x}$, i.e. $\|P_S \bar{x}\| = \|\bar{x}\|$, i.e. $\frac{\|P_S \bar{x}\|}{\|\bar{x}\|} = 1$. Then $\|P_S\| = \sup_{x \in \mathcal{H},\, x \neq 0} \frac{\|P_S x\|}{\|x\|} = 1$.

7) $\boxed{\mathrm{Im}(P_S) = S}$. This is obvious, by definition of the projection operator.

8) $\boxed{\ker P_S = S^\perp}$. We must show the double inclusion:

$\underline{x \in \ker P_S} \implies P_S(x) = 0$, so, for all $y \in S$ it holds that $0 = \langle P_S(x), y \rangle = \langle x, P_S^\dagger(y) \rangle = \langle x, y \rangle$ since $P_S^\dagger = P_S$ and $P_S(y) = y$, thus $\underline{x \in S^\perp}$.

$\underline{x \in S^\perp} \implies x = 0 + x = P_S(x) + P_{S^\perp}(x)$, by uniqueness of the orthogonal decomposition, thus $P_S(x) = 0$ and then $\underline{x \in \ker P_S}$.

9) One immediate consequence of the two previous properties and the projection theorem is that:

$$\boxed{\mathcal{H} = \mathrm{Im}(P_S) \oplus \ker(P_S)} \qquad \forall S \text{ closed subspace of } \mathcal{H}.$$

10) $\boxed{P_S + P_{S^\perp} = id_{\mathcal{H}}}$ (*decomposition of the identity*). For any $x \in \mathcal{H}$, we always have the decomposition $x = x' + x''$ with $x' = P_S(x)$ and $x'' = P_{S^\perp}(x)$. We thus have $P_S(x) + P_{S^\perp}(x) = (P_S + P_{S^\perp})(x) = x' + x'' = x$, $\forall x \in \mathcal{H}$. An immediate consequence is that:

$$\boxed{P_{S^\perp} = id_{\mathcal{H}} - P_S} \qquad \forall S \text{ closed subspace of } \mathcal{H}.$$

$P_{S^\perp}$ is also called *complementary projector* and it is denoted with ${P_S}^\perp$.

For all $x \in \mathcal{H}$, the residual vector of the projection of x on S is obtained *via* $P_{S^\perp} = (id_{\mathcal{H}} - P_S)(x) = x - P_S(x)$;

11) *Characterization of the projection subspace*:

$$\boxed{S = \{x \in \mathcal{H} : P_S(x) = x\} = \{x \in \mathcal{H} : \|P_S(x)\| = \|x\|\}}$$

that is, the elements of S are the fixed points of P_S in $\mathcal{H}$, which may themselves be characterized as elements of $\mathcal{H}$ which have a norm equal to that of their projection on S. Let us prove that $S = \{x \in \mathcal{H} : P_S(x) = x\}$: an element of S is a point of $\mathcal{H}$ on which P_S acts as the identity, vice-versa, if $x \in \mathcal{H}$ satisfies $P_S(x) = x$ then, by applying P_S to both sides we get $P_S^2(x) = P_S(x)$, but then, thanks to the idempotence of P_S, $x = P_S(x) \in S$. Let us now check that $P_S(x) = x \iff \|P_S(x)\| = \|x\|$:

$\boxed{\implies}$: evidently, $P_S(x) = x \implies \|P_S(x)\| = \|x\|$;

$\boxed{\impliedby}$: again, taking $\mathcal{H} \ni x = x' + x''$ with $P_S(x) = x'$, we have:

$$\|P_S(x)\| = \|x\| \implies \|P_S(x)\|^2 = \|x\|^2 \implies \|x'\|^2 = \|x' + x''\|^2$$
$$\implies \|x'\|^2 = \|x'\|^2 + \|x''\|^2 \text{ (since } x' \perp x'')$$

$$\iff \|x''\|^2 = 0$$
$$\iff x'' = 0 \text{ (property of the norm)}$$
$$\iff x = x',$$

we thus have $\|P_S(x)\| = \|x\| \implies x = x' = P_S(x)$.

12) $\boxed{\langle P_S x, x\rangle = \langle x, P_S x\rangle = \|P_S x\|^2}$ for all $x \in \mathcal{H}$. The first equality is simply a consequence of the fact that P_S is self-adjoint, then, by idempotence, $P_S^2 = P_S P_S = P_S$, so $\langle P_S^2 x, x\rangle = \langle P_S x, P_S x\rangle = \|P_S x\|^2$.

Two of the properties proven above characterize bounded linear operators as orthogonal projectors.

THEOREM 6.31 ("Algebraic" characterization of orthogonal projectors).– Taking $A \in \mathcal{B}(\mathcal{H})$, the following statements are equivalent:

1) A is an orthogonal projector;

2) $A^\dagger A = A$;

3) $A^\dagger A = A^\dagger$;

4) A is self-adjoint and idempotent, i.e. $A^\dagger = A$ and $A^2 = A$.

PROOF.– The theorem will be proven by the logical loop $1) \implies 2) \implies 3) \implies 4) \implies 1)$.

$\boxed{1) \implies 2)}$: A is an orthogonal projector, hence $A^\dagger = A$ and $A^2 = A$, then $A^\dagger A = AA = A^2 = A$.

$\boxed{2) \implies 3)}$: if $A^\dagger A = A$, then $(A^\dagger A)^\dagger = A^\dagger$, but we know that $A^\dagger A$ is self-adjoint, thus $(A^\dagger A)^\dagger = A^\dagger A = A^\dagger$.

$\boxed{3) \implies 4)}$: if $A^\dagger A = A^\dagger$, then $(A^\dagger A)^\dagger = A^{\dagger\dagger} = A$; moreover, $A^\dagger A$ is self-adjoint, hence $(A^\dagger A)^\dagger = A^\dagger A = A$. By hypothesis, $A^\dagger A = A^\dagger$, so $A^\dagger = A$, that is, A is self-adjoint. Reusing the starting hypothesis $A^\dagger A = A^\dagger$, the fact that A is self-adjoint implies that $A^2 = A$, that is, A is idempotent.

$\boxed{4) \implies 1)}$: let $A \in \mathcal{B}(\mathcal{H})$ be self-adjoint and idempotent. We wish to show that A is an orthogonal projector. By definition, an orthogonal projector projects onto a closed vector subspace, so we first need to show that $\text{Im}(A)$, the subspace which is intended to be the "site" of the projection, is closed, given the hypotheses of the theorem.

We can show that *the continuity and idempotence of a linear operator A imply the closure of its image*; this is remarkable, since the relationship between the concept of closure of a vector subspace and idempotence is far from obvious.

Let $(x_n)_{n\in\mathbb{N}} \subset \mathrm{Im}(A)$ be a sequence converging to $x_0 \in \mathcal{H}$. We wish to show that $x_0 \in \mathrm{Im}(A)$. Since each $x_n \in \mathrm{Im}(A)$, then, $\forall n \in \mathbb{N}$, there exists $\xi_n \in \mathcal{H}$ such that $x_n = A\xi_n$ and we can thus rewrite $x_n \xrightarrow[n\to+\infty]{} x_0$ as $A\xi_n \xrightarrow[n\to+\infty]{} x_0$. The continuity of A implies that $A^2\xi_n \xrightarrow[n\to+\infty]{} Ax_0$, but $A^2 = A$, hence $A\xi_n \xrightarrow[n\to+\infty]{} Ax_0$. We then have $A\xi_n \xrightarrow[n\to+\infty]{} Ax_0$ and $A\xi_n \xrightarrow[n\to+\infty]{} x_0$, and the uniqueness of the limit implies that $Ax_0 = x_0$, i.e. $x_0 \in \mathrm{Im}(A)$, thus $\mathrm{Im}(A)$ is closed.

In this case, property $A^\dagger = A$ is used alongside idempotence to show that A projects in an orthogonal manner. First, let us write an *orthogonal* decomposition of $\mathcal{H}$ with respect to $\mathrm{Im}(A)$: for all $x \in \mathcal{H}$, let us consider $x = Ax + (x - Ax)$, then $Ax \in \mathrm{Im}(A)$ by definition. We need to show that $x - Ax$ is orthogonal to any vector of the form $A\xi$, $\xi \in \mathcal{H}$:

$$\langle A\xi, x - Ax\rangle = \langle \xi, A^\dagger(x - Ax)\rangle \underset{A\text{ s.a.}}{=} \langle \xi, A(x - Ax)\rangle$$

$$= \langle \xi, Ax - A^2x\rangle \underset{A^2=A}{=} \langle \xi, Ax - Ax\rangle = 0$$

thus $x - Ax \in \mathrm{Im}(A)^\perp$ and then $A = P_{\mathrm{Im}(A)}$ by the orthogonal projection theorem. □

Exercise 6.5

Let $E \subset \ell^2(\mathbb{N}, \mathbb{C})$ be the set:

$$E = \{x = (x_n)_{n\in\mathbb{N}} \in \ell^2(\mathbb{N}, \mathbb{C}) \ : \ x_0 + x_1 = 0\}$$

1) Show that E is a closed vector subspace of $\ell^2(\mathbb{N}, \mathbb{C})$.

2) Provide an explicit description of $E^\perp$ and determine the orthogonal projection operator $P_E : \ell^2(\mathbb{N}, \mathbb{C}) \to E$ on E. Determine $\|P_E\|$ and $P_E^\dagger$.

3) Let $x = (x_n)_{n\in\mathbb{N}} \in \ell^2(\mathbb{N}, \mathbb{C})$ such that:

$$x_n = \begin{cases} 1 & \text{if } n = 0 \\ 0 & \text{otherwise} \end{cases}$$

Calculate the distance between x and the subspace E, i.e. $\delta = \inf_{y\in E} \{\|x - y\|\}$.

4) Let $A : \ell^2(\mathbb{N}, \mathbb{C}) \to \ell^2(\mathbb{N}, \mathbb{C})$ be the operator defined by:

$$(A(x_0, x_1, x_2, \dots))_n = \begin{cases} -x_1 & \text{if } n = 0 \\ x_n & \text{otherwise} \end{cases}$$

Determine $\|A\|$ and $A^\dagger$. (Hint: calculate $A(0, 1, 0, 0, ...)$).

5) Show that $A^2 = A$ and determine $\operatorname{Im}(A)$. Is A an orthogonal projector?

Solution to Exercise 6.5

1) We begin by showing that E is a vector subspace of $\ell^2(\mathbb{N}, \mathbb{C})$: let us consider any $\lambda \in \mathbb{C}$ and arbitrary $x, y \in E$. Then, given that the linear structure of $\ell^2(\mathbb{N}, \mathbb{C})$ is defined pointwise, [6.24] tells us that:

$$z := \lambda x + y = (\lambda x_n)_{n\in\mathbb{N}} + (y_n)_{n\in\mathbb{N}} = (\lambda x_n + y_n)_{n\in\mathbb{N}}$$

thus $z_0 = \lambda x_0 + y_0$ and $z_1 = \lambda x_1 + y_1$, and then $z_0 + z_1 = \lambda(x_0 + x_1) + (y_0 + y_1) = \lambda \cdot 0 + 0 = 0$ since $x, y \in E$, showing that E is stable with respect to the linear combinations of its elements.

We can show that E is closed using a technique which is particularly useful in the context of constraints as $x_0 + x_1 = 0$. This approach consists of *establishing an identity between the constraint and the condition defining the kernel of a continuous linear operator between normed vector spaces*, which we know from Theorem 6.8 to be a closed vector subspace of the operator domain. In our case, it is easy to identify the sum of the projection operators on the first and second components, that is, $A := P_0 + P_1 : \ell^2(\mathbb{N}, \mathbb{C}) \to \mathbb{C}$, $A(x) := P_0(x) + P_1(x) = x_0 + x_1$, with the continuous linear operator (insofar as it is a sum of continuous linear operators) between two Hilbert spaces such that $\ker(A) = \{x \in \ell^2(\mathbb{N}, \mathbb{C}) \;:\; Ax = 0 \iff x_0 + x_1 = 0\} = E$, demonstrating the closure of E.

2) Using the constraint $x_0 + x_1 = 0$, a sequence $x \in E$ can be written as $(x_0, -x_0, x_2, x_3, \dots)$, of course by respecting the fact that $x \in \ell^2(\mathbb{N}, \mathbb{C})$. This implies that the canonical Hilbert basis of $\ell^2(\mathbb{N}, \mathbb{C})$, that is, $e = ((1, 0, 0, \dots), (0, 1, 0, \dots), \dots)$, can be used to construct a Hilbert basis of E as:

$$\tilde{e} := ((1, -1, 0, \dots), (-1, 1, 0, \dots), (0, 0, 1, 0, \dots), \dots)$$

thus: $E^\perp = \{y \in \ell^2(\mathbb{N}, \mathbb{C}) \;:\; \langle y, \tilde{e}_n \rangle = 0 \; \forall n \in \mathbb{N}\}$.

Taking $y = (y_0, y_1, y_2, \dots) \in \ell^2(\mathbb{N}, \mathbb{C})$, then:

- $n = 0$: $\langle y, \tilde{e}_0 \rangle = y_0 - y_1 + 0 + \cdots = y_0 - y_1$ null if and only if $y_0 = y_1$;

- $n = 1$: $\langle y, \tilde{e}_1 \rangle = y_0 - y_1 + 0 + \cdots = -y_0 + y_1$ null if and only if $y_0 = y_1$, as in the case where $n = 0$;

- $n = 2$: $\langle y, \tilde{e}_2 \rangle = 0 + 0 + y_2 + 0 + \cdots = y_2$ null if and only if $y_2 = 0$.

Evidently, for all $n \geqslant 2$, $\langle y, \tilde{e}_n \rangle = y_n$, which is null if and only if $y_n = 0$. Thus, the only vector $y \in \ell^2(\mathbb{N}, \mathbb{C})$ which is orthogonal to all elements in the Hilbert basis $\tilde{e}$ of E is $y = (y_0, y_0, 0, \dots)$, that is:

$$E^\perp = \{(y, y, 0, 0, \dots),\ y \in \mathbb{C}\}$$

The orthogonal projection operator on E can be determined using the projection theorem: $\ell^2(\mathbb{N}, \mathbb{C}) = E \oplus E^\perp$. We decompose the arbitrary vector $z = (z_n)_{n \in \mathbb{N}} \in \ell^2(\mathbb{N}, \mathbb{C})$ into a sum of two vectors, one belonging to E and the other to $E^\perp$. This is done by noting that, given $z = (z_0, z_1, z_2, \dots)$, $z \in E$ if the first two components are the inverse of one another, and $z \in E^\perp$ if the first two components are equal and are null from the third position onward, then:

$$\begin{aligned} z &= (z_0, z_1, z_2, z_3, \dots) = (a, -a, z_2, z_3, \dots) + (b, b, 0, 0, \dots) \\ &= (a + b, b - a, z_2, z_3, \dots) \end{aligned}$$

which implies the system of constraints:

$$\begin{cases} a + b = z_0 \\ b - a = z_1 \end{cases}$$

solved by $a = (z_0 - z_1)/2$ and $b = (z_0 + z_1)/2$, that is:

$$(z_0, z_1, z_2, \dots) = \left(\frac{z_0 - z_1}{2}, -\frac{z_0 - z_1}{2}, z_2, \dots\right) + \left(\frac{z_0 + z_1}{2}, \frac{z_0 + z_1}{2}, 0, 0, \dots\right)$$

with the first vector in E and the second in $E^\perp$; thus:

$$P_E(z_0, z_1, z_2, \dots) = \left(\frac{z_0 - z_1}{2}, -\frac{z_0 - z_1}{2}, z_2, \dots\right)$$

is the explicit expression of the orthogonal projector on E. Finally, without carrying out a single calculation, we can state that P_E has unit norm, $\|P_E\| = 1$, given that it is a non-trivial orthogonal projector, and also that $P_E^\dagger = P_E$, as orthogonal projectors are self-adjoint.

3) Let $x = (x_n)_{n \in \mathbb{N}}$ be the element in $\ell^2(\mathbb{N}, \mathbb{C})$ such that:

$$x_n = \begin{cases} 1 & \text{if } n = 0 \\ 0 & \text{otherwise} \end{cases}$$

Since E is a closed vector subspace of $\ell^2(\mathbb{N}, \mathbb{C})$, the distance between x and E is well defined thanks to the projection theorem. $P_E(x)$ represents the vector in E which is the closest to x; therefore; this distance is equal to $\delta = \|x - P_E(x)\|_{\ell^2}$:

$$P_E(x) = P_E((1, 0, \dots)) = \left(\frac{1 - 0}{2}, -\frac{1 - 0}{2}, 0, \dots\right) = \left(\frac{1}{2}, -\frac{1}{2}, 0, \dots\right)$$

Then :

$$\delta = \left\|(1,0,\dots) - \left(\frac{1}{2},-\frac{1}{2},0,\dots\right)\right\|_{\ell^2} = \left\|\left(\frac{1}{2},\frac{1}{2},0,\dots\right)\right\|_{\ell^2}$$
$$= \sqrt{\left(\frac{1}{2}\right)^2+\left(\frac{1}{2}\right)^2} = \frac{1}{\sqrt{2}}$$

4) First, we note that x_0 plays no part in the action of A, thus:

$$A(x_0,x_1,x_2,\dots) = A(y_0,x_1,x_2,\dots) = \begin{cases} -x_1 & \text{if } n=0 \\ x_n & \text{otherwise} \end{cases} \qquad \forall y_0 \in \mathbb{C}$$

Notably, this holds true for $y_0 = 0$, so we can limit the action of A on the elements of $\ell^2(\mathbb{N},\mathbb{C})$ of the form $x = (0,x_1,x_2,\dots)$. Using this specification, by direct calculation, we obtain:

$$\begin{aligned}\|Ax\|_{\ell^2} &= \sqrt{(-x_1)^2+x_1^2+x_2^2+\dots} = \sqrt{2x_1^2+x_2^2+\dots} \leqslant \sqrt{2x_1^2+2x_2^2+\dots} \\ &\leqslant \sqrt{2}\sqrt{0^2+x_1^2+x_2^2+\dots} = \sqrt{2}\|x\|_{\ell^2}\end{aligned}$$

With this majorization, the definition of the operator norm from equation [6.3] becomes:

$$\|A\| = \inf\{0 < c \leqslant \sqrt{2} \ : \ \|Ax\|_{\ell^2} \leqslant c\|x\|_{\ell^2} \ \ \forall x = (0,x_1,x_2,\dots) \in \ell^2(\mathbb{N},\mathbb{C})\}$$

The inf is the sup of the minimizer set; thus, if we can identify a vector $x \in \ell^2(\mathbb{N},\mathbb{C})$ for which $\|Ax\|_{\ell^2} = \sqrt{2}$, then the norm of A must be $\sqrt{2}$. Taking the hint given in the question, we calculate:

$$\|A(0,1,0,\dots)\|_{\ell^2} = \|(-1,1,0,\dots)\|_{\ell^2} = \sqrt{(-1)^2+1^2+0^2+\dots} = \sqrt{2}$$

and then $\|A\| = \sqrt{2}$.

Now, let us determine $A^\dagger$. For all $x,y \in \ell^2(\mathbb{N},\mathbb{C})$ (in this case, x is not necessarily of the form $(0,x_1,x_2,\dots)$) it holds that:

$$\begin{aligned}&\langle Ax,y\rangle_{\ell^2} = \langle(-x_1,x_1,x_2,\dots),(y_0,y_1,y_2,\dots)\rangle_{\ell^2} = -x_1\overline{y_0}+x_1\overline{y_1}+x_2\overline{y_2}+\cdots \\ &= x_0\cdot 0 + x_1(\overline{y_1-y_0}) + x_2\overline{y_2}+\cdots = \langle x,(0,y_1-y_0,y_2,\dots)\rangle = \langle x,A^\dagger y\rangle\end{aligned}$$

and then the adjoint operator of A is:

$$A^\dagger(y) = (0,y_1-y_0,y_2,\dots) \qquad \forall y \in \ell^2(\mathbb{N},\mathbb{C})$$

5) We have $A^2x = AAx = A(-x_1,x_1,x_2,\dots) = (-x_1,x_1,x_2,\dots) = Ax$ for all $x \in \ell^2(\mathbb{N},\mathbb{C})$, thus A is idempotent. Moreover, we clearly see that $\text{Im}(A) = E$, where E is the subspace defined at the beginning of the exercise. Thus A is a projection operator on E, but since $\|A\| = \sqrt{2} \neq 1$, it cannot be an orthogonal projector. A is

therefore an oblique projection operator on E. The difference between the actions of A and P_E is:

$$Ax = (-x_1, x_1, x_2, \dots) \qquad \text{oblique projector on } E$$

$$P_E\, x = \left(\frac{x_0 - x_1}{2}, -\frac{x_0 - x_1}{2}, x_2, \dots\right) \text{ orthogonal projector on } E \qquad \square$$

6.7.1. *Bounded multiplication operators and their relation to orthogonal projectors*

In this section, we shall present a concrete application of the last theorem, while taking the opportunity to introduce a new category of highly useful linear operators.

DEFINITION 6.17.– *Let $\mathcal{H} = L^2(X, \mathcal{A}, \mu)$ and take $g \in L^\infty(X, \mathcal{A}, \mu)$. The multiplication operator by g is defined by:*

$$\begin{aligned} M_g : L^2(X, \mathcal{A}, \mu) &\longrightarrow L^2(X, \mathcal{A}, \mu) \\ f &\longmapsto M_g f = f \cdot g \end{aligned}$$

% where $f \cdot g(x) = f(x)g(x)\ \forall x \in X$ (pointwise multiplication).

Now, let us examine the properties of M_g.

– M_g is *bounded* $\forall g \in L^\infty(X, \mathcal{A}, \mu)$:

$$\|M_g f\|_2^2 = \int_X |f(x)g(x)|^2 d\mu(x) \leqslant \int_X \left(\sup_{x \in X} |g(x)|^2\right) |f(x)|^2 d\mu(x)$$

$$= \|g\|_\infty^2 \, \|f\|_2^2 < +\infty$$

thus[5] $\|M_g\|_2 \leqslant \|g\|_\infty$ then $M_g \in \mathcal{B}(L^2(X, \mathcal{A}, \mu))\ \forall g \in L^\infty(X, \mathcal{A}, \mu)$.

– $\forall g, h \in L^\infty(X, \mathcal{A}, \mu)$, by the commutativity of the pointwise product, *multiplication operators commute*, that is, $M_g M_h = M_h M_g$, $[M_g, M_h] = 0$.

– $\ker(M_g) = \{f \in L^2(X, \mathcal{A}, \mu) \ :\ M_g(f) = f \cdot g = 0_{L^2(X,\mathcal{A},\mu)}\}$. Defining the set:

$$N_g = \{x \in X \ :\ g(x) = 0\},$$

it is clear that $(f \cdot g)(x) = 0\ \forall x \in N_g$. Thus, since $g(x) \neq 0\ \forall x \in {N_g}^c = X \backslash N_g$, to obtain the zero function on X *via* the product $f \cdot g$, we must simply impose the

5 It is possible to show that if $(X, \mathcal{A}, \mu)$ is a measure space with a σ-finite measure, i.e. $X = \bigcup_{k \in \mathbb{N}} A_k$, where $\mu(A_k) < +\infty$, then $\|M_g\|_2 = \|g\|_\infty$.

condition that f must be null on ${N_g}^c$ (remember that f is an equivalence class of functions which are equal a.e.). In short:

$$\boxed{\ker(M_g) = \{f \in L^2(X, \mathcal{A}, \mu) \ : \ f(x) = 0 \ \forall x \in {N_g}^c\}}$$

– Now, let us consider the invertibility of M_g. For the kernel of M_g to be trivial, the only element in $\ker(M_g)$ must be the equivalence class in which the identically zero function appears. This corresponds to requiring that $\mu(N_g) = 0$, since in this case $\mu({N_g}^c) = \mu(X) - \mu(N_g) = \mu(X)$ thus $\ker(M_g) = \{f \in L^2(X, \mathcal{A}, \mu) \ : \ f(x) = 0 \text{ a.e.}\}$.

– If $\mu(N_g) = 0$, then there exists an inverse operator of M_g: ${M_g}^{-1} : \mathrm{Im}(M_g) \to L^2(X, \mathcal{A}, \mu)$ which can be characterized using the function $\frac{1}{g} : X \to \mathbb{K}$, $\frac{1}{g}(x) = \begin{cases} \frac{1}{g(x)} & \text{if } g(x) \neq 0 \\ 0 & \text{otherwise} \end{cases}$.

By definition, $\mathrm{Im}(M_g) = \{h \in L^2(X, \mathcal{A}, \mu) \ : \ \exists f \in L^2(X, \mathcal{A}, \mu) \ : \ h = M_g(f) = f \cdot g\}$; it is thus clear that $\frac{1}{g} \cdot h = \frac{1}{g(x)} \cdot f \cdot g = f$. This simple observation allows us to characterize both $\mathrm{Im}(M_g)$ and the action of ${M_g}^{-1}$:

$$\boxed{\mathrm{Im}(M_g) = \{h \in L^2(X, \mathcal{A}, \mu) \ : \ \frac{1}{g} \cdot h \in L^2(X, \mathcal{A}, \mu)\}} \text{ and } \boxed{{M_g}^{-1} = M_{\frac{1}{g}}}$$

– Let us determine the adjoint of M_g: $\forall f, h \in L^2(X, \mathcal{A}, \mu)$, $g \in L^\infty(X, \mathcal{A}, \mu)$:

$$\begin{aligned} \langle M_g^\dagger f, h\rangle &= \langle f, M_g h\rangle = \langle f, gh\rangle = \int_X f(x)\overline{g(x)}\,\overline{h(x)}d\mu(x) \\ &= \int_X \left(\overline{g(x)}f(x)\right)\overline{h(x)}d\mu(x) \\ &= \langle \bar{g} \cdot f, h\rangle = \langle M_{\bar{g}} f, h\rangle \end{aligned}$$

that is, $\boxed{M_g^\dagger = M_{\bar{g}}}$, by Theorem 6.10.

– Now, we calculate M_g^2:

$$\begin{aligned} &M_g^2 f = M_g(M_g f) = M_g(fg) = fg^2 = M_{g^2} f \\ &\forall f \in L^2(X, \mathcal{A}, \mu), \ g \in L^\infty(X, \mathcal{A}, \mu) \end{aligned}$$

Thus, the bounded linear operator M_g is self-adjoint and idempotent if and only if $\bar{g} = g$ and $g^2 = g$. The first condition means that g must be a real-valued function, but the only function with real values which is equal to its own square is a function which only takes values of 0 and 1, that is, the indicator function of a measurable subset of $\mathbb{R}^n$, which is clearly an element of $L^\infty(X, \mathcal{A}, \mu)$.

In summary, the multiplication operator M_g *is an orthogonal projection operator if and only if* $g = \chi_E$, *with* $E \subseteq X$ *measurable*. M_{χ_E} is invertible if and only if $\mu(E^c) \neq 0$, i.e. $\mu(E) \neq \mu(X)$.

Leaving aside invertibility, let us calculate the image of M_{χ_E}: the condition which determines this subspace is $\frac{1}{\chi_E} \cdot h \in L^2(X, \mathcal{A}, \mu)$, but, by definition, $\frac{1}{\chi_E}(x) = 0 \; \forall x \in X$ such that $\chi_E(x) = 0$, that is, $\forall x \in E^c$ and, in this case, $\frac{1}{\chi_E} \cdot h \in L^2(X, \mathcal{A}, \mu)$. When $x \in E$, $\frac{1}{\chi_E}(x) = 1$, so the defining condition of $\mathrm{Im}(M_g)$ becomes $h \in L^2(E, \mathcal{A}, \mu)$.

In conclusion, for any measurable set $E \subseteq X$, the orthogonal projector and multiplication operator M_{χ_E} is, explicitly:

$$\begin{aligned} M_{\chi_E} : L^2(X, \mathcal{A}, \mu) &\longrightarrow L^2(E, \mathcal{A}, \mu) \\ f &\longmapsto M_{\chi_E} f = \begin{cases} f(x) & x \in E \\ 0 & \text{otherwise} \end{cases} \end{aligned}$$

6.7.2. *Geometric realization of orthogonal projection operators* via *orthonormal systems*

We now have the means of proving another important analogy between Hilbert spaces and finite-dimensional Euclidean spaces related to the geometric realization of orthogonal projectors on a vector subspace generated by an orthogonal family, which we have already discussed in Chapter 1.

We recall that the orthogonal projector of an inner product vector space V of finite dimension n on a vector subspace S of dimension s can be written as:

$$P_S(x) = \sum_{i=1}^{s} \langle x, u_i \rangle u_i$$

where $(u_i)_{i=1}^s$ is any orthonormal basis of S.

In a Hilbert space, we have the following result.

THEOREM 6.32.– Take $A \in \mathcal{B}(\mathcal{H})$, $A \neq \mathbf{0}$. The following statements are equivalent:

1) A is an orthogonal projector;

2) there exists an orthonormal system[6] $(u_n)_{n\in\mathbb{N}}$ in $\mathcal{H}$ such that:

$$Ax = \sum_{n\in\mathbb{N}} \langle x, u_n\rangle u_n \qquad \forall x \in \mathcal{H}$$

Where applicable, A projects onto the closed subspace $\overline{span(u_n,\ n \in \mathbb{N})}$.

PROOF.–

$\boxed{1) \implies 2)}$: First, we note that an orthogonal projector A is surjective if and only if it is the identity operator. The condition $\mathrm{Im}(A) = \mathcal{H}$ implies, by properties 7 and 8 of orthogonal projectors, that $\mathrm{Im}(A)^{\perp} = \ker(A) = \{0_{\mathcal{H}}\}$; thus, by property 9, it holds that $Ax = x\ \forall x \in \mathcal{H}$, i.e. $A = id_{\mathcal{H}}$. In this case, any complete orthonormal system $(u_n)_{n\in\mathbb{N}}$ in $\mathcal{H}$ realizes A since, on the one hand, $Ax = x$, and on the other hand, by Theorem 5.11 regarding the characterization of complete orthonormal systems, we can write $x = \sum_{n\in\mathbb{N}} \langle x, u_n\rangle u_n$. Given that a complete orthonormal system is a special instance of an orthonormal system, the implication $1) \implies 2)$ when $\mathrm{Im}(A) = \mathcal{H}$ is true.

Now, let A be an orthogonal projector such that $\mathrm{Im}(A) \subset \mathcal{H}$, that is, $\mathrm{Im}(A)$ is a closed vector subspace (by definition of an orthogonal projector) and proper in $\mathcal{H}$; thus, it is a Hilbert space itself, properly included in $\mathcal{H}$.

Let $(u_n)_{n\in\mathbb{N}}$ be any complete orthonormal system in Im(A). Given our hypotheses, $(u_n)_{n\in\mathbb{N}}$ is only an orthonormal system (and not, generally, a complete orthonormal system) of $\mathcal{H}$. For all $y \in \mathrm{Im}(A)$, we have the following decomposition:

$$y = \sum_{n\in\mathbb{N}} \langle y, u_n\rangle u_n$$

Moreover, $\mathrm{Im}(A) = \{Ax,\ x \in \mathcal{H}\}$, so, using the fact that A, as an orthogonal projector, is self-adjoint:

$$Ax = \sum_{n\in\mathbb{N}} \langle Ax, u_n\rangle u_n \underset{(A \text{ s.a.})}{=} \sum_{n\in\mathbb{N}} \langle x, Au_n\rangle u_n, \quad \forall x \in \mathcal{H}$$

Since A is the identity on $\mathrm{Im}(A)$ and $u_n \in \mathrm{Im}(A)\ \forall n \in \mathbb{N}$, then $Au_n = u_n$, hence:

$$Ax = \sum_{n\in\mathbb{N}} \langle x, u_n\rangle u_n, \quad \forall x \in \mathcal{H}$$

that is, the orthogonal projector A is realized on the orthonormal system $(u_n)_{n\in\mathbb{N}}$ of $\mathcal{H}$ as described in point 2.

6 Note that although we write $(u_n)_{n\in\mathbb{N}}$, the orthonormal system may be finite, i.e. it may include a finite number of $u_n \neq 0$.

$\boxed{2) \implies 1)}$: for any pair $x, y \in \mathcal{H}$, let $(u_n)_{n\in\mathbb{N}}$ be an orthonormal system of $\mathcal{H}$ such that:

$$Ax = \sum_{m\in\mathbb{N}} \langle x, u_m\rangle u_m, \quad Ay = \sum_{n\in\mathbb{N}} \langle y, u_n\rangle u_n$$

then, by the continuity of the inner product:

$$\langle x, Ay\rangle = \langle x, \sum_{n\in\mathbb{N}} \langle y, u_n\rangle u_n\rangle = \sum_{n\in\mathbb{N}} \overline{\langle y, u_n\rangle}\langle x, u_n\rangle = \sum_{n\in\mathbb{N}} \langle x, u_n\rangle\langle u_n, y\rangle$$

Again, using the continuity of the inner product, as we saw when proving Parseval's identity (Theorem 5.11):

$$\begin{aligned}\langle x, A^\dagger Ay\rangle &= \langle Ax, Ay\rangle = \langle \sum_{m\in\mathbb{N}} \langle x, u_m\rangle u_m, \sum_{n\in\mathbb{N}} \langle y, u_n\rangle u_n\rangle \\ &= \sum_{m\in\mathbb{N}} \sum_{n\in\mathbb{N}} \langle x, u_m\rangle\overline{\langle y, u_n\rangle}\langle u_m, u_n\rangle \\ &= \sum_{m\in\mathbb{N}} \sum_{n\in\mathbb{N}} \langle x, u_m\rangle \langle u_n, y\rangle \, \delta_{n,m} = \sum_{n\in\mathbb{N}} \langle x, u_n\rangle \langle u_n, y\rangle\end{aligned}$$

so $\langle x, Ay\rangle = \langle x, A^\dagger Ay\rangle$ $\forall x, y \in \mathcal{H}$, that is, $A = A^\dagger A$ by Theorem 6.10. Using the algebraic characterization of orthogonal projectors, Theorem 6.31, we can therefore state that A is an orthogonal projector.

Supposing that 1 and 2 are verified, then:

1) $\implies$ $\text{Im}(A) = \ker(A)^\perp$ and $\ker(A) = \text{Im}(A)^\perp$;

2) $\implies$ $\text{Im}(A) \subseteq \overline{\text{span}(u_n,\ n \in \mathbb{N})}$, which is obvious, and $\ker(A) \subseteq \left(\overline{\text{span}(u_n,\ n \in \mathbb{N})}\right)^\perp$, which is not quite so obvious. For $x = 0_{\mathcal{H}}$ this is true; taking $x \in \ker(A)$, $x \neq 0_{\mathcal{H}}$, then $Ax = 0 = \sum_{n\in\mathbb{N}} \langle x, u_n\rangle u_n = \lim_{N \implies +\infty} \sum_{n=1}^{N} \langle x, u_n\rangle u_n$. The vectors u_n are linearly independent since they are mutually orthogonal, so, for all $N \in \mathbb{N}$, the linear combination $\sum_{n=1}^{N} \langle x, u_n\rangle u_n$ is zero if and only if the coefficients $\langle x, u_n\rangle$ are zero, that is, $x \in (u_n,\ n \in \mathbb{N})^\perp = (\text{span}(u_n,\ n \in \mathbb{N}))^\perp = \left(\overline{\text{span}(u_n,\ n \in \mathbb{N})}\right)^\perp$, by the properties of the orthogonal complement.

To summarize: on one side, $\text{Im}(A) \subseteq \overline{\text{span}(u_n,\ n \in \mathbb{N})}$, while, on the other side $\ker(A) \subseteq \left(\overline{\text{span}(u_n,\ n \in \mathbb{N})}\right)^\perp$, thus $\overline{\text{span}(u_n,\ n \in \mathbb{N})} \subseteq \ker(A)^\perp = \text{Im}(A)$, that is, $\text{Im}(A) = \overline{\text{span}(u_n,\ n \in \mathbb{N})}$. □

REMARK.– We see from the proof of this theorem that any orthonormal system $(u_n)_{n\in\mathbb{N}}$ in $\text{Im}(A)$ can be used to realize a projector in the sense defined by the theorem.

This means that, although each term in the summation may be different, the overall action of the operator will be the same for any orthonormal system $(u_n)_{n\in\mathbb{N}}$ in $\text{Im}(A)$.

A remarkable application of this result is shown in Exercise 6.6, which illustrates the way in which the best linear approximation of a parabola on a real interval may be found using the orthogonal projection theory on a Hilbert space.

Exercise 6.6

Let $f(x) \equiv 1$ (th constant function equal to 1) and $g(x) = x$, the identity function, seen as two elements of $L^2[0,1]$. Calculate:

1) the angle ϑ between f and g;

2) their distance in $L^2[0,1]$;

3) the projection $P_W h$ of the function $h \in L^2[0,1]$, $h(x) = x^2$, on the vector subspace $W = \text{span}(f,g)$. Interpret your findings.

Solution to Exercise 6.6

1) The angle between f and g is obtained using the definition of inner product: $\langle f,g\rangle = \|f\|\,\|g\|\cos(\vartheta)$, so we need to calculate $\langle f,g\rangle$, $\|f\|$, $\|g\|$:

$$\langle f,g\rangle = \int_0^1 x dx = \left[\frac{x^2}{2}\right]_0^1 = \frac{1}{2}$$

$$\|f\| = \left(\int_0^1 dx\right)^{1/2} = 1, \quad \|g\| = \left(\int_0^1 x^2 dx\right)^{1/2} = \left(\left[\frac{x^3}{3}\right]_0^1\right)^{1/2} = \frac{1}{\sqrt{3}}$$

In conclusion, $\cos(\vartheta) = \frac{\langle f,g\rangle}{\|f\|\|g\|} = \frac{\sqrt{3}}{2}$, thus $\vartheta = \frac{\pi}{6}$.

2) Distance:

$$\begin{aligned} d(f,g) &= \|f-g\| = \left(\textstyle\int_0^1 (f(x)-g(x))^2 dx\right)^{1/2} = \left(\textstyle\int_0^1 (1-x)^2 dx\right)^{1/2} \\ &= \left(-\left[\tfrac{(1-x)^3}{3}\right]_0^1\right)^{1/2} = \tfrac{1}{\sqrt{3}} = \tfrac{\sqrt{3}}{3} \end{aligned}$$

3) Projection on W: We use the characterization of projection given by the previous theorem. We need to construct an orthonormal basis of W, which can be done by using the Gram-Schmidt procedure. A wise choice is to begin with the function f which is a generator of W and, furthermore, has a unitary norm. The second (and final) element in the orthonormal basis of W is then:

$$\tilde{g}(x) = \frac{g(x) - \langle f,g\rangle f(x)}{\|g(x) - \langle f,g\rangle f(x)\|} = \frac{x - \frac{1}{2}}{\|x - \frac{1}{2}\|},$$

with

$$\left\|x - \frac{1}{2}\right\| = \left(\int_0^1 \left(x - \frac{1}{2}\right)^2 dx\right)^{1/2} = \left(\frac{1}{3}\left[\left(x - \frac{1}{2}\right)^3\right]_0^1\right)^{1/2} = \frac{1}{2\sqrt{3}}$$

so $\tilde{g}(x) = 2\sqrt{3}\left(x - \frac{1}{2}\right)$ and the desired orthonormal basis is $B = (1, \sqrt{3}\,(2x-1))$. The orthogonal projection $h(x) = x^2$ on W is thus:

$$P_W h = \langle h, f\rangle f + \langle h, \tilde{g}\rangle \tilde{g}$$

By direct calculation, $\langle h, f\rangle = \frac{1}{3}$ and $\langle h, \tilde{g}\rangle = \frac{\sqrt{3}}{6}$, hence:

$$P_W h(x) = \frac{1}{3} + \frac{\sqrt{3}}{6}\sqrt{3}\,(2x-1) = x - \frac{1}{6}$$

The interpretation of this result is as follows: The functions $[0,1] \underset{f}{\mapsto} 1$ and $[0,1] \underset{g}{\mapsto} x$ are the generators of the space W of linear functions (straight lines) defined on the interval $[0,1]$. In fact, any linear function $\ell : [0,1] \to \mathbb{K}$ may be written as $\ell(x) = \alpha + \beta x$, $x \in [0,1]$ with $\alpha, \beta \in \mathbb{K}$; since $\alpha + \beta x = \alpha f(x) + \beta g(x)$ $\forall x \in [0,1]$, then $\ell = \alpha f + \beta g$.

The function $[0,1] \underset{h}{\mapsto} x^2$ is a parabola defined on the same interval. By definition of orthogonal projection:

$$P_W h = \underset{w \in W}{\arg\min}\ \|h - w\|$$

that is, $P_W h$ is the element in W which minimizes the L^2 distance between h and the straight lines. So, the straight line with equation $y = x - \frac{1}{6}$ is the best approximation of the parabola with equation $y = x^2$, in the sense of the norm L^2, on the interval $[0,1]$. □

Figure 6.1 shows a graphical representation of this approximation.

A list of properties of orthogonal projectors follows (for the proofs of these properties, see, for example, Abbati and Cirelli 1997). For all $A, B \in \mathcal{B}(\mathcal{H})$, we recall that:

$$[A, B] = AB - BA$$

$[A, B]$ is said to be the *commutator* of A and B. If $[A, B] = 0$, the zero operator, that is, $AB = BA$, then A and B are said to *commute*.

Let R and S be two closed vector subspaces in the Hilbert space $\mathcal{H}$ and let P_R, P_S be the orthogonal projectors on R and S, respectively.

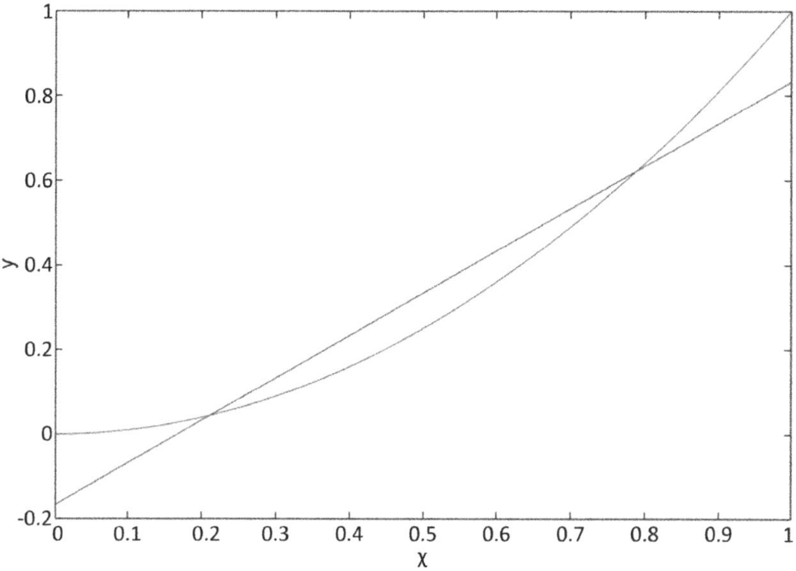

Figure 6.1. *The line of equation $y = x - \frac{1}{6}$ (shown in blue) is the best approximation of the parabola with equation $y = x^2$ (in red) with respect to the Hilbert norm of $L^2[0,1]$. For a color version of this figure, see www.iste.co.uk/provenzi/spaces.zip*

THEOREM 6.33 (Sum of orthogonal projectors).– The following statements are equivalent:

1) $P_R + P_S$ is an orthogonal projector;

2) $P_R P_S = P_S P_R = 0$;

3) $P_R(x) = 0 \ \forall x \in S$ and $P_S(x) = 0 \ \forall x \in R$;

4) $R \perp S$.

Moreover, if $P_R + P_S$ is an orthogonal projector, then it projects on $R + S$.

THEOREM 6.34 (Product of orthogonal projectors).– The following statements are equivalent:

1) $P_R P_S$ is an orthogonal projector;

2) $P_S P_R$ is an orthogonal projector;

3) $[P_R, P_S] = 0$;

4) $R = (R \cap S) \oplus (R \cap S^{\perp})$;

5) $S = (R \cap S) \oplus (R^{\perp} \cap S)$.

If $P_R P_S$ and $P_S P_R$ are orthogonal projectors, then they project on $R \cap S$.

THEOREM 6.35 (Difference between orthogonal projectors).– The following statements are equivalent:

1) $P_R - P_S$ is a projector;

2) $P_R P_S = P_S P_R = P_S$;

3) $R \cap S = S$, i.e. $S \subset R$.

If $P_R - P_S$ is an orthogonal projector, then it projects on $R \cap S^{\perp}$.

THEOREM 6.36 (Mixing projector sum, difference and product).– If

$$[P_S, P_R] = 0,$$

then $P_R + P_S - P_R P_S$ is an orthogonal projector which projects on $span(R \cup S)$.

6.8. Isometric and unitary operators

In this section, we shall determine the properties of isometric and unitary operators in a Hilbert space of infinite dimension, and provide an algebraic and geometric characterization of these operators. Once again, the adjoint operator plays a fundamental role in algebraic characterization, while orthonormal systems and Hilbert bases are crucial for the geometric characterization.

In finite-dimensional vector spaces V, a linear operator which preserves the inner product, that is, $A : V \to V$, $\langle Ax, Ay \rangle = \langle x, y \rangle$, $\forall x, y \in V$, also preserves the norm of the vectors (simply by considering $x = y$), that is, $\|Ax\| = \|x\| \ \forall x \in V$. To prove the converse, it is sufficient to consider the polarization formula[7] [1.7], $\forall x, y \in V$:

$$\langle x, y \rangle = \frac{1}{4}\left(\|x+y\|^2 - \|x-y\|^2 + i\,\|x+iy\|^2 - i\,\|x-iy\|^2\right)$$

If we replace x, y with Ax, Ay and use the linearity of A:

$$\langle Ax, Ay \rangle = \frac{1}{4}\Big(\|A(x+y)\|^2 - \|A(x-y)\|^2$$

$$+i\,\|A(x+iy)\|^2 - i\,\|A(x-iy)\|^2\Big)$$

7 The complex case is considered here; the real case is even simpler.

Assuming that A preserves the norm, we have:

$$\langle Ax, Ay\rangle = \frac{1}{4}\left(\|x+y\|^2 - \|x-y\|^2 + i\,\|x+iy\|^2 - i\,\|x-iy\|^2\right) = \langle x, y\rangle$$

As we know, the norm canonically generates a metric *via* $d(x,y) = \|x-y\|$. For this reason, an operator which preserves the inner product or norm is said to be *isometric*. The only vector which has a norm of zero is the vector 0_V, thus an isometric operator A never transforms a non-zero vector (whose norm is > 0) into the null vector, that is, $\ker(A) = \{0_V\}$. Hence, $\dim(\ker(A)) = 0$ and then, by the rank theorem, $\dim(\mathrm{Im}(A)) = \dim(V)$. In other words, *an isometric endomorphism in finite dimensions is automatically surjective.*

In an infinite-dimensional Hilbert space, it is still true that a bounded linear operator preserves the scalar product if and only if it is isometric. However, the statement that an isometric operator $A : \mathcal{H} \to \mathcal{H}$ is always surjective is no longer true. One counter-example is provided by the operator $A \in \mathcal{B}(\mathcal{H})$ defined by $Au_n = u_{2n}$, where $(u_n)_{n\in\mathbb{N}}$ is an arbitrary Hilbert basis of $\mathcal{H}$. Evidently, A is isometric, but, as we will see in Theorem 6.39, $\mathrm{Im}(A) = \overline{\mathrm{span}(u_k,\ k \in \mathbb{N},\ \text{even})} \subset \mathcal{H}$; the inclusion is strict, since $(u_k,\ k \in \mathbb{N},\ k \text{ even})$ is not a complete orthonormal system, as it is a proper subset of $(u_n)_{n\in\mathbb{N}}$.

These considerations lead to Definition 6.18.

DEFINITION 6.18.– *The operator $A : \mathcal{H} \to \mathcal{H}$ is said to be:*

– isometric, *if $\langle Ax, Ay\rangle = \langle x, y\rangle$, $\forall x, y \in \mathcal{H}$, or, in an equivalent manner, if $\|Ax\| = \|x\|$, $\forall x \in \mathcal{H}$;*

– unitary, *if A is isometric and surjective.*

Let us calculate the norm of an isometric operator:

$$\|A\| = \sup_{\|x\|=1} \|Ax\| = \sup_{\|x\|=1} \|x\| = 1$$

Since a unitary operator is also isometric, we have that *the norm of isometric and unitary operators is 1.*

BASIC EXAMPLES OF UNITARY OPERATORS.–

Let us consider $\mathbb{R}^n$ with the Borel σ-algebra and the Lebesgue measure. Given a fixed element $a \in \mathbb{R}^n$, *any translation operator*:

$$\begin{array}{rccl} T_a : & L^2(\mathbb{R}^n) & \longrightarrow & L^2(\mathbb{R}^n) \\ & f & \longmapsto & T_a f, \text{ where } T_a f(x) = f(x-a),\ \forall x \in \mathbb{R}^n \end{array}$$

is unitary. In fact, we know that it is well defined, linear and isometric due to the shift invariance of the Lebesgue measure. It is also surjective, since, for any element $g \in L^2(\mathbb{R}^n)$, we simply need to consider $f \in L^2(\mathbb{R}^n)$, $f(x) = g(x+a)$ $\forall x \in \mathbb{R}^n$ to obtain $T_a f = g$.

Now, let $R \in \mathrm{O}(n)$ be a rotation matrix of $\mathbb{R}^n$, where $\mathrm{O}(n)$ is the orthogonal group of dimension n, that is, the group of square matrices R of dimension n which are orthogonal, that is, such that $R^t = R^{-1}$. *Any rotation operator*:

$$\begin{aligned} T_R : L^2(\mathbb{R}^n) &\longrightarrow L^2(\mathbb{R}^n) \\ f &\longmapsto T_R f, \text{ where } T_R f(x) = f(Rx), \ \forall x \in \mathbb{R}^n \end{aligned}$$

is unitary, due to the fact that the Jacobian of the transformation, that is, the determinant of R, has an absolute value of 1 and thus the integrals used to calculate the norm of $T_R f$ and of f are equal. It is also surjective, since for any element $g \in L^2(\mathbb{R}^n)$, we simply need to consider $f \in L^2(\mathbb{R}^n)$, $f(x) = g(R^t x)$ $\forall x \in \mathbb{R}^n$ to obtain $T_R f = g$.

A special case of the rotation operator is the inverse identity matrix: $P = -I$ such that $T_P f = f_P$, with $f_P(x) = f(-x)$. T_P is known as the *parity operator*.

6.8.1. *Characterizations of isometric and unitary operators*

The following results establish a useful characterization of isometric and unitary operators.

THEOREM 6.37 (Algebraic characterization of isometric operators).– $A \in \mathcal{B}(\mathcal{H})$ is an isometric operator if and only if $A^\dagger A = id_{\mathcal{H}}$.

PROOF.– Let A be isometric, then $\forall x \in \mathcal{H}$:

$$\langle A^\dagger Ax, x\rangle = \langle Ax, Ax\rangle = \|Ax\|^2 \underset{A \text{ isometric}}{=} \|x\|^2 = \langle x, x\rangle$$

that is, $A^\dagger A = id_{\mathcal{H}}$.

Conversely, if $A^\dagger A = id_{\mathcal{H}}$, then $\forall x, y \in \mathcal{H}$:

$$\langle Ax, Ay\rangle = \langle A^\dagger Ax, y\rangle = \langle x, y\rangle$$

that is, A conserves the inner product, and thus it is isometric. □

The following result is particularly useful in optimization theory and in quantum mechanics.

THEOREM 6.38.– Let $A \in \mathcal{B}(\mathcal{H})$ be isometric, then $AA^\dagger$ is an orthogonal projector.

PROOF.– We will use the algebraic characterization of orthogonal projectors: since we already know that $AA^\dagger$ is self-adjoint for all $A \in \mathcal{B}(\mathcal{H})$, we simply need to verify that $AA^\dagger$ is bounded and idempotent.

1) $AA^\dagger$ is bounded: $\forall x \in \mathcal{H}$ it holds that:

$$\|AA^\dagger x\| = \|A(A^\dagger x)\| \underset{A \text{ isometric}}{=} \|A^\dagger x\| \leqslant \|A^\dagger\| \, \|x\| = \|A\| \, \|x\| \underset{\|A\|=1}{=} \|x\|$$

2) $AA^\dagger$ is idempotent : $(AA^\dagger)^2 = AA^\dagger(AA^\dagger) = A(A^\dagger A)A^\dagger \underset{A^\dagger A = id_{\mathcal{H}}}{=} AA^\dagger$. □

$AA^\dagger$ projects onto its image,which can be characterized as follows.

THEOREM 6.39.– Let $A \in \mathcal{B}(\mathcal{H})$ be isometric. The image of the orthogonal projector $AA^\dagger$ is $Im(A)$, so the image of an isometric $A \in \mathcal{B}(\mathcal{H})$ is a closed vector subspace of $\mathcal{H}$.

PROOF.– We wish to show that $\text{Im}(AA^\dagger) = \text{Im}(A)$. We begin by observing that in general, $\forall A, B \in \mathcal{B}(\mathcal{H})$, it holds that $\text{Im}(AB) \subseteq \text{Im}(A)$ and $\text{Im}(AB) = \text{Im}(A)$ if and only if B is surjective:

$$\text{Im}(A) = \{y \in \mathcal{H} \;:\; \exists x \in \mathcal{H} \;:\; Ax = y\}, \ \text{Im}(AB)$$
$$= \{z \in \mathcal{H} \;:\; \exists x \in \text{Im}(B) \;:\; Ax = y\}$$

This tells us that $\text{Im}(AB) = \text{Im}(A)$ if and only if B is surjective, otherwise the images of A and those of AB would not agree.

Taking $B = A^\dagger$, then $\underline{\text{Im}(AA^\dagger) \subseteq \text{Im}(A)}$ $\forall A \in \mathcal{B}(\mathcal{H})$. Now, let A be isometric, that is, $A^\dagger A = id_{\mathcal{H}}$ and $y \in \text{Im}(A)$, then $\exists x \in \mathcal{H}$ such that:

$$y = Ax = A(A^\dagger A)x = AA^\dagger(Ax)$$

that is, $y \in \text{Im}(AA^\dagger)$, hence $\underline{\text{Im}A \subseteq \text{Im}(AA^\dagger)}$. Thus $\text{Im}(AA^\dagger)$=$\text{Im}(A)$.

Since $AA^\dagger$ is an orthogonal projector, we know that its image is closed; hence, $\text{Im}(A)$ is closed for any isometric operator. □

The fact that an isometric operator has a closed image can be shown directly, using a proof very similar to that of Theorem 6.13.

If A is unitary, that is, isometric and surjective, then $\text{Im}(AA^\dagger) = \text{Im}(A) = \mathcal{H}$ and then $AA^\dagger = id_{\mathcal{H}}$.

The fact that $\ker(A^\dagger) = \text{Im}(A)^\perp$ gives us immediately sufficient conditions to guarantee the invertibility or non-invertibility of the adjoint of an operator in $\mathcal{B}(\mathcal{H})$.

THEOREM 6.40.– Taking $A \in \mathcal{B}(\mathcal{H})$:

– if A is isometric and not surjective, then $\ker(A^\dagger) = Im(A)^\perp \neq \{0_\mathcal{H}\}$, i.e. $A^\dagger$ is not invertible;

– if A is unitary, then $\ker(A^\dagger) = \mathcal{H}^\perp = \{0\}$, i.e. $A^\dagger$ is invertible.

Now, let us apply these results to the case of the operator $Au_n = u_{2n}$, where $(u_n)_{n\in\mathbb{N}}$ is a complete orthonormal system in $\mathcal{H}$. As noted before, since $\|Au_n\| = \|u_{2n}\| = 1 = \|u_n\|$, A is isometric, but it is not unitary, as $\text{Im}(A) = \overline{\text{span}(u_k,\ k \text{ even})} \subset \mathcal{H}$.

Let us determine $A^\dagger$: $\langle Au_n, u_m\rangle = \langle u_n, A^\dagger u_m\rangle$, furthermore, $\langle Au_n, u_m\rangle = \langle u_{2n}, u_m\rangle = \delta_{2n,m}$, then we can write $\langle u_n, A^\dagger u_m\rangle = \delta_{2n,m}$, that is:

$$A^\dagger u_m = \begin{cases} u_{\frac{m}{2}} & \text{if } m = 2n \\ 0 & \text{if } m \neq 2n \end{cases}$$

We see that $\ker(A^\dagger) = \overline{\text{span}(u_m,\ m \text{ odd})} = \text{Im}(A)^\perp$, confirming our results.

The following theorem gives a complete algebraic characterization of unitary operators.

THEOREM 6.41 (algebraic characterization of unitary operators).– $A \in \mathcal{B}(\mathcal{H})$, the following statements are equivalent:

1) A is unitary;

2) A is invertible and $A^{-1} = A^\dagger \in \mathcal{B}(\mathcal{H})$;

3) $A^\dagger A = AA^\dagger = id_\mathcal{H}$;

4) $A^\dagger$ is unitary.

PROOF.–

$\boxed{1) \implies 2)}$: We know that if A is unitary, then A is injective, that is, invertible on its image; by definition, A is surjective; therefore, it is bijective and invertible on all $\mathcal{H}$. To show that $A^{-1} = A^\dagger$, we write:

$$\langle A^\dagger Ax, y\rangle = \langle Ax, Ay\rangle = \langle x, y\rangle, \quad \forall x, y \in \mathcal{H}$$

hence $A^\dagger A = id_\mathcal{H}$ (showing that $A^\dagger$ is the left inverse of A) and then:

$$A^\dagger = A^\dagger(AA^{-1}) = (A^\dagger A)A^{-1} = id_\mathcal{H} A^{-1} = A^{-1}$$

that is, $A^\dagger = A^{-1}$. Now, we need only to prove that $A^\dagger = A^{-1}$ is bounded: Since A is surjective, $\exists y \in \mathcal{H}$ such that $x = Ay$, and, by unitarity: $\|x\| = \|Ay\| = \|y\|$, that is, $\|x\| = \|y\|$ and then:

$$\|A^\dagger x\| = \|A^{-1}x\| = \|A^{-1}Ay\| = \|y\| = \|x\| \quad \forall x \in \mathcal{H}$$

which implies that $\|A^{-1}\| = \|A^\dagger\| = \sup_{\|x\|=1} \|x\| = 1$.

$\boxed{2) \implies 3)}$:

$$A^\dagger = A^{-1} \implies A^\dagger A = A^{-1}A = id_{\mathcal{H}}$$

and:

$$A^\dagger = A^{-1} \implies AA^\dagger = AA^{-1} = id_{\mathcal{H}}$$

$\boxed{3) \implies 4)}$: From $AA^\dagger = id_{\mathcal{H}}$, we obtain: $\langle x, AA^\dagger y\rangle = \langle x, y\rangle \; \forall x, y \in \mathcal{H}$; furthermore, $\langle x, AA^\dagger y\rangle = \langle A^\dagger x, A^\dagger y\rangle$, thus $\langle A^\dagger x, A^\dagger y\rangle = \langle x, y\rangle \; \forall x, y \in \mathcal{H}$, that is, $A^\dagger$ is isometric. Now, we only need to prove that $A^\dagger$ is surjective. We do this using the other identity, $A^\dagger A = id_{\mathcal{H}}$, which implies:

$$A^\dagger(Ay) = (A^\dagger A)y = id_{\mathcal{H}}(y) = y, \forall y \in \mathcal{H}$$

that is, $\forall y \in \mathcal{H} \; \exists \xi = Ay \in \mathcal{H}$ such that $A^\dagger \xi = y$, i.e. $A^\dagger$ is surjective.

$\boxed{1) \implies 4) \implies 1)}$: As we have seen, given an arbitrary unitary operator, its adjoint is also unitary. Using the hypothesis that $A^\dagger$ is a unitary operator, then $A^{\dagger\dagger}$ is unitary, and, since $A^{\dagger\dagger} = A$, then unitary $A^\dagger$ implies unitary A. □

One consequence of this result is that we can study the unitarity of an operator A by considering that of its adjoint, which can be simpler.

Corollary 6.5 shows that the norm of a unitary operator is invariant with respect to adjunction and inversion.

COROLLARY 6.5.– If $A \in \mathcal{B}(\mathcal{H})$ is unitary, then $\|A\| = \|A^\dagger\| = \|A^{-1}\| = 1$.

Let $\mathcal{U}(\mathcal{H})$ be the set of unitary operators on a Hilbert space $\mathcal{H}$. If $A, B \in \mathcal{U}(\mathcal{H})$, then, by direct calculation, we can verify that $AB \in \mathcal{U}(\mathcal{H})$. The theorem proved above tells us that if $A \in \mathcal{U}(\mathcal{H})$ then $A^{-1} = A^\dagger \in \mathcal{U}(\mathcal{H})$, that is, $\mathcal{U}(\mathcal{H})$ verifies the group axioms with respect to composition.

DEFINITION 6.19.– *$\mathcal{U}(\mathcal{H})$ denotes the unitary group of $\mathcal{H}$. $\mathcal{U}(\mathcal{H})$ coincides with the group of automorphisms of $\mathcal{H}$: Aut($\mathcal{H}$).*

Some applications of the characterization of unitary operators are shown below.

Taking $\mathcal{H} = L^2(X, \mathcal{A}, \mu)$ and $g \in L^\infty(X, \mathcal{A}, \mu)$, we have seen that the multiplication operator by g defined by:

$$\begin{aligned} M_g : L^2(X, \mathcal{A}, \mu) &\longrightarrow L^2(X, \mathcal{A}, \mu) \\ f &\longmapsto M_g f = f \cdot g \end{aligned}$$

where $f \cdot g(x) = f(x)g(x)\ \forall x \in X$ is linear and bounded. Moreover, we know that $M_g^\dagger = M_{\bar{g}}$, thus $M_g M_g^\dagger = M_{g\bar{g}} = M_{|g|^2}$, and then $M_g M_g^\dagger = id_{\mathcal{H}}$ if and only if $M_{|g|^2} = id_{\mathcal{H}}$, but this is, equivalent to requiring that $|g|^2 = 1$, that is, the equivalence class of g must contain at least one representative, also noted g for simplicity's sake, of the form $g(x) = e^{ih(x)}$, with $h : X \to \mathbb{R}$ measurable.

Let us apply the last theorem that we proven to verify that for all complete orthonormal system $(u_n)_{n\in\mathbb{N}}$ of $\mathcal{H}$, the operator U defined as:

$$Uu_n = (-i)^n u_n$$

is a unitary operator. We will use the algebraic characterization $U^\dagger U = UU^\dagger = id_{\mathcal{H}}$. On one side, by definition:

$$\langle Uu_n, u_n \rangle = \langle u_n, U^\dagger u_n \rangle \qquad [6.25]$$

and, on the other side:

$$\langle Uu_n, u_n \rangle = \langle (-i)^n u_n, u_n \rangle = (-i)^n \langle u_n, u_n \rangle = \langle u_n, \overline{(-i)^n} u_n \rangle$$

$$\underset{[6.25]}{=} \langle u_n, U^\dagger u_n \rangle \quad \forall n \in \mathbb{N}$$

hence $U^\dagger u_n = \overline{(-i)^n} u_n$, and then:

$$U^\dagger U u_n = U^\dagger (U u_n) = \overline{(-i)^n} U u_n = \overline{(-i)^n} (-i)^n u_n = |(-i)^n|^2 u_n = u_n.$$

Moreover:

$$UU^\dagger u_n = U(U^\dagger u_n) = U\overline{(-i)^n} u_n = (-i)^n \overline{(-i)^n} u_n = |(-i)^n|^2 u_n = u_n$$

Since $(u_n)_{n\in\mathbb{N}}$ is a complete orthonormal system, the fact that $U^\dagger U = UU^\dagger$ is the identity on any element u_n can be extended to all $\mathcal{H}$. In fact, for all $x \in \mathcal{H}$, it holds that $x = \sum_{n\in\mathbb{N}} \langle x, u_n \rangle u_n$ and, by the linearity and continuity of $U^\dagger U$ and $U^\dagger U$, we can write:

$$U^\dagger U x = \sum_{n\in\mathbb{N}} \langle x, u_n \rangle U^\dagger U u_n = \sum_{n\in\mathbb{N}} \langle x, u_n \rangle u_n = x.$$

The same is true for $UU^\dagger x$, i.e. $U^\dagger U = UU^\dagger = id_{\mathcal{H}}$, proving that U is unitary.

As in finite dimensions, unitary operators allow us to establish an equivalence relation between operators, as formalized in the following definition.

DEFINITION 6.20.– *Two operators $A, B \in \mathcal{B}(\mathcal{H})$ are unitarily equivalent if there exists a unitary operator $U \in \mathcal{B}(\mathcal{H})$ such that $A = UBU^{-1}$.*

We do not have the space to go into greater detail regarding the properties of unitary equivalence here. We simply note that unitary equivalence preserves operator properties, such as continuity, invertibility and self-adjointness.

6.8.2. *Relationship between isometric and unitary operators and orthonormal systems*

The final property of isometric and unitary operators that we wish to discuss here is their interaction with orthonormal systems and complete orthonormal systems in Hilbert spaces. In finite dimension, isometric and unitary operators coincide, and they transform orthonormal bases into orthonormal bases. In infinite dimension, this remains true only for unitary operators.

THEOREM 6.42 (Geometric characterization of isometric operators).– $A \in \mathcal{B}(\mathcal{H})$ is isometric if and only if it transforms complete orthonormal systems $(u_n)_{n\in\mathbb{N}}$ in $\mathcal{H}$ into orthonormal systems $(Au_n)_{n\in\mathbb{N}}$.

PROOF.–

$\boxed{\Longrightarrow}$: for any complete orthonormal system $(u_n)_{n\in\mathbb{N}}$ in $\mathcal{H}$, by the isometry of A, we can write:

$$\langle Au_n, Au_m\rangle = \langle u_n, u_m\rangle = \delta_{n,m} \qquad \forall n, m \in \mathbb{N}$$

thus $(Au_n)_{n\in\mathbb{N}}$ is an orthonormal system of $\mathcal{H}$.

$\boxed{\Leftarrow}$: let $A \in \mathcal{B}(\mathcal{H})$, $(u_n)_{n\in\mathbb{N}}$ be the complete orthonormal system of $\mathcal{H}$ and $(Au_n)_{n\in\mathbb{N}}$ an orthonormal system of $\mathcal{H}$. We wish to prove that A is isometric.

On one side, the fact that $(u_n)_{n\in\mathbb{N}}$ is a complete orthonormal system implies that, for all $x \in \mathcal{H}$, we have the decomposition into a generalized Fourier series $x = \sum\limits_{n\in\mathbb{N}} \langle x, u_n\rangle u_n$ and Plancherel's identity $\|x\|^2 = \sum\limits_{n\in\mathbb{N}} |\langle x, u_n\rangle|^2$.

On the other side, by the continuity of A, we can write $Ax = \sum_{n\in\mathbb{N}} \langle x, u_n\rangle\, Au_n$; furthermore, the hypothesis that $(Au_n)_{n\in\mathbb{N}}$ is an orthonormal system of $\mathcal{H}$ allows us to use the second part of the Riesz-Fischer theorem (Theorem 5.10) to state that[8]:

$$\|Ax\|^2 = \sum_{n\in\mathbb{N}} |\langle x, u_n\rangle|^2$$

that is, $\|Ax\|^2 = \|x\|^2$, therefore A is isometric. □

THEOREM 6.43 (Geometric characterization of unitary operators).– $A \in \mathcal{B}(\mathcal{H})$ is unitary if and only if it transforms complete orthonormal systems $(u_n)_{n\in\mathbb{N}}$ in $\mathcal{H}$ into complete orthonormal systems $(Au_n)_{n\in\mathbb{N}}$.

PROOF.–

$\boxed{\Longrightarrow}$: a unitary operator A is isometric, thus by Theorem 6.43, $(Au_n)_{n\in\mathbb{N}}$ is an orthonormal system and we simply need to show that $(Au_n)_{n\in\mathbb{N}}$ is complete. We do this using one of the characteristic properties of a Hilbert basis: As we saw in point 2 of Theorem 5.11, if $\langle x, Au_n\rangle = 0\ \forall n \in \mathbb{N}$ implies $x = 0_{\mathcal{H}}$, then $(Au_n)_{n\in\mathbb{N}}$ is a complete orthonormal system. Since A is surjective, there exists $y \in \mathcal{H}$ such that $x = Ay$; hence, the condition $\langle x, Au_n\rangle = 0\ \forall n \in \mathbb{N}$ becomes:

$$\forall n \in \mathbb{N} : \langle Ay, Au_n\rangle \underset{A \text{ unitary}}{=} \langle y, u_n\rangle = 0$$

and since $(u_n)_{n\in\mathbb{N}}$ is a complete orthonormal system of $\mathcal{H}$, $y = 0_{\mathcal{H}}$, implying $x = A0_{\mathcal{H}} = 0_{\mathcal{H}}$, then $(Au_n)_{n\in\mathbb{N}}$ is a complete orthonormal system of $\mathcal{H}$.

$\boxed{\Longleftarrow}$: by Theorem 6.43, we can guarantee that, if A transforms complete orthonormal systems $(u_n)_{n\in\mathbb{N}}$ of $\mathcal{H}$ into complete orthonormal system $(Au_n)_{n\in\mathbb{N}}$ in $\mathcal{H}$, then A is at least isometric; thus, we only need to demonstrate its surjectivity.

We have seen that the image of an isometric operator is closed, that is, by linearity, $\mathrm{Im}(A) = \overline{\mathrm{span}(Au_n,\ n \in \mathbb{N})} = \mathcal{H}$, since $(Au_n)_{n\in\mathbb{N}}$ is a complete orthonormal system by hypothesis, thus A is surjective, implying that A is unitary. □

We end this section with a simple exercise involving both unitary operators and orthogonal projectors.

---

8 Explicitly, the second part of the Riesz-Fischer theorem tells us that, given an orthonormal system $(v_n)_{n\in\mathbb{N}}$ in a Hilbert space $\mathcal{H}$, if the series $\sum_{n\in\mathbb{N}} k_n v_n$ converges to $y \in \mathcal{H}$, then it holds that $\|y\|^2 = \sum_{n\in\mathbb{N}} |k_n|^2$; in our case, $y = Ax$, $v_n = Au_n$ and $k_n = \langle x, u_n\rangle$.

Exercise 6.7

Let $\mathcal{H}$ be a Hilbert space. Show that the following properties are equivalent.

1) $A \in \mathcal{B}(\mathcal{H})$ is self-adjoint and unitary.

2) The operator $P = \frac{1}{2}(A + id_{\mathcal{H}})$ is an orthogonal projector.

3) There exist two mutually orthogonal closed subspaces $\mathcal{H}_1$ and $\mathcal{H}_2$ in $\mathcal{H}$ such that $\mathcal{H} = \mathcal{H}_1 \oplus \mathcal{H}_2$ and there exists an operator $A \in \mathcal{B}(\mathcal{H})$ such that, for all $x = x_1 + x_2$, $x_i \in \mathcal{H}_i$, it holds that $Ax = x_1 - x_2$.

Suggestion: show that 1) $\iff$ 2) and 2) $\iff$ 3).

Solution to Exercise 6.7

We begin with the equivalence 1) $\iff$ 2)

$\boxed{1) \implies 2)}$: By hypothesis, A is self-adjoint, that is, $A = A^\dagger$, and unitary, that is, $A^\dagger = A^{-1}$. Then $A = A^{-1}$ and thus $A^2 = AA = AA^{-1} = id_{\mathcal{H}}$. We can use this fact to show that $P = \frac{1}{2}(A + id_{\mathcal{H}})$ is self-adjoint and idempotent, implying that it is an orthogonal projector:

$$P^\dagger = \frac{1}{2}(A + id_{\mathcal{H}})^\dagger \underset{\text{linearity of }\dagger}{=} \frac{1}{2}(A^\dagger + id_{\mathcal{H}}^\dagger) = \frac{1}{2}(A + id_{\mathcal{H}}) = P$$

$$P^2 = \frac{1}{4}(A^2 + 2A + id_{\mathcal{H}}) = \frac{1}{4}(id_{\mathcal{H}} + 2A + id_{\mathcal{H}}) = \frac{1}{2}(A + id_{\mathcal{H}}) = P$$

$\boxed{2) \implies 1)}$: If property 2 holds, then we write $A = 2P - id_{\mathcal{H}}$, where P is an orthogonal projector, and we prove that A is self-adjoint and unitary:

$$A^\dagger = 2P^\dagger - id_{\mathcal{H}}^\dagger = 2P - id_{\mathcal{H}} = A$$

$$\begin{aligned} A^\dagger A &= A^\dagger A = (2P - id_{\mathcal{H}})^2 = 4P^2 - 4P + id_{\mathcal{H}} = 4P - 4P + id_{\mathcal{H}} \\ &= id_{\mathcal{H}} \implies A^\dagger = A^{-1} \end{aligned}$$

The next step is to analyze the equivalence 2) $\iff$ 3).

$\boxed{2) \implies 3)}$: If property 2 holds, then we know that $\mathcal{H} = \mathrm{Im}(P) \oplus \ker(P)$, hence $\mathcal{H}_1 = \mathrm{Im}(P)$ and $\mathcal{H}_2 = \ker(P)$. Furthermore, if we write $\mathcal{H} \ni x = x_1 + x_2$, $x_1 \in \mathrm{Im}(P)$ and $x_2 \in \ker(P)$:

$$Px = x_1 \text{ and } P(x) = \frac{1}{2}(A + id_{\mathcal{H}})(x) = \frac{1}{2}(Ax + x_1 + x_2)$$

that is, $x_1 = \frac{1}{2}(Ax + x_1 + x_2)$, and then $Ax = 2x_1 - x_1 - x_2 = x_1 - x_2$.

$\boxed{3) \implies 2)}$: Assuming that property 3 is verified, $P(x) = \frac{1}{2}(Ax + x) = \frac{1}{2}(x_1 - x_2 + x_1 + x_2) = x_1$ for all $x \in \mathcal{H}$, thus, by definition, P is the orthogonal projector $P_{\mathcal{H}_1}$ by the hypothesis that $\mathcal{H}_1$ and $\mathcal{H}_2$ are orthogonal and closed. □

6.9. The Fourier transform on $\mathcal{S}(\mathbb{R}^n)$, $L^1(\mathbb{R}^n)$ and $L^2(\mathbb{R}^n)$

The Fourier transform on $L^2(\mathbb{R}^n)$ is the most important example of a unitary operator on $L^2(\mathbb{R}^n)$ in terms of its applications to theoretical physics, differential equation theory and signal processing, among others.

Nonetheless, this operator is not simple to construct, as $L^2(\mathbb{R}^n)$ is not the most natural space for the Fourier transform; the most suitable environment for the Fourier transform is, in fact, the Schwartz space.

Several constructions of the Fourier transform on $L^2(\mathbb{R}^n)$ can be found in the literature; the most widespread, which shall be used here, consists of defining the Fourier transform on the Schwartz space to highlight its remarkable properties, and then operating an extension to $L^2(\mathbb{R}^n)$ using a limit procedure. In addition to this result, we shall present an explicit formula which makes use of the Hermite basis of $L^2(\mathbb{R}^n)$.

6.9.1. *The invariance of the Schwartz space with respect to the Fourier transform*

Let us begin by defining the Fourier transform on the Schwartz space $\mathcal{S}(\mathbb{R}^n)$ for $n = 1$. We will then generalize this definition for an arbitrary (finite) n.

DEFINITION 6.21.– *The Fourier transform on $\mathcal{S}(\mathbb{R})$ is the following linear operator:*

$$\begin{aligned} \hat{F} : \mathcal{S}(\mathbb{R}) &\longrightarrow \mathcal{S}(\mathbb{R}) \\ f &\longmapsto \hat{F}(f) = \hat{f}, \text{ where: } \hat{f}(\omega) = \frac{1}{\sqrt{2\pi}} \int_{\mathbb{R}} f(x) e^{-i\omega x} dm(x) \end{aligned}$$

where m is the Lebesgue measure on $\mathbb{R}$ and $\omega \in \mathbb{R}$. The inverse Fourier transform on $\mathcal{S}(\mathbb{R})$ is the following linear operator:

$$\begin{aligned} \check{F} : \mathcal{S}(\mathbb{R}) &\longrightarrow \mathcal{S}(\mathbb{R}) \\ f &\longmapsto \check{F}(f) = \check{f}, \text{ where: } \check{f}(x) = \frac{1}{\sqrt{2\pi}} \int_{\mathbb{R}} f(\omega) e^{i\omega x} dm(\omega) \end{aligned}$$

More generally, the Fourier transform on $\mathcal{S}(\mathbb{R}^n)$ is the following linear operator:

$$\begin{aligned} \hat{F} : \mathcal{S}(\mathbb{R}^n) &\longrightarrow \mathcal{S}(\mathbb{R}^n) \\ f &\longmapsto \hat{F}(f) = \hat{f}, \text{ where: } \hat{f}(\omega) = \frac{1}{(2\pi)^{n/2}} \int_{\mathbb{R}^n} f(x) e^{-i\langle \omega, x \rangle} dm(x) \end{aligned}$$

where m is the Lebesgue measure on $\mathbb{R}^n$, $\omega \in \mathbb{R}^n$ and $\langle \omega, x \rangle = \sum_{k=1}^{n} \omega_1 x_i$ is the Euclidean inner product in $\mathbb{R}^n$. The inverse Fourier transform on $\mathcal{S}(\mathbb{R}^n)$ is the following linear operator:

$$\begin{aligned} \check{F} : \mathcal{S}(\mathbb{R}^n) &\longrightarrow \mathcal{S}(\mathbb{R}^n) \\ f &\longmapsto \check{F}(f) = \check{f}, \text{ where: } \check{f}(x) = \frac{1}{(2\pi)^{n/2}} \int_{\mathbb{R}^n} f(\omega) e^{i\langle \omega, x \rangle} dm(\omega) \end{aligned}$$

To verify that these definitions are well posed, we must ensure that the integrals exist and that $\hat{f}$ and $\check{f}$ are rapidly decreasing functions. The existence of the integrals is evident if we consider that $\mathcal{S}(\mathbb{R}^n) \subset L^1(\mathbb{R}^n)$, thus:

$$\int_{\mathbb{R}^n} \left| f(x) e^{-i\langle \omega, x \rangle} \right| dm(x) = \int_{\mathbb{R}^n} |f(x)| \, dm(x) < +\infty.$$

The same is true for the inverse Fourier transform. The fact that $\hat{f}$ and $\check{f}$ are rapidly decreasing functions can be verified by iterating the derivation under the integral sign and by integrating by parts.

A summary of the most important properties of the Fourier transform for a function $f \in \mathcal{S}(\mathbb{R})$, $a, b, c \in \mathbb{R}$, $a \neq 0$ is given in Table 6.4.

IMPORTANT OBSERVATIONS.–

– $\hat{F}$ transforms the product by a constant into a division by the same constant (up to a coefficient).

– $\hat{F}$, like the DFT, transforms the shift of the initial variable into the product by a complex exponential.

– $\hat{F}$ transforms the n-th derivation into the product by a power of $i\omega$. This property is crucial for transforming differential equations into algebraic equations.

– $\hat{F}$ transforms a Gaussian with unit standard deviation into a Gaussian with unit standard deviation. More generally, $\hat{F}$ *inverts the standard deviation*: a Gaussian with a small standard deviation, that is, with values located in close proximity to its mean, is transformed by $\hat{F}$ into a Gaussian with a large standard deviation, that is, with values which are spread away from the mean, and *vice versa*.

Original function $f \in \mathcal{S}(\mathbb{R})$	Fourier transform $\hat{f} \in \mathcal{S}(\mathbb{R})$
$f(ax)$	$\frac{1}{\lvert a \rvert} \hat{f}\left(\frac{\omega}{a}\right)$
$f(x-b)$	$e^{-i\omega b} \hat{f}(\omega)$
$f(ax-b)$	$\frac{e^{-i\omega \frac{b}{a}}}{\lvert a \rvert} \hat{f}\left(\frac{\omega}{a}\right)$
$e^{icx} f(x)$	$\hat{f}(\omega - c)$
$f'(x)$	$i\omega \hat{f}(\omega)$
$f''(x)$	$-\omega^2 \hat{f}(\omega)$
$\frac{d^n f}{dx^n}$	$(i\omega)^n \hat{f}(\omega)$
$(-ix)^n f(x)$	$\frac{d^n \hat{f}}{d\omega^n}(\omega)$
$e^{\frac{x^2}{2}}$	$e^{-\frac{\omega^2}{2}}$
$e^{-c^2 x^2}$	$\frac{1}{c\sqrt{2}} e^{-\frac{\omega^2}{4c^2}}$

Table 6.1. *Properties of the Fourier transform on $\mathcal{S}(\mathbb{R})$*

We wish to prove the property $\widehat{f'}(\omega) = i\omega\hat{f}(\omega)$.

PROOF.– We begin by observing that for $f : \mathbb{R} \to \mathbb{C}$, $f \in \mathcal{S}(\mathbb{R})$, then $f(x) \underset{|x|\to+\infty}{\longrightarrow} 0$. We write the Fourier transform of f':

$$\begin{aligned}\widehat{f'}(\omega) &= \frac{1}{\sqrt{2\pi}}\int_{-\infty}^{+\infty} f'(x)e^{-i\omega x}dx = \text{(int. by parts)}\\ &= \frac{1}{\sqrt{2\pi}}\left[f(x)e^{-i\omega x}\right]_{-\infty}^{+\infty} - \frac{1}{\sqrt{2\pi}}\int_{-\infty}^{+\infty} f(x)(-i\omega)e^{-i\omega x}dx\\ &= 0 - (-i\omega)\frac{1}{\sqrt{2\pi}}\int_{-\infty}^{+\infty} f(x)e^{-i\omega x}dx\\ &= i\omega\hat{f}(\omega)\end{aligned}$$

□

The fact that the Gaussian with unit standard deviation is invariant with respect to the Fourier transform is not immediately evident, so a proof is helpful. For that, we need two lemmas.

LEMMA 6.1.– It holds that:

$$\int_{-\infty}^{+\infty} e^{-\frac{x^2}{2}}dx = \sqrt{2\pi}$$

PROOF.– We write $I = \int_{-\infty}^{+\infty} e^{-\frac{x^2}{2}}dx$, then:

$$I^2 = \int_{-\infty}^{+\infty} e^{-\frac{x^2}{2}}dx \cdot \int_{-\infty}^{+\infty} e^{-\frac{y^2}{2}}dy \underset{\text{(th. Fubini)}}{=} \int_{-\infty}^{+\infty}\int_{-\infty}^{+\infty} e^{-\frac{1}{2}(x^2+y^2)}dxdy$$

Switching to polar coordinates (ρ, ϑ), $\rho \in [0, +\infty)$, $\vartheta \in [0, 2\pi)$ and recalling that the Jacobian in polar coordinates is ρ, we obtain:

$$\begin{aligned}I^2 &= \int_0^{+\infty}\int_0^{2\pi} e^{-\frac{\rho^2}{2}}\rho\, d\rho d\vartheta = \int_0^{2\pi} d\vartheta \int_0^{+\infty} e^{-\frac{\rho^2}{2}}\rho\, d\rho\\ &= 2\pi\left[-e^{-\frac{\rho^2}{2}}\right]_0^{+\infty} = 2\pi\left[-e^{-\infty} + e^0\right] = 2\pi\end{aligned}$$

Thus $I = \sqrt{2\pi}$. □

LEMMA 6.2.– It holds that:

$$\int_{-\infty}^{+\infty} e^{-\frac{(x+i\omega)^2}{2}}dx = \int_{-\infty}^{+\infty} e^{-\frac{x^2}{2}}dx$$

The proof uses the calculus of residues of complex analysis.

We can now prove that:

$$\widehat{e^{-\frac{x^2}{2}}}(\omega) = e^{-\frac{\omega^2}{2}}$$

PROOF.–

$$\begin{aligned}
\widehat{e^{-\frac{x^2}{2}}}(\omega) &= \frac{1}{\sqrt{2\pi}} \int_{-\infty}^{+\infty} e^{-\frac{x^2}{2}} e^{-i\omega x} dx \\
&\underset{\cdot e^{-\frac{\omega^2}{2}} \cdot e^{\frac{\omega^2}{2}}}{=} \frac{e^{-\frac{\omega^2}{2}}}{\sqrt{2\pi}} \int_{-\infty}^{+\infty} e^{-\frac{x^2}{2}} e^{-i\omega x} e^{\frac{\omega^2}{2}} dx \\
&= \frac{e^{-\frac{\omega^2}{2}}}{\sqrt{2\pi}} \int_{-\infty}^{+\infty} e^{-\frac{x^2+2i\omega x-\omega^2}{2}} dx \\
&= \frac{e^{-\frac{\omega^2}{2}}}{\sqrt{2\pi}} \int_{-\infty}^{+\infty} e^{-\frac{(x+i\omega)^2}{2}} dx \\
&\underset{\text{Lemma 6.2}}{=} \frac{e^{-\frac{\omega^2}{2}}}{\sqrt{2\pi}} \int_{-\infty}^{+\infty} e^{-\frac{x^2}{2}} dx \\
&\underset{\text{Lemma 6.1}}{=} \frac{e^{-\frac{\omega^2}{2}}}{\sqrt{2\pi}} \sqrt{2\pi} = e^{-\frac{\omega^2}{2}}
\end{aligned}$$

□

The inversion of the standard deviation, i.e. the fact that $e^{-c^2x^2} \underset{\hat{F}}{\mapsto} \frac{1}{c\sqrt{2}} e^{-\frac{\omega^2}{4c^2}}$, can be proven using an alternative technique (evidently, the technique presented earlier is also an option).

This technique is based on solving a differential equation. If $f(x) = e^{-c^2x^2}$, then $f'(x) = -2c^2xf(x)$, thus $f' + 2c^2xf = 0$ and, given the properties $f'(x) \underset{\hat{F}}{\mapsto} i\omega\hat{f}(\omega)$, $-ixf(x) \underset{\hat{F}}{\mapsto} \hat{f}'(\omega)$ and the fact that $2c^2xf = i2c^2(-ixf)$, by applying $\hat{F}$ to both sides of the previous differential equation we can write:

$$i\omega\hat{f}(\omega) + i2c^2\hat{f}'(\omega) = 0 \iff \omega\hat{f}(\omega) + 2c^2\hat{f}'(\omega) = 0 \qquad [6.26]$$

This gives us a *separable differential equation*[9] *with respect to* $\hat{f}$. The canonical technique for solving this type of differential equation is to first search for constant solutions $\hat{f}(\omega) = C \in \mathbb{R}\ \forall\omega \in \mathbb{R}$, implying $\hat{f}'(\omega) = 0\ \forall\omega \in \mathbb{R}$, thus [6.26] becomes $\omega\hat{f}(\omega) = 0$ which may only be verified for all $\omega \in R$ when $\hat{f}(\omega) = 0$; hence, the only

9 We recall that a differential equation with respect to a function $y(t)$ is said to be separable if it can be written as $y'(t) = f(y(t)) \cdot g(t)$, where f and g are two continuous functions.

constant solution to the differential equation [6.26] is the identically zero function. However, this function is not coherent with the fact that $\hat{f}(0) \neq 0$:

$$\begin{aligned}\hat{f}(0) &\underset{\text{def. of } \hat{f}(0)\,!}{=} \frac{1}{\sqrt{2\pi}} \int_{\mathbb{R}} f(x) e^{-i0x} dx \\ &= \frac{1}{\sqrt{2\pi}} \int_{\mathbb{R}} f(x) dx = \frac{1}{\sqrt{2\pi}} \int_{\mathbb{R}} e^{-c^2x^2} dx \\ &= \frac{1}{\sqrt{2\pi} c\sqrt{2}} \int_{\mathbb{R}} e^{-y^2/2} dy \underset{\text{Lemma (6.1)}}{=} \frac{1}{\sqrt{2\pi} c\sqrt{2}} \sqrt{2\pi} = \frac{1}{c\sqrt{2}}\end{aligned}$$

Hence, $\hat{f}(0) \equiv 0$ is not a solution to [6.26]. Now, let us suppose that $\hat{f}(\omega) \neq 0$ and look for non-constant solutions to [6.26] using the variable separation technique. We write the equation as follows:

$$\frac{\hat{f}'(\omega)}{\hat{f}(\omega)} = -\frac{\omega}{2c^2}$$

Integrating both sides we obtain: $\log |\hat{f}(\omega)| = -\frac{\omega^2}{4c^2} + \log C$, $C > 0$, where $\log C$ is the arbitrary constant resulting from integration. It is written in this way because, taking the exponential of both sides, we obtain:

$$|\hat{f}(\omega)| = e^{-\frac{\omega^2}{4c^2} + \log C} = Ce^{-\frac{\omega^2}{4c^2}} \qquad \hat{f}(\omega) = \pm Ce^{-\frac{\omega^2}{4c^2}}$$

that is, $\hat{f}(\omega) = Ke^{-\frac{\omega^2}{4c^2}}$, $K \in \mathbb{R} \backslash \{0\}$. Now, we simply observe that $K = \hat{f}(0) = \frac{1}{c\sqrt{2}}$, as before, which gives us the solution $\hat{f}(\omega) = \frac{1}{c\sqrt{2}} e^{-\frac{\omega^2}{4c^2}}$.

The properties of the Fourier transform defined on $\mathcal{S}(\mathbb{R}^n)$, summarized in Table 6.2 (where $c \in \mathbb{R}$, $c \neq 0$, $a, b \in \mathbb{R}^n$, $k \in \{1, \ldots, n\}$), follow directly from those obtained in the case where $n = 1$, with relatively straightforward changes to the demonstration technique, notably involving the use of Fubini's theorem to calculate multiple integrals.

We end this section by presenting the result which makes the Schwartz space so interesting for Fourier transform theory (and which justifies the name of $\check{F}$).

THEOREM 6.11. The transform $\hat{F}$ is a linear isomorphism of $\mathcal{S}(\mathbb{R}^n)$ in itself, and its inverse transformation is $\check{F}$: $\check{F} = \hat{F}^{-1}$. Furthermore, if $f \in \mathcal{S}(\mathbb{R}^n)$ is interpreted as a function of $L^2(\mathbb{R}^n)$, then: $\|f\| = \|\hat{f}\|\ \forall f \in \mathcal{S}(\mathbb{R}^n) \subset L^2(\mathbb{R}^n)$.

The Schwartz space is thus invariant with respect to the application of the Fourier transform $\hat{F}$, which possesses an explicit integral formula and an explicit inverse given by $\check{F}$ and conserves the norm of rapidly decreasing functions when these are interpreted as elements of $L^2(\mathbb{R}^n)$. There is no other infinite-dimensional functional space in which the Fourier transform possesses all of these properties simultaneously.

Original function $f \in \mathcal{S}(\mathbb{R}^n)$	Fourier transform $\hat{f} \in \mathcal{S}(\mathbb{R}^n)$
$f(cx)$	$\frac{1}{\lvert c\rvert}\hat{f}\left(\frac{\omega}{c}\right)$
$f(x-b)$	$e^{-i\langle\omega,b\rangle}\hat{f}(\omega)$
$f(cx-b)$	$\frac{e^{-i\omega\frac{b}{c}}}{\lvert c\rvert}\hat{f}\left(\frac{\omega}{c}\right)$
$e^{i\langle a,x\rangle}f(x)$	$\hat{f}(\omega-a)$
$\partial_{x_k}f(x)$	$i\omega_k\hat{f}(\omega)$
$\partial^2_{x_k}f(x)$	$-\omega_k^2\hat{f}(\omega)$
$\partial^n_{x_k}f(x)$	$(i\omega_k)^n\hat{f}(\omega)$
$(-ix_k)^n f(x)$	$\partial^n_{x_k}\hat{f}(\omega)$
$e^{-\frac{\lVert x\rVert^2}{2}}$	$e^{-\frac{\lVert\omega\rVert^2}{2}}$
$e^{-c^2\lVert x\rVert^2}$	$\frac{1}{c\sqrt{2}}e^{-\frac{\lVert\omega\rVert^2}{4c^2}}$

Table 6.2. *Properties of the Fourier transform on $\mathcal{S}(\mathbb{R}^n)$*

As we shall see, $L^1(\mathbb{R})$ is not invariant under Fourier transform, while in $L^2(\mathbb{R})$ we loose the explicit integral formula.

6.9.2. *Extension of the Fourier transform of $\mathcal{S}(\mathbb{R}^n)$ to $L^1(\mathbb{R}^n)$: the Riemann-Lebesgue theorem*

The functions which constitute the elements of the Schwartz are too regular to be exhaustive, particularly with respect to applications.

It is thus important to consider the extension of the Fourier transform to less regular function spaces, such as $L^1(\mathbb{R})$ and $L^2(\mathbb{R})$. In this section, we shall consider $L^1(\mathbb{R})$, for which we have a particularly famous result.

THEOREM 6.45 (Riemann-Lebesgue theorem).– The operator $\hat{F}$ from section 6.9.1 can be extended in a unique manner to the *injective* and *continuous* linear operator defined as follows:

$$\begin{aligned} \hat{F}_1 : L^1(\mathbb{R}^n) &\longrightarrow C_\infty(\mathbb{R}^n) \\ f &\longmapsto \hat{F}_1(f) =, \text{ where: } \hat{F}_1 f(\omega) = \frac{1}{(2\pi)^{n/2}}\int_{\mathbb{R}^n} f(x)e^{-i\langle\omega,x\rangle}dm(x) \end{aligned}$$

The same statement holds for the extension of $\check{F}$ to $L^1(\mathbb{R}^n)$ with the corresponding integral function, that is:

$$\begin{aligned} \check{F}_1 : L^1(\mathbb{R}^n) &\longrightarrow \mathcal{C}_\infty(\mathbb{R}^n) \\ f &\longmapsto \check{F}_1(f) =, \text{ or}: \check{F}_1 f(x) = \tfrac{1}{(2\pi)^{n/2}} \textstyle\int_{\mathbb{R}^n} f(\omega) e^{i\langle \omega, x\rangle} dm(\omega) \end{aligned}$$

We recall that $\mathcal{C}_\infty(\mathbb{R}^n)$ is the space of defined and continuous functions on $\mathbb{R}^n$ which tend toward 0 as we approach infinity, equipped with the norm $\|f\|_\infty = \sup_{x\in\mathbb{R}^n} |f(x)|$.

OBSERVATIONS.–

– The Riemann-Lebesgue theorem tells us that the integral formula of the Fourier transform remains valid for the elements of $L^1(\mathbb{R}^n)$; this is very important, since functions which are absolutely integrable in the Lebesgue sense are much more widespread than rapidly decreasing functions in practical applications.

– The injectivity of F_1 means that it can be inverted on the image $F_1(L^1(\mathbb{R}^n)) \subset \mathcal{C}_\infty(\mathbb{R}^n)$ but not on $L^1(\mathbb{R}^n)$. A classic counter-example for the case where $n = 1$ is the indicator function for the interval $[-1, 1]$ in $\mathbb{R}$, that is, $\chi_{[-1,1]}$; it belongs to $L^1(\mathbb{R})$, but by direct calculation we obtain:

$$\hat{F}_1(\chi_{[-1,1]})(\omega) = \sqrt{\frac{2}{\pi}} \frac{\sin \omega}{\omega} \tag{6.27}$$

This evidently belongs to $\mathcal{C}_\infty(\mathbb{R})$ but not to $L^1(\mathbb{R})$; it actually belongs to $L^2(\mathbb{R})$. Thus, $\check{F}_1$, which is defined on all $L^1(\mathbb{R})$, is not the inverse of $\hat{F}_1$.

6.9.3. *Extension of the Fourier transform to a unitary operator on $L^2(\mathbb{R}^n)$: the Fourier-Plancherel transform*

The technique which is classically used to extend $\hat{F}$ to $L^2(\mathbb{R}^n)$ consists of using a theorem that is of fundamental importance in functional analysis, which will be presented and proved below. First, however, we must establish a definition of the extension of a linear operator.

DEFINITION 6.22 (bounded extension of bounded linear operators).– *Let E, V, W be vector spaces on the same field $\mathbb{K}$ and let E be a vector subspace of V. Let $A : E \to W$ be a linear operator. The linear operator $B : V \to W$ is an extension of A on V if the restriction of B to E coincides with A, that is, if $Ax = Bx\ \forall x \in E$.*

THEOREM 6.46 (Theorem of extension of a bounded linear operator).– Let E and F be two normed vector spaces, with F a Banach space. Let $A : D_A \subseteq E \to F$ be a bounded linear operator, where D_A is a vector subspace of E. Then, there exists only one linear operator $\overline{A}$ with the following properties:

1) the domain of $\overline{A}$ is the closure of D_A in E: $D_{\overline{A}} = \overline{D_A}$;

2) $\overline{A}$ is continuous: $A \in \mathcal{B}(\overline{D_A}, F)$;

3) $\|\overline{A}\| = \|A\|$.

This operator is defined as follows. Let $(x_n)_{n\in\mathbb{N}} \subset D_A$ be an *arbitrary* sequence which converges to $x \in \overline{D_A}$, then:

$$\begin{aligned} \overline{A} : \overline{D_A} \subseteq E &\longrightarrow F \\ x &\longmapsto \boxed{\overline{A}x = \lim_{n\to+\infty} Ax_n} \end{aligned}$$

PROOF.– Let x be an arbitrary element in $D_{\overline{A}} = \overline{D_A}$, then, by definition, there exists a sequence $(x_n)_{n\in\mathbb{N}} \subset D_A$ such that $x = \lim_{n\to+\infty} x_n$. $(x_n)_{n\in\mathbb{N}}$. Being convergent, $(x_n)_{n\in\mathbb{N}}$ is a Cauchy sequence and, since A is continuous, the sequence $(Ax_n)_{n\in\mathbb{N}} \subset F$ is also a Cauchy sequence, by Theorem 6.9.

Since F is a Banach space, there exists $y = \lim_{n\to+\infty} Ax_n$; thus, the operator $\overline{A}$: $\overline{D_A} \to F$, $\overline{A}x = \lim_{n\to+\infty} Ax_n$ is well defined and linear, as it is defined *via* the limit operation, which is linear.

Furthermore, $\overline{A}$ does not depend on the sequence which converges to x; in fact, if $(x'_n)_{n\in\mathbb{N}} \subset E$ is another sequence such that $x = \lim_{n\to+\infty} x'_n$, then:

$$\begin{aligned} &\|\lim_{n\to+\infty} Ax_n - \lim_{n\to+\infty} Ax'_n\| = \lim_{n\to+\infty} \|Ax_n - Ax'_n\| \quad \text{(Continuity of } \| \ \|) \\ &= \lim_{n\to+\infty} \|A(x_n - x'_n)\| \leqslant \lim_{n\to+\infty} \|A\|\|x_n - x'_n\| \quad (A \text{ bounded}) \\ &= \|A\| \lim_{n\to+\infty} \|x_n - x'_n\| = \|A\| \, \|\lim_{n\to+\infty}(x_n - x'_n)\| \quad \text{(Continuity of } \| \ \|) \\ &= \|A\| \, \|\lim_{n\to+\infty} x_n - \lim_{n\to+\infty} x'_n\| = \|A\|\|x - x\| = 0 \end{aligned}$$

Evidently, any $x \in D_A$ may be identified as the limit of the constant sequence $x_n = x \ \forall n \in \mathbb{N}$; hence, given that the definition of $\overline{A}$ is independent with respect to the chosen sequence, $\overline{A}x = Ax \ \forall x \in D_A$, that is, the restriction of $\overline{A}$ to D_A is A and, inversely, $\overline{A}$ is an extension of A on $\overline{D_A}$.

The fact that $\overline{A}$ is a bounded operator on $\overline{D_A}$ can be verified by considering the limit of the inequality $\|Ax_n\| \leqslant \|A\|\|x_n\|$. The limit conserves the order relation, that is:

$$\lim_{n\to+\infty} \|Ax_n\| \leqslant \lim_{n\to+\infty} \|A\|\|x_n\|$$

and, by the continuity of the norm, we have:

$$\| \lim_{n \to +\infty} Ax_n \| \leqslant \|A\| \, \| \lim_{n \to +\infty} x_n \| \iff \|\overline{A}x\| \leqslant \|A\| \|x\|$$

for all $x \in \overline{D_A}$, that is, $\overline{A}$ is bounded, and thus continuous.

Now, let us prove that any other extension of A to $\overline{D_A}$ must coincide with $\overline{A}$. Let B be another bounded extension of A on $\overline{D_A}$, then, for all $x \in \overline{D_A}$, there exists a sequence $(x_n)_{n \in \mathbb{N}} \subset D_A$, such that $x = \lim_{n \to +\infty} x_n$ and by the *definition* of $\overline{A}$ and the *continuity* of B we have:

$$\overline{A}x - Bx = \lim_{n \to +\infty} Ax_n - B\left(\lim_{n \to +\infty} x_n \right) = \lim_{n \to +\infty} Ax_n - \lim_{n \to +\infty} Bx_n$$

$$= \lim_{n \to +\infty} (Ax_n - Bx_n)$$

For all $n \in \mathbb{N}$, $x_n \in D_A$ and, since B is an extension of A, by definition $Bx_n = Ax_n \; \forall n \in \mathbb{N}$, then $Ax_n - Bx_n = 0 \; \forall n \in \mathbb{N}$ and thus $\overline{A}x - Bx = \lim_{n \to +\infty} (Ax_n - Bx_n) = \lim_{n \to +\infty} 0 = 0$, i.e. $\overline{A} = B$.

Finally, we need to show that the extension is isometric, that is, $\|\overline{A}\| = \|A\|$. We have already seen that $\left\|\overline{A}x\right\| \leqslant \|A\| \, \|x\|$ for all $x \in \overline{D_A}$, thus:

$$\left\|\overline{A}\right\| = \sup_{\|x\|=1} \left\|\overline{A}x\right\| \leqslant \sup_{\|x\|=1} \|A\| \, \|x\| = \|A\|$$

then, if we can show that $\left\|\overline{A}\right\| \geqslant \|A\|$, this will prove the isometry of the extension. The proof is straightforward if we consider the definition of the following operator norm:

$$\|\overline{A}\| = \sup \left\{ \frac{\|\overline{A}x\|}{\|x\|}, \; x \in \overline{D_A} \backslash \{0_E\} \right\} \geqslant \sup \left\{ \frac{\|Ax\|}{\|x\|}, \; x \in D_A \backslash \{0_E\} \right\} = \|A\|$$

since $\overline{A}x = Ax \; \forall x \in D_A$ and $D_A \subseteq \overline{D_A}$, hence $\left\|\overline{A}\right\| = \|A\|$ and the theorem is fully proven. □

Using the extension theorem and the fact that $\overline{\mathcal{S}(\mathbb{R}^n)} = L^2(\mathbb{R}^n)$, the Fourier transform of the Schwartz space can be extended to $L^2(\mathbb{R}^n)$ *via* the limit formula of the extension theorem, as formalized as follows.

THEOREM 6.47.– The operators $\hat{F}$ and $\check{F}$ which define the Fourier transform and the inverse Fourier transform on $\mathcal{S}(\mathbb{R}^n)$, respectively, can be extended in a unique manner to two *unitary* operators F and $\tilde{F}$ on $L^2(\mathbb{R}^n)$; furthermore, $\tilde{F} = F^{-1}$.

The operator F is known as the *Fourier-Plancherel transform* and it is defined as follows: let $(f_n)_{n\in\mathbb{N}} \subset \mathcal{S}(\mathbb{R}^n)$ be an arbitrary sequence of elements in $\mathcal{S}(\mathbb{R}^n)$ which converge to $f \in L^2(\mathbb{R}^n)$, then:

$$\begin{aligned} F : L^2(\mathbb{R}^n) &\longrightarrow L^2(\mathbb{R}^n) \\ f &\longmapsto F(f) = \lim_{n\to+\infty} \hat{f}_n. \end{aligned}$$

Analogously:

$$\begin{aligned} F^{-1} : L^2(\mathbb{R}^n) &\longrightarrow L^2(\mathbb{R}^n) \\ f &\longmapsto F(f) = \lim_{n\to+\infty} \check{f}_n \end{aligned}$$

Thus, *the Fourier-Plancherel transform F on $L^2(\mathbb{R}^n)$ has the vital properties of being a unitary operator* with inverse given by the unitary operator F^{-1}.

One reason $L^2(\mathbb{R}^n)$ is a less natural space than $\mathcal{S}(\mathbb{R}^n)$ for studying the Fourier transform is the lack of a valid integral formula for all elements of $L^2(\mathbb{R}^n)$. Theorem 6.48 provides a partial solution to this problem.

THEOREM 6.48.– If $f \in L^1(\mathbb{R}^n) \cap L^2(\mathbb{R}^n)$, then $F = \hat{F}_1$ and, for functions belonging to $L^1(\mathbb{R}^n) \cap L^2(\mathbb{R}^n)$, the integral formula of the Fourier transform remains valid.

Thankfully, as we saw in section 4.4.4, the functions of $L^1(\mathbb{R}^n) \cap L^2(\mathbb{R}^n)$ include the bounded functions of $L^1(\mathbb{R}^n)$ and those of $L^2(\mathbb{R}^n)$ which cancel outside of a compact subspace, often encountered in practical applications.

6.9.4. *Relationship between the Fourier-Plancherel transform and the Hermitian Hilbert basis*

One very important Hilbert basis in $L^2(\mathbb{R})$ is the Hermite basis, defined as:

$$u_n(x) = \frac{(-1)^n}{\sqrt{2^n n! \sqrt{\pi}}} e^{\frac{1}{2}x^2} \frac{d^n}{dx^n} e^{-x^2}, \quad x \in \mathbb{R},\ n \in \mathbb{N}$$

The functions u_n can be shown to decay rapidly, so their Fourier transform is obtained by applying the integral formula, that is:

$$\begin{aligned} Fu_n(\omega) &= \frac{1}{\sqrt{2\pi}} \int_{\mathbb{R}} u_n(x) e^{-i\omega x} dm(x) \\ &= \frac{1}{\sqrt{2\pi}} \frac{(-1)^n}{\sqrt{2^n n! \sqrt{\pi}}} \int_{\mathbb{R}} e^{-i\omega x + \frac{1}{2}x^2} \frac{d^n}{dx^n} e^{-x^2} dm(x). \end{aligned}$$

By means of some simple algebraic manipulations, we can show that:

$$Fu_n = (-i)^n u_n, \qquad n \in \mathbb{N}$$

and thus F coincides with the unitary operator introduced in section 6.8.1. By the continuity of F, we can write:

$$\boxed{Ff = \sum_{n \in \mathbb{N}} (-i)^n \langle f, u_n \rangle u_n}, \qquad \forall f \in L^2(\mathbb{R}),\ (u_n)_{n\in\mathbb{N}} \ : \ \text{Hermite basis}$$

6.9.5. *The Fourier transform and convolution*

The properties of the Fourier transform with respect to convolution merit a separate discussion, given their importance and usefulness in both theoretical and applied mathematics. Readers wishing to study this subject in greater detail are advised to consult Gasquet and Witomski (2013).

We shall begin by defining convolution and discussing its basic properties, before proving the best-known and most important property of the Fourier transform in $L^1(\mathbb{R}^n)$ with respect to convolution: the convolution product is transformed into the pointwise product of the Fourier transforms (to within a coefficient).

DEFINITION 6.23.– *Taking $f, g : \mathbb{R}^n \to \mathbb{R}$, the convolution between f and g is the function $f * g$ defined by:*

$$(f * g)(x) = \int_{\mathbb{R}^n} f(x-y)g(y)dm(y), \qquad \forall x \in \mathbb{R}^n$$

as long as the integral exists in the Lebesgue sense.

THEOREM 6.49 (Basic properties of convolution).– The following properties hold:

1) if $f \in L^1(\mathbb{R}^n)$ and $g \in L^\infty(\mathbb{R}^n)$ or *vice versa* , then the convolution is well defined;

2) if $f, g \in L^2(\mathbb{R}^n)$, then the convolution is well defined and, in general, is an element of $L^\infty(\mathbb{R}^n)$;

3) if $f, g \in L^1(\mathbb{R}^n)$, then the convolution is well defined and belongs to $L^1(\mathbb{R}^n)$, which becomes a Banach algebra with respect to the convolution;

4) if convolution is well defined, then:

- $f * (\alpha g + \beta h) = \alpha f * g + \beta f * h$ (linearity);
- $f * g = g * f$ (commutativity) ;
- $f * (g * h) = (f * g) * h$ (associativity).

PROOF.– Only the first two properties will be proved here. Proof of the remaining properties is left to the reader as an exercise.

1) If $f \in L^1(\mathbb{R}^n)$ and $g \in L^\infty(\mathbb{R}^n)$, then:

$$\int_{\mathbb{R}^n} |f(x-y)g(y)|dm(y) \leqslant \|g\|_\infty \int_{\mathbb{R}^n} |f(x-y)|dm(y) = \|g\|_\infty \|f\|_1$$

by the shift invariance of the Lebesgue measure.

2) If $f, g \in L^2(\mathbb{R}^n)$, then, by the Hölder inequality [4.19]:

$$\int_{\mathbb{R}^n} |f(x-y)g(y)|dm(y) \leqslant \left(\int_{\mathbb{R}^n} |f(x-y)|^2 dm(y)\right)^{1/2} \left(\int_{\mathbb{R}^n} |g(y)|^2 dm(y)\right)^{1/2}$$
$$= \|f\|_2 \|g\|_2$$

again by the shift invariance of the Lebesgue measure. □

THEOREM 6.50 (Convolution and Fourier transform in L^1).– Taking $f, g \in L^1(\mathbb{R}^n)$, then the Fourier transform verifies the following property:

$$\widehat{f * g} = (2\pi)^{n/2} \hat{f} \cdot \hat{g}$$

PROOF.– We simply write the definition of convolution and of the Fourier transform, then apply the Fubini theorem twice, with a minor algebraic manipulation in between:

$$\begin{aligned}
(\widehat{f * g})(\omega) &= \frac{1}{(2\pi)^{n/2}} \int_{\mathbb{R}^n} \left(\int_{\mathbb{R}^n} f(x-y)g(y)dm(y)\right) e^{-i\langle\omega,x\rangle} dm(x) \\
&= \frac{1}{(2\pi)^{n/2}} \int_{\mathbb{R}^n}\int_{\mathbb{R}^n} f(x-y)g(y)e^{-i\langle\omega,x\rangle} dm(x)dm(y) \quad \text{(Fubini)} \\
&= \frac{1}{(2\pi)^{n/2}} \int_{\mathbb{R}^n}\int_{\mathbb{R}^n} f(x-y)g(y)e^{-i\langle\omega,x-y+y\rangle} dm(x)dm(y) \\
&= \frac{1}{(2\pi)^{n/2}} \int_{\mathbb{R}^n}\int_{\mathbb{R}^n} f(x-y)e^{-i\langle\omega,x-y\rangle} g(y)e^{-i\langle\omega,y\rangle} dm(x)dm(y) \\
&= \frac{1}{(2\pi)^{n/2}} \int_{\mathbb{R}^n} f(x-y)e^{-i\langle\omega,x-y\rangle} dm(x) \int_{\mathbb{R}^n} g(y)e^{-i\langle\omega,y\rangle} dm(y) \quad \text{(Fubini)} \\
&= (t = x-y,\ dm(t) = dm(x-y)) \\
&= (2\pi)^{n/2} \frac{1}{(2\pi)^{n/2}} \int_{\mathbb{R}^n} f(t)e^{-i\langle\omega,t\rangle} dm(t) \frac{1}{(2\pi)^{n/2}} \int_{\mathbb{R}^n} g(y)e^{-i\langle\omega,y\rangle} dm(y) \\
&= (2\pi)^{n/2} \hat{f}(\omega) \cdot \hat{g}(\omega), \qquad \forall \omega \in \mathbb{R}^n.
\end{aligned}$$

□

If we inverse the Fourier transform on the image $F_1(L^1(\mathbb{R}^n)) \subset C_\infty(\mathbb{R}^n)$, then we obtain $f * g = (2\pi)^{n/2}(\hat{f} \cdot \hat{g})^\vee$, which is often written in the form:

$$(f \cdot g)^\vee = (2\pi)^{-n/2} \check{f} * \check{g} \qquad [6.28]$$

This formula will be used in section 6.11.

Convolution is a stationary operation, that is, it commutes with translation, as in the discrete case. Fixing $s \in \mathbb{R}$ and $g \in L^1(\mathbb{R})$, then we can define the right translation operator R_s and the convolution operator with g, T_g, as follows:

$$R_s f(t) = f(t-s), \quad T_g f(t) = (f * g)(t) = \int_{\mathbb{R}} f(t-x)g(x)dx$$

then, for all $t \in \mathbb{R}$:

$$R_s T_g f(t) = T_g f(t-s) = \int_{\mathbb{R}} f(t-s-x)g(x)dx = \int_{\mathbb{R}} R_s f(t-x)g(x)dx = T_g R_s f(t)$$

As we saw in the discrete case (see section 2.9.6), the convolution operation with the Gaussian function results in blurring of a signal. This result can be understood from a different perspective, using the following impulse function:

$$I_\varepsilon(t) = \begin{cases} \frac{1}{\varepsilon} & 0 < t < \varepsilon \\ 0 & \text{otherwise} \end{cases}$$

If $f \in L^1(\mathbb{R})$, then:

$$(f * I_\varepsilon)(t) = \int_{\mathbb{R}} f(t-x)I_\varepsilon(x)dx = \frac{1}{\varepsilon}\int_0^\varepsilon f(t-x)dx$$

Now, applying the variable substitution $u = t - x$, we obtain $du = -dx$ and the lower and upper extrema of the integral with respect to the new variable u become t and $t - \varepsilon$. Then:

$$(f * I_\varepsilon)(t) = -\frac{1}{\varepsilon}\int_t^{t-\varepsilon} f(u)du = \frac{1}{\varepsilon}\int_{t-\varepsilon}^{t} f(u)du = \langle f \rangle_{[t,t+\varepsilon]},$$

that is, the mean of f in the interval $[t - \varepsilon, t]$, of size ε.

A Gaussian $G_{\mu,\sigma}$ with mean μ and standard deviation σ is a "smooth" version of the pulse I_ε, which rapidly tends toward 0 outside of the interval $[\mu - \sigma, \mu + \sigma]$, thus:

$$\boxed{f * G_{\mu,\sigma} \simeq \text{local mean of } f \text{ in } [\mu - \sigma, \mu + \sigma]}$$

In section 2.9.6, we saw that blurring, in the frequency domain, results from the fact that the Fourier multiplier corresponding to the convolution with the Gaussian constitutes a low-pass filter. Here, we find the explanation for the blurring effect in the original domain of a signal f: following convolution with a Gaussian, each value of f in t is replaced by an approximation of the local mean value of f, with a locality parameter determined by the standard deviation of the Gaussian.

A further property of convolution, which is crucial for applications to the theory of differential equations, is discussed in Theorem 6.51.

THEOREM 6.51.– Taking $f \in \mathcal{C}(\mathbb{R}^n)$ with bounded partial derivatives and $g \in L^1(\mathbb{R}^n)$, then $f * g \in \mathcal{C}(\mathbb{R}^n)$ and:

$$\partial_{x_k}(f * g) = (\partial_{x_k} f) * g, \qquad \forall k = 1, \dots, n$$

In the same way, if $g \in \mathcal{C}(\mathbb{R}^n)$ with bounded parial derivatives and $f \in L^1(\mathbb{R}^n)$, then:

$$\partial_{x_k}(f * g) = f * (\partial_{x_k} g), \qquad \forall k = 1, \dots, n$$

PROOF.– The hypotheses of the theorem ensure that the derivation can be passed under the integral sign, thus $\forall k = 1, \dots, n$:

$$\partial_{x_k}\left(\int_{\mathbb{R}^n} f(x-y)g(y)dm(y)\right) = \int_{\mathbb{R}^n} \partial_{x_k}\left(f(x-y)g(y)\right)dm(y)$$
$$= \int_{\mathbb{R}^n} \left(\partial_{x_k} f(x-y)\right) g(y)dm(y)$$

since f is the only element which depends on x, that is, $\partial_{x_k}(f * g) = (\partial_{x_k} f) * g$. The second formula is a consequence of the commutative property of convolution, which allows us to switch the roles of f and g. □

6.9.6. *Convolution and Fourier transforms in L^2: localization of the Fourier transform*

A generalization of equation [6.27] allows us to highlight a significant limitation of the Fourier transform. The formalization of this statement relies on the following result, taken from Gasquet and Witomski (2013), which shows that the Fourier transform of the product of the elements in $L^2(\mathbb{R}^n)$ is proportional to the convolution of their Fourier transforms.

THEOREM 6.52.– If $f, g \in L^2(\mathbb{R}^n)$, then:

$$\boxed{\widehat{f \cdot g} = (2\pi)^{-n/2} \hat{f} * \hat{g}}$$

Let us consider the spectrum of $f \in L^2(\mathbb{R})$, but only in the neighborhood of a value of t_0. Using translation, it is always possible to consider $t_0 = 0$. The simplest, but *incorrect* (for reasons which we shall see later) approach to localizing the analysis of the spectrum of $f(t)$ consists of truncating it, that is, multiplying it by the step function of size $2T$:

$$\chi(t) = \begin{cases} 1 & \text{if } |t| \leqslant T \\ 0 & \text{otherwise} \end{cases},$$

where $2T$ is the size of the neighborhood that we wish to consider. Since $\chi \in L^2(\mathbb{R})$, by Theorem 6.52[10], the Fourier transform of the *truncated signal* $\tilde{f}(t) = f(t)\chi(t)$ is $\widehat{\tilde{f}}(\omega) = 1/\sqrt{2\pi}\hat{f}(\omega) * \hat{\chi}(\omega)$, where:

$$\hat{\chi}(\omega) = \frac{1}{\sqrt{2\pi}}\sqrt{\frac{2}{\pi}}T\frac{\sin(\omega T)}{\omega T} = \frac{T}{\pi}\text{sinc}(\omega T)$$

where the function $\mathbb{R} \ni t \mapsto \text{sinc}(t) := \frac{\sin t}{t}$. Thus:

$$\widehat{\tilde{f}}(\omega) = \frac{T}{\pi}\left(\hat{f}(\omega) * \text{sinc}(\omega T)\right)$$

that is, *the spectrum of the truncated signal is proportional to the convolution between the spectrum of the original signal and the sinc function of* ωT.

We thus see that precise localized information concerning the original signal cannot be obtained by truncation alone. This is one of the difficulties inherent in localizing frequency analysis of a signal within the context of Fourier analysis. Wavelet theory (Frazier 2001), developed to a significant extent in the late 1980s, offers powerful tools for handling this phenomenon.

6.10. The Nyquist-Shannon sampling theorem

The Nyquist-Shannon theorem[11] is one of the most important theorems in signal theory. It states that, when a function f possesses a bounded spectrum as specified in Definition 6.24, this function can be reconstructed using a discrete set of samples.

DEFINITION 6.24.– *The function* $f : \mathbb{R} \to \mathbb{C}$ *is said to be a* continuous signal of finite bandwidth *if there exists* $\Omega \in \mathbb{R}^+$ *such that:*

$$\hat{f}(\omega) = 0 \qquad \forall |\omega| > \Omega$$

The human visual system is incapable of perceiving an electromagnetic wave as light when the oscillating frequency of the wave is lower than 400 THz or higher than 800 THz, where T = Tera = 10^{12}. Moreover, humans are able to hear sounds as variations in air pressure only at frequencies between 20 Hz and 20 KHz, where K = Kilo = 10^3.

Visual and auditory signals, which are transmitted to the brain for interpretation, are two key examples of finite-bandwidth signals.

10 This argument is not valid if $f \in L^1(\mathbb{R})$, as, in this case, the formula from Theorem 6.52 would only be valid if $\hat{f}$ and $\hat{\chi}$ belong to $L^1(\mathbb{R})$; however, as we saw in section 6.9.2, $\hat{\chi} \notin L^1(\mathbb{R})$.

11 This theorem is known by several different names, sometimes including the names of Whittaker and Kotelnikov, two other mathematicians who independently discovered it.

THEOREM 6.53 (Shannon-Nyquist sampling theorem[12]).– Let:

– $f : \mathbb{R} \to \mathbb{C}$ be a signal of finite bandwidth: $\exists \Omega \in \mathbb{R}^+$ such that $\hat{f}(\omega) = 0 \quad \forall |\omega| > \Omega$;

– $\hat{f}$ be continuous and $C^1(\mathbb{R})$ piecewise.

Thus, f is fully determined by its samples at points $t_n = \frac{\pi}{\Omega} n$, $n \in \mathbb{Z}$, and the following formula holds:

$$\boxed{f(t) = \sum_{n \in \mathbb{Z}} f\left(\frac{\pi}{\Omega} n\right) sinc(\Omega t - \pi n)} \qquad [6.29]$$

where the convergence of the series is uniform.

There are several proofs of the sampling theorem, including a notable example which uses Poisson's summation formula (1781, Pithiviers-1840, Paris); here, we have chosen to present an alternative proof, found in Boggess and Narcowich (2015, p. 118).

PROOF.– We shall use the series and Fourier transform of $\hat{f}$. To do this, we interpret $\hat{f}$ as a 2Ω-periodic function when we write its Fourier series and as a function with support bounded in $[-\Omega, \Omega]$ when we calculate its Fourier transform.

Thanks to our hypotheses, $\hat{f} \in L^2[-\Omega, \Omega]$ and thus we can develop $\hat{f}$ into a Fourier series:

$$\hat{f}(\omega) = \sum_{k \in \mathbb{Z}} c_k e^{i \frac{2\pi \omega k}{2\Omega}} = \sum_{k \in \mathbb{Z}} c_k e^{i \frac{\pi \omega k}{\Omega}} \qquad [6.30]$$

with:

$$\begin{aligned} c_k &= \frac{1}{2\Omega} \int_{-\Omega}^{\Omega} \hat{f}(\omega) e^{-i \frac{\pi \omega k}{\Omega}} d\omega \\ &\underset{\hat{f}(\omega)=0 \ \forall |\omega| > \Omega}{=} \frac{\sqrt{2\pi}}{2\Omega} \frac{1}{\sqrt{2\pi}} \int_{\mathbb{R}} \hat{f}(\omega) e^{i\left(\frac{-\pi}{\Omega} k\right)\omega} d\omega \\ &= \frac{\sqrt{2\pi}}{2\Omega} \check{\hat{f}}\left(-\frac{\pi}{\Omega} k\right) = \frac{\sqrt{2\pi}}{2\Omega} f\left(-\frac{\pi}{\Omega} k\right), \end{aligned}$$

where in the final step of the previous computation we have used the definition of the inverse Fourier transform of $\hat{f}$, i.e. f, calculated in $-\frac{\pi}{\Omega} k$, and we included the normalization factor of the series in c_k.

12 Shannon (b. 1916, Petoskey; d. 2001, Medford), Nyquist (b. 1889, Stora Kil; d. 1976, Harlingen)

The Fourier series [6.30] can thus be rewritten as follows:

$$\hat{f}(\omega) = \sum_{k\in\mathbb{Z}} \frac{\sqrt{2\pi}}{2\Omega} f\left(-\frac{\pi}{\Omega}k\right) e^{i\frac{\pi\omega k}{\Omega}} \underset{(n=-k \iff k=-n)}{=} \sum_{n\in\mathbb{Z}} \frac{\sqrt{2\pi}}{2\Omega} f\left(\frac{\pi}{\Omega}n\right) e^{-i\frac{\pi\omega n}{\Omega}}$$

and this series is *uniformly convergent* since $\hat{f}$ is continuous and C^1 piecewise.

We calculate $f(t)$ *via* the inverse Fourier transform of $\hat{f}(\omega)$:

$$\begin{aligned} f(t) &= \frac{1}{\sqrt{2\pi}} \int_{\mathbb{R}} \hat{f}(\omega) e^{i\omega t} d\omega \\ &\underset{\hat{f}(\omega)=0\ \forall|\omega|>\Omega}{=} \frac{1}{\sqrt{2\pi}} \int_{-\Omega}^{\Omega} \hat{f}(\omega) e^{i\omega t} d\omega \\ &= \frac{1}{\sqrt{2\pi}} \int_{-\Omega}^{\Omega} \sum_{n\in\mathbb{Z}} \frac{\sqrt{2\pi}}{2\Omega} f\left(\frac{\pi}{\Omega}n\right) e^{-i\frac{\pi\omega n}{\Omega}} e^{i\omega t} d\omega \\ &= \sum_{n\in\mathbb{Z}} \frac{1}{2\Omega} f\left(\frac{\pi}{\Omega}n\right) \int_{-\Omega}^{\Omega} e^{i\omega\frac{t\Omega-\pi n}{\Omega}} d\omega \end{aligned} \qquad [6.31]$$

In the final step of the previous calculation, the series and the integral can be switched thanks to the fact that the series is uniformly convergent. Now, let us analyze the integral:

$$\int_{-\Omega}^{\Omega} e^{i\omega\frac{t\Omega-\pi n}{\Omega}} d\omega = \int_{-\Omega}^{\Omega} \cos\left(\omega\frac{t\Omega-\pi n}{\Omega}\right) d\omega + i \int_{-\Omega}^{\Omega} \sin\left(\omega\frac{t\Omega-\pi n}{\Omega}\right) d\omega$$

The second integral is zero, as the sine function is odd and the domain of integration is symmetric; on the other hand, the cosine function is even, so we obtain:

$$\begin{aligned} \int_{-\Omega}^{\Omega} e^{i\omega\left(\frac{t\Omega-\pi n}{\Omega}\right)} d\omega &= 2\int_{0}^{\Omega} \cos\left(\omega\frac{t\Omega-\pi n}{\Omega}\right) d\omega = 2\left[\frac{\sin\left(\omega\frac{t\Omega-\pi n}{\Omega}\right)}{\frac{t\Omega-\pi n}{\Omega}}\right]_0^{\Omega} \\ &= 2\Omega\frac{\sin\left(\Omega\frac{t\Omega-\pi n}{\Omega}\right)}{t\Omega-\pi n} - 0 = 2\Omega\frac{\sin(t\Omega-\pi n)}{t\Omega-\pi n} \end{aligned}$$

Inserting this result in [6.31], we obtain:

$$f(t) = \sum_{n\in\mathbb{Z}} \frac{2\Omega}{2\Omega} f\left(\frac{\pi}{\Omega}n\right) \frac{\sin(t\Omega-\pi n)}{t\Omega-\pi n} = \sum_{n\in\mathbb{Z}} f\left(\frac{\pi}{\Omega}n\right) \operatorname{sinc}(t\Omega-\pi n)$$

and, as underlined before, the series is uniformly convergent. □

6.10.1. *The Nyquist frequency: aliasing and oversampling*

Since the sinc function is fixed, the signal f is unequivocally characterized by the sequence of samples $f\left(\frac{\pi}{\Omega}n\right)$.

The sampling period used in the theorem is $T = \frac{\pi}{\Omega}$, so the sampling frequency, known as the *Nyquist frequency* and noted ν_N, is $\nu_N = \frac{1}{T} = \frac{\Omega}{\pi}$.

We now wish to compare the Nyquist frequency with the maximal frequency present in the signal f. Remember that we started with the hypothesis that f is a finite-bandwidth signal with maximum pulse Ω. Then the maximum frequency $\nu_{\max}$ of f is defined by the relation $\Omega = 2\pi\nu_{\max}$, i.e. $\nu_{\max} = \frac{\Omega}{2\pi}$.

Comparing the Nyquist sampling frequency ν_N with the maximal frequency $\nu_{\max}$ of signal f, we obtain $\nu_N = 2\nu_{\max}$, which tells us that *the sampling theorem holds if and only if the sampling frequency is at least twice the maximal frequency present in the signal f*. This is coherent with the results of the discrete Fourier transform, where we have seen that the highest frequency of a discrete signal given by N periodic samples is $\frac{N}{2}$ if N is even, or the integer part of $\frac{N}{2}$ if N is odd.

If the sampling frequency is lower than the Nyquist frequency, then a phenomenon known as *aliasing* occurs; this corresponds to errors in signal reconstruction. These errors result from the fact that, as we saw in our proof, we need to consider a periodic extension of the spectrum of f; the Nyquist frequency ν_N is the minimum frequency which allows f to be reconstructed without "overflowing" into the next period of the spectrum. A lower sampling frequency results in the inclusion of parasite information from the adjacent spectrum periods on each side.

Finally, we note that the general term of the series in the theorem converges to 0 with the same speed as $\frac{1}{n}$ when $n \to +\infty$; this is a relatively slow convergence. The convergence speed can be increased, for example to $\frac{1}{n^2}$, by increasing the sampling frequency: this technique is known as *oversampling*.

6.11. Application of the Fourier transform to solve ordinary and partial differential equations

The way the Fourier transform behaves with respect to derivatives makes it particularly helpful for solving certain types of differential equations. The general idea is illustrated below in the case of an ordinary differential equation (ODE).

6.11.1. *Solving an ordinary differential equation using the Fourier transform*

Taking $y, g : \mathbb{R} \to \mathbb{R}$, $y, g \in L^1(\mathbb{R})$, y twice differentiable, consider the following ODE:

$$y''(t) - y(t) = -g(t) \qquad \forall t \in \mathbb{R}$$

Applying the Fourier transform to both sides, by the property of linearity, we can write:

$$\widehat{y''}(\omega) - \hat{y}(\omega) = -\hat{g}(\omega)$$

that is:

$$-\omega^2\hat{y}(\omega) - \hat{y}(\omega) = -\hat{g}(\omega) \quad \Longleftrightarrow \quad (1+\omega^2)\hat{y}(\omega) = \hat{g}(\omega)$$

that is:

$$\hat{y}(\omega) = \frac{1}{1+\omega^2} \cdot \hat{g}(\omega) \qquad \text{(Solution in the frequency domain)}$$

We see that the properties of the Fourier transform allowed us to transform the ODE into an algebraic equation in the frequency domain. If we know the Fourier transform of g, then the ODE is solved in the Fourier space.

However, as the original ODE was formulated in terms of the variable t, we must return to the original representation by applying the inverse Fourier transform to both sides of the final equation, using property [6.28] we have:

$$(\hat{y}(\omega))^\vee(t) = y(t) = \left[\frac{1}{1+\omega^2} \cdot \hat{g}(\omega)\right]^\vee(t) = \frac{1}{\sqrt{2\pi}}\left(\left(\frac{1}{1+\omega^2}\right)^\vee(t) * g(t)\right) \tag{6.32}$$

We can verify by direct calculation that:

$$\widehat{e^{-a|t|}}(\omega) = \sqrt{\frac{2}{\pi}}\frac{a}{a^2+\omega^2}$$

so, considering $a = 1$:

$$\sqrt{\frac{\pi}{2}}\widehat{e^{-|t|}}(\omega) = \frac{1}{1+\omega^2}$$

and then:

$$y(t) = \frac{1}{\sqrt{2\pi}}\sqrt{\frac{\pi}{2}}e^{-|t|} * g(t)$$

that is:

$$y(t) = \frac{1}{2}\int_{-\infty}^{+\infty} e^{-|t-s|}g(s)ds = \frac{1}{2}\int_{-\infty}^{+\infty} g(t-s)e^{-|s|}ds$$

If we are able to calculate the integral (this depends on the analytical expression of g), then $y(t)$ can be determined explicitly; otherwise, the value must be approximated.

To solve an ODE *via* the Fourier transform, we thus need to perform the following operations:

1) transform the ODE in the frequency domain, applying the Fourier transform to both sides of the equation;

2) solve the algebraic ODE in the Fourier space;

3) apply the inverse Fourier transform to obtain the solution to the ODE in its original representation;

4) typically, the solution in the Fourier space is given by a product; hence, the solution in the original representation is given by a convolution.

This technique can only be used if the coefficients of the derivatives are constant, and if the functions are integrable.

6.11.2. *The Fourier transform and partial differential equations*

The Fourier transform is even more effective when applied to partial differential equations. For the purposes of our presentation, we shall only consider functions of the type $u = u(t,x)$ or $u = u(t,x,y,z)$, where t is the time coordinate and x or (x,y,z) are one-dimensional (1D) or three-dimensional (3D) coordinates, respectively. It is implicitly considered that $u \in L^1(\mathbb{R}^2)$ or $u \in L^1(\mathbb{R}^4)$, respectively, and that u can be derived enough times so that the corresponding PDE is well defined.

For simplicity's sake, we write:

$$\frac{\partial u}{\partial x} = u_x, \ \frac{\partial^2 u}{\partial x^2} = u_{xx}, \ \frac{\partial u}{\partial t} = u_t, \ \ldots$$

The properties of the Fourier transform with respect to the partial derivatives are as follows:

– if the integration variable of the Fourier transform is x, then:

$$\widehat{u_x}(t,\omega) = i\omega\hat{u}(t,\omega), \ \ \widehat{u_{xx}}(t,\omega) = -\omega^2\hat{u}(t,\omega)$$

$$\widehat{u_t}(t,\omega) = \frac{\partial}{\partial t}\hat{u}(t,\omega), \ \ \widehat{u_{tt}}(t,\omega) = \frac{\partial^2}{\partial t^2}\hat{u}(t,\omega)$$

The first two formulas are straightforward; to obtain the remaining two, we note that, since $u \in L^1(\mathbb{R}^2)$, the order of derivation and integration can be modified:

$$\frac{1}{\sqrt{2\pi}}\int_{-\infty}^{+\infty}\frac{\partial u(t,x)}{\partial t}e^{-i\omega x}dx = \frac{\partial}{\partial t}\frac{1}{\sqrt{2\pi}}\int_{-\infty}^{+\infty}u(t,x)\,e^{-i\omega x}dx = \frac{\partial}{\partial t}\hat{u}(t,\omega)$$

The same is true for u_{tt};

– if the integration variable of the Fourier transform is t, then:

$$\widehat{u_t}(x,\omega) = i\omega\hat{u}(x,\omega), \quad \widehat{u_{tt}}(x,\omega) = -\omega^2\hat{u}(x,\omega)$$

$$\widehat{u_x}(x,\omega) = \frac{\partial}{\partial x}\hat{u}(x,\omega), \quad \widehat{u_{xx}}(x,\omega) = \frac{\partial^2}{\partial x^2}\hat{u}(x,\omega)$$

– these considerations can be extended to $u(t,x,y,z)$.

6.11.3. *Solving the partial differential equation for heat propagation using the Fourier transform*

Consider the Cauchy problem for $u \in \mathcal{C}^2(\mathbb{R}^2) \cap L^1(\mathbb{R}^2)$ and $\varphi \in \mathcal{C}^2(\mathbb{R}) \cap L^1(\mathbb{R})$ defined by:

$$\begin{cases} u_t = \alpha^2 u_{xx} & \forall x \in (-\infty,+\infty), \forall t \in (0,+\infty), \ \alpha \in \mathbb{R}^+ \\ u(0,x) = \varphi(x) & \forall x \in (-\infty,+\infty), t = 0 \end{cases}$$

where

– $u(t,x)$ is the temperature of a 1D bar at time t and at the point x;

– $u_t(t,x)$ is the rate of temperature change at time t and at the point x;

– $u_{xx}(t,x)$ is the concavity of the temperature profile at time t and x (note that the second derivative is with respect to the spatial variable, thus it would be wrong to interpret u_{xx} as an acceleration);

– $\varphi(x)$ is the initial concavity of the temperature profile at the point x.

If we write the second discrete derivative (with step Δx) with respect to x, we see that it defines the comparison of the temperature at point x at time t with that of its neighbors at the same instant:

$$\begin{aligned} u_{xx}(t,x) &\simeq \tfrac{u(t,x+\Delta x)-2u(t,x)+u(t,x-\Delta x)}{(\Delta x)^2} \\ &= \tfrac{2}{(\Delta x)^2}\left[\underbrace{\frac{u(t,x+\Delta x)+u(t,x-\Delta x)}{2}}_{\text{mean temperature of neighboring points}} - u(t,x)\right] \end{aligned}$$

Thus, the equation $u_t = \alpha^2 u_{xx}$ tells us that:

– if $u(t,x)$ is less than the mean temperature of its neighbors, then $u_{xx} > 0$ and thus $u_t(t,x)(= \alpha^2 u_{xx}) > 0$, meaning that the temperature at the point x will increase over time: the neighboring points lose some of their heat in favor of x in order to attain thermal equilibrium;

– in the opposite case, $u_t(t,x)(=\alpha^2 u_{xx}) < 0$ and so the temperature at point x decreases over time: x loses heat to its neighbors in order to attain thermal equilibrium;

– the positive constant α^2 is a characteristic of the material, known as the thermal diffusion coefficient. The higher the value of α^2, the faster the bar will reach thermal equilibrium.

The heat equation is used in many other domains: for instance, in image processing, it is used to smooth out imperfections, and in the field of economics, it plays an important role in the Black-Scholes-Merton model of financial markets.

The heat equation is solved by calculating the Fourier transform (integrating with respect to variable x) on both sides:

$$u_t(t,x) = \alpha^2 u_{xx}(t,x) \quad \overset{\wedge}{\longrightarrow} \quad \frac{\partial}{\partial t}\widehat{u}(t,\omega) = -\alpha^2\omega^2\widehat{u}(t,\omega)$$

The initial condition in the Fourier space becomes $\widehat{u}(0,\omega) = \widehat{\varphi}(\omega)$. The PDE has thus been transformed into an ODE:

$$\begin{cases} u_t(t,x) = \alpha u_{xx}(t,x) \\ u(0,x) = \varphi(x) \end{cases} \quad \overset{\wedge}{\longrightarrow} \begin{cases} \frac{\partial}{\partial t}\widehat{u}(t,\omega) = -\alpha^2\omega^2\widehat{u}(t,\omega) \\ \widehat{u}(0,\omega) = \widehat{\varphi}(\omega) \end{cases}$$

because ω is a constant with respect to variable t, thus the equation $\frac{\partial}{\partial t}\widehat{u}(t,\omega) = -\alpha^2\omega^2\widehat{u}(t,\omega)$ is ordinary. We recall that the solution of the Cauchy problem:

$$\begin{cases} y' = -ky \\ y(0) = y_0 \end{cases}$$

is $y(t) = y_0 e^{-kt}$ and thus, in the present case:

$$\widehat{u}\,(t,\omega) = \widehat{\varphi}(\omega)\cdot e^{-\alpha^2\omega^2 t} = \widehat{\varphi}(\omega)\cdot e^{-(\alpha^2 t)\omega^2} \qquad \text{(Solution in the Fourier space)}$$

The inverse Fourier transform is then applied to obtain the solution in the original representation. Using equation [6.28], we obtain:

$$\begin{aligned} u(t,x) &= \left(\widehat{\varphi}(\omega)\cdot e^{-(\alpha^2 t)\omega^2}\right)^{\vee}(t,x) \\ &= \frac{1}{\sqrt{2\pi}}\check{\widehat{\varphi}}(x) * \left(e^{-(\alpha^2 t)\omega^2}\right)^{\vee}(t,x) \end{aligned}$$

Furthermore, $\check{\widehat{\varphi}}(x) = \varphi(x)$, and $e^{-(\alpha^2 t)\omega^2}$ is a Gaussian with respect to ω, so we can use the following property:

$$\widehat{e^{-c^2x^2}}(\omega) = \frac{1}{c\sqrt{2}}e^{-\frac{\omega^2}{4c^2}} \iff c\sqrt{2}\widehat{e^{-c^2x^2}}(\omega) = e^{-\frac{1}{4c^2}\omega^2}$$

In our case, this gives us $\frac{1}{4c^2} = \alpha^2 t$; moreover, $c^2 = \frac{1}{4\alpha^2 t}$ and then $c = \frac{1}{2\alpha\sqrt{t}}$ (in physical terms, only the positive determination of the root is relevant). Finally, we can write:

$$\left(e^{-(\alpha^2 t)\omega^2}\right)^{\vee}(t,x) = \frac{\sqrt{2}}{2\alpha\sqrt{t}} e^{-\frac{x^2}{4\alpha^2 t}} = \frac{1}{\alpha\sqrt{2t}} e^{-\frac{x^2}{4\alpha^2 t}}$$

and the solution of the heat equation is thus:

$$\boxed{u(t,x) = \frac{1}{\alpha\sqrt{4\pi t}} \int_{-\infty}^{+\infty} e^{-\frac{(x-y)^2}{4\alpha^2 t}} \varphi(y)dy}$$

Certain expressions of $\varphi(x)$ permit exact integration, and an analytical expression of $u(t,x)$ is thus possible. Generally, however, it is only possible to approximate $u(t,x)$.

It is interesting to note that, as the standard expression of a Gaussian is:

$$\frac{1}{\sigma\sqrt{2\pi}} e^{-\frac{(x-\mu)^2}{2\sigma^2}}$$

then $\sigma = \alpha\sqrt{2t}$, i.e. $\sigma^2 = 2\alpha^2 t$: the variance of the Gaussian featured in the solution of the heat equation is not fixed, but increases linearly as the time t increases.

This tells us that the support of the Gaussian widens over time; this is perfectly coherent with common experience, given that as $t \to +\infty$, the bar reaches thermal equilibrium and thus the temperature is uniform across the whole bar.

The observations above provide a deeper insight into the technique of convolution with a Gaussian, widely used in signal processing, for example to blur digital images. Taking $\varphi(y)$ to represent the original intensity of any given pixel y in a digital image, and interpreting $u(t,x)$ as the intensity of the blurred image at time t and in a fixed pixel at position x, the convolution of an image with a Gaussian may be considered as the exchange of intensity ("heat") between x and its neighbors.

Furthermore, just as heat propagation is an *irreversible process*, the blurring effect obtained by convolution with a Gaussian cannot be directly inverted.

One final observation linked to the spatial dimension of the problem is that the application of the technique described above requires x to be variable between $-\infty$ and $+\infty$. Other techniques are used to solve problems where x varies within a bounded interval, including the sine and cosine Fourier transforms and the Laplace transform.

6.12. Summary

Linear operators between normed vector spaces are continuous at a given point if and only if they are continuous everywhere, and if and only if they are bounded. All linear operators defined on a finite-dimensional vector space are continuous (and thus bounded); this ceases to be true, in general, when the space in which the operator is defined is not of finite dimension. A classic example is provided by the derivation operation.

For bounded linear operators, we can define a norm, with four equivalent definitions, which makes the set $\mathcal{B}(V, W)$ of bounded linear operators between two normed vector spaces V and W a normed vector space in its own right. In the specific case where $V = W$, the composition of operators defines a product in $\mathcal{B}(V)$ with respect to which $\mathcal{B}(V)$ becomes a unital normed associative algebra. Furthermore, if W is complete, $\mathcal{B}(V, W)$ is complete; in the specific case where $V = W = \mathcal{H}$, a Hilbert space, $\mathcal{B}(\mathcal{H})$ is a unital Banach algebra, that is, a complete associative normed algebra such that $\|AB\| \leqslant \|A\| \, \|B\| \; \forall A, B \in \mathcal{B}(\mathcal{H})$.

The kernel of a bounded operator is always a closed vector subspace in the domain of the operator. If the kernel consists solely of the zero vector, then the operator is inversible, but its inverse will not necessarily be bounded. The existence of $\mu > 0$ such that $\|Ax\| \geqslant \mu\|x\|$ gives a simple and useful characterization of the bounded invertibility of an operator $A : V \to W$. If V is a Banach space and this condition is verified, then $\text{Im}(A)$, the image space of A, is closed. In practical applications, the closure of $\ker(A)$ (where A is continuous) and of $\text{Im}(A)$ (in the hypotheses given above) may be used to characterize a closed subspace: we must simply show that this coincides with the kernel or image of a linear operator which satisfies those hypotheses.

The dual of an arbitrary vector space V on the field $\mathbb{K} = \mathbb{R}$ or $\mathbb{C}$ is the vector space V^* of linear functionals defined on the vector space itself. If the space is normed, then it is natural to require compatibility with the topological structure generated by the norm, that is, the functionals are continuous, that is, we define $V^* = \mathcal{B}(V, \mathbb{K})$. Given that $\mathbb{K}$ is complete, V^* is always complete, even when V is not. In the case of a Hilbert space $\mathcal{H}$, the Riesz representation theorem tells us that $\mathcal{H}$ and $\mathcal{H}^*$ are isomorphic by the transformation which associates each $x \in \mathcal{H}$ with the functional T_x which implements the inner product, that is, $T_x(y) = \langle y, x\rangle \; \forall y \in \mathcal{H}$. This theorem makes it possible to define the adjoint $A^\dagger$ of any operator $A \in \mathcal{B}(\mathcal{H})$ *via* the relationship $\langle A^\dagger x, y\rangle = \langle x, Ay\rangle \; \forall x, y \in \mathcal{H}$. If $A = A^\dagger$, then A is said to be self-adjoint. Two examples of self-adjoint operators are $A^\dagger A$ and $AA^\dagger$.

The adjoint of a bounded linear operator is a particularly important operator in both theory and practice. An idea of its importance can be seen in the theorem used to characterize an orthogonal projection operator on a Hilbert space: $A \in \mathcal{B}(\mathcal{H})$ is an orthogonal projector on $\mathrm{Im}(A)$ if and only if A is self-adjoint $A = A^\dagger$ and idempotent $A^2 = A$. This result can be used, for example, to show that multiplication operators on $L^2(\mathbb{R}^n)$ are orthogonal projectors if and only if they multiply by the indicator function of a measurable subset of $\mathbb{R}^n$. There is also a highly important geometric representation of orthogonal projectors: $A \in \mathcal{B}(\mathcal{H})$ is an orthogonal projector if and only if there exists an orthonormal system $(u_n)_{n\in\mathbb{N}}$ in $\mathcal{H}$ such that $Ax = \sum_{n\in\mathbb{N}} \langle x, u_n\rangle u_n$, $\forall x \in \mathcal{H}$. This realization of the projector is the extension, in infinite dimensions, of the analogous formula valid in finite dimension.

The adjoint also plays a role in the analysis of isometric and unitary operators. An operator $A \in \mathcal{B}(\mathcal{H})$ is isometric if it conserves the norm (or, in an equivalent manner, the inner product); a unitary operator is isometric and surjective. The two categories of operators have unit norm. The relationship between isometric operators and orthogonal projectors is given by the following result: if $A \in \mathcal{B}(\mathcal{H})$ is isometric, then $AA^\dagger$ is an orthogonal projector. If $A \in \mathcal{B}(\mathcal{H})$ is isometric, then $\mathrm{Im}(A) = \mathrm{Im}(AA^\dagger)$ and, given that $AA^\dagger$ is an orthogonal projector (since A is taken to be isometric), $\mathrm{Im}(AA^\dagger)$ is closed; thus the image space of an isometric operator is always closed. Since $\ker(A^\dagger) = \mathrm{Im}\,(A)^\perp$, if A is isometric but not surjective, then $\mathrm{Im}(A) \neq \mathcal{H}$; hence, $\mathrm{Im}\,(A)^\perp \neq \{0_\mathcal{H}\}$ and then $A^\dagger$ is not invertible. Using the same argument, we also see that if A is unitary, then $A^\dagger$ is invertible.

As in the case of orthogonal projection operators, an algebraic characterization of isometric and unitary operators can be obtained *via* the adjoint: $A \in \mathcal{B}(\mathcal{H})$ is isometric if and only if $A^\dagger A = id_\mathcal{H}$, while $A \in \mathcal{B}(\mathcal{H})$ is unitary if and only if $A^\dagger A = AA^\dagger = id_\mathcal{H}$; in this final case, A is invertible and $A^{-1} = A^\dagger$. Moreover, we can show that A is unitary if and only if $A^\dagger$ is unitary. One consequence of this result is that the unitary nature of an operator A can be studied by examining that of its adjoint, which, in some cases, is simpler. Regarding the geometric realization of isometric and unitary operators, $A \in \mathcal{B}(\mathcal{H})$ is isometric if and only if it transforms Hilbert bases into orthonormal systems, while $A \in \mathcal{B}(\mathcal{H})$ is unitary if and only if it transforms Hilbert bases into Hilbert bases. This is an important difference with respect to the finite dimensional case.

The Fourier transform $f(x) \mapsto \hat{f}(\omega) = \frac{1}{(2\pi)^{n/2}} \int_{\mathbb{R}^n} f(x) e^{-i\langle\omega,x\rangle}\,dx$ is widely used in both pure and applied mathematics. The most "natural" space in which to define this transform is the Schwartz space; in this space, the Fourier transform has the integral formula given above, and is an isometric isomorphism with respect to the norm inherited by $L^2(\mathbb{R}^n)$. If we wish to extend the transform to a space with less regular functions, for example $L^1(\mathbb{R}^n)$ or $L^2(\mathbb{R}^n)$, certain properties must be sacrificed. On $L^1(\mathbb{R}^n)$, the image is $\mathcal{C}_\infty(\mathbb{R}^n)$, but the integral formula is preserved. On $L^2(\mathbb{R}^n)$, the integral formula must be replaced by a limit formula, but the isomorphic character of the transform is retained; the extension of the Fourier transform on $L^2(\mathbb{R}^n)$ defines a unitary operator $F \in \mathcal{B}(L^2(\mathbb{R}^n))$. An explicit formula for this unitary operator can be obtained by means of the Hermite basis: $Ff = \sum_{n\in\mathbb{N}} (-i)^n \langle f, u_n \rangle u_n$. Finally, we note that – to within a constant – the Fourier transform of the convolution of two functions in $L^1(\mathbb{R}^n)$ is the pointwise product of the transforms.

Finally, we presented the Nyquist-Shannon sampling theorem, which enables the reconstruction of a signal with bounded bandwidth using a sufficiently dense, but finite, set of samples of this signal. We also described applications of the Fourier transform in solving differential equations, notably the heat equation, which played a crucial role in the development of Fourier's theory.

Appendix 1

Quotient Space

The concept of quotient of a vector space is essential in mathematics, and, in our opinion, does not always receive the attention it deserves in works on linear algebra. For this reason, we have chosen to devote an appendix to the definition and interpretation of this concept.

DEFINITION A1.1.– *An equivalence relation* $\sim$ *defined on a vector space* V *(of arbitrary dimension) on the field* $\mathbb{K}$ *is said to be* compatible *with the linear structure of* V *if:*

$$v \sim v', \ w \sim w' \quad \Longrightarrow \quad \alpha v + \beta w \sim \alpha v' + \beta w', \qquad \forall \alpha, \beta \in \mathbb{K}$$

The equivalence class of 0 in V *is a vector space* Z *(since it is stable with respect to linear combinations, and contains the neutral element) known as the* kernel *of the equivalence relationship.*

One special case of this definition is when $w = w' = v'$ and $\alpha = 1, \beta = -1$, which implies:

$$v \sim v' \quad \Longrightarrow \quad v - v' \sim 0 \iff v - v' \in Z$$

Conversely, if $v - v' \in Z$, i.e. $v - v' \sim 0 \iff v - v' \sim v' - v'$, and by the fact that $v' \sim v'$, and since $\sim$ is compatible with the linear structure of V, we obtain: $v - v' + v' \sim v' - v' + v'$, that is, $v \sim v'$.

In short: $v \sim v' \iff v - v' \in Z$, which tells us that an equivalence relationship compatible with the linear structure of a vector space is univocally determined by its kernel, which is a vector subspace of V. This observation allows us to reverse the process. Given an arbitrary vector subspace W in V, if we define:

$$\boxed{v \sim_W v' \iff v - v' \in W} \qquad \forall v, v' \in V$$

then $\sim$ *is an equivalence relationship in V which is compatible with its linear structure and with W as kernel.*

By symmetry, $v \sim_W v' \iff v' \sim_W v \iff v' - v \in W$, and this means that there exists $w \in W$ such that $v' - v = w$, that is, $v' = v + w$; thus, the equivalence class $[v]_W$ containing $v \in V$ is the subset of V given by:

$$\boxed{[v]_W = v + W = \{v + w \ : \ w \in W\}}$$

This is referred to as a *linear subvariety* and interpreted geometrically as the shift of the subspace W by the vector v.

We observe that if $v \in W$, then, by linearity, the shift of W through v does not modify W. As a vector subspace of V, W contains the 0, thus if $v \notin W$, then the equivalence class $v + W$ does not contain the 0 and cannot, therefore, be a vector subspace of V.

Lemma A1.1 is essential for defining a quotient space, and will be used extensively in the rest of this appendix.

LEMMA A1.1.– Let V be an arbitrary vector space, let $v_1, v_2 \in V$ and let W be a vector subspace of V. Then the equality $v_1 + W = v_2 + W$ holds if and only if $W_1 = W_2 \equiv W$ and $v_1 - v_2 \in W$.

PROOF.–

$\boxed{\Leftarrow}$: taking $v_1 - v_2 \equiv w_0 \in W = W_1 = W_2$, then $v_2 = v_1 - w_0$ and:

$$v_1 + W = \{v_1 + w \ : \ w \in W\}, \quad v_2 + W = \{v_1 + w - w_0 \ : \ w \in W\}$$

but evidently $W = \{w \ : \ w \in W\} = \{w - w_0 \ : \ w \in W\}$, hence $v_1 + W = v_2 + W$.

$\boxed{\Rightarrow}$: inversely, taking $v_1 + W_1 = v_2 + W_2$, then, by the definition of a linear subvariety, $v_1 - v_2 + W_1 = v_2 - v_2 + W_2 = W_2$, thus, if $w_0 = v_1 - v_2$, we obtain $w_0 + W_1 = W_2$. Since W_2 is a vector subspace of V, it contains 0. $w_0 + W_1$ also contains 0, i.e. $-w_0 \in W_1$ and thus $w_0 \in W_1$ since W_1 is also a vector subspace. Shifting the vectors of W_1 using w_0, which is a vector in W_1, does not change the subspace, i.e. $w_0 + W_1 = W_1$; however, since $w_0 + W_1 = W_2$, we obtain $W_1 = W_2 = W$ and $w_0 = v_1 - v_2 \in W_1 = W$. □

This lemma implies that *every linear subvariety is uniquely determined by a single subspace W, of which the subvariety is the shift*. Moreover, *the vector which induces the shift is uniquely determined*, up to the sum with a vector in W.

It is now possible to establish the definition of quotient space and prove that this definition is well posed.

DEFINITION A1.2 (quotient (vector) space).– *Let V be any vector space and W a vector subspace of V. The* quotient vector space *V/W is the set of all linear subvarieties of V which are shifts of W, equipped with the following linear operations:*

$$(v_1 + W) + (v_2 + W) = (v_1 + v_2) + W, \qquad \forall v_1, v_2 \in V$$

$$\alpha(v + W) = \alpha v + W, \qquad \forall v \in V, \forall \alpha \in \mathbb{K}$$

Let us verify that these operations are well defined and that V/W is a vector space on $\mathbb{K}$.

The easy proof that the vector space axioms for V/W are directly induced by the vector space properties of V is left to the reader. Let us just underline the following properties:

a) if $v_1 + W = v_1' + W$ and $v_2 + W = v_2' + W$, then $v_1 + v_2 + W = v_1' + v_2' + W$;
b) if $v_1 + W = v_1' + W$, then $\alpha v_1 + W = \alpha v_1' + W$;

$\forall v_1, v_2, v_1', v_2' \in V$ and $\forall \alpha \in \mathbb{K}$.

We begin by proving the validity of property a. Lemma A1.1 tells us that $v_1 - v_1' \equiv w_1 \in W$ and $v_2 - v_2' \equiv w_2 \in W$, thus:

$$(v_1 + v_2) + W = (v_1' + v_2') + \underbrace{(w_1 + w_2) + W}_{=W} = (v_1' + v_2') + W$$

To prove the validity of property b, we simply note that if we write $v_1 - v_1' \equiv w \in W$, then $\alpha(v_1 - v_1') = \alpha w \in W$, and thus by Lemma A1.1, $\alpha v_1 + W = \alpha v_1' + W$.

It is natural to wonder what the dimension of V/W is, and whether it is linked or not to the dimensions of V and W. To answer this question we need the following preliminary result.

LEMMA A1.2.– Let V be an arbitrary vector space, W a vector subspace of V and H a subspace of V which is supplementary to W, i.e. such that $W \cap H = \{0\}$ and $V = W \oplus H$. Then, for any vector $v \in V$ which implements a translation of W, there exists only one vector $h_v \in H \cap (v + W)$. This vector is used to write v in a unique manner in the direct sum $v = w_v + h_v$.

PROOF.– Let us begin by proving the existence of a vector h_v belonging to H and to $v + W$. By the hypothesis $V = W \oplus H$, any vector $v \in V$ may be written in a unique manner as $v = w_v + h_v$, $w_v \in W$ and $h_v \in H$; we must prove that $h_v \in v + W$.

To do this, let us now consider a vector $v' = w_{v'} + h_{v'} \in V$, $w_{v'} \in W$ and $h_{v'} \in H$, which belongs to the same equivalence class as v, that is, which is such that $v + W =$

$v' + W$. Again, using Lemma A1.1, $v' - v \in W$, that is, $w_{v'} + h_{v'} - w_v - h_v \in W$, that is, $w_{v'} - w_v + h_{v'} - h_v \in W$. Moreover, since $w_{v'} - w_v \in W$ and $h_{v'} - h_v \in H$, the only case in which their sum remains within W is where $h_{v'} - h_v = 0$ (given that $W \cap H = \{0\}$).

Hence, $h_{v'} = h_v$ and then $v+W \ni v' = w_{v'}+h_v$, that is, $w_{v'}+h_v \in v+W$. Using Lemma A1.1 once more, we know that the sum of a vector belonging to $v + W$ and a vector in W does not take us outside of the equivalence class $v+W$, thus $h_v \in v+W$. Since $h_v \in H$ and $h_v \in v + W$, then $h_v \in H \cap (v + W)$.

Inversely, if $h \in H \cap (v+W)$, then, in particular, $h \in v+W$, that is, $h \sim_W v$, that is, $\exists v \in V$ and $\tilde{w}_v \in W$ such that $h = \tilde{w}_v + v$, that is, $v = w_v + h$, where $w_v = -\tilde{w}_v$, that is, $h = h_v$. □

THEOREM A1.1.– If W is a subspace of the vector space V which admits a supplement H in V, then H is isomorphic to V/W:

$$V/W \simeq H, \qquad V = W \oplus H$$

PROOF.– The uniqueness of vector h_v, as established by Lemma A1.2, allows us to construct the bijective and intrinsically linear correspondence which associates an arbitrary linear subvariety $v + W$ in V with the component in H of an arbitrary representative $v \in v + W$, that is:

$$\begin{aligned} V/W &\longrightarrow H \\ v + W &\longmapsto h_v, \text{ such that: } v = w_v + h_v \end{aligned}$$

is a linear isomorphism. □

Note that, given a *closed vector subspace* W of a Hilbert space $\mathcal{H}$, the orthogonal projection theorem 5.7 tells us that a supplementary space always exists in the form of the orthogonal complement $W^\perp$; hence, in this case:

$$\boxed{\mathcal{H}/W \simeq W^\perp}$$

that is, the quotient vector space of a Hilbert space on a closed vector subspace W is isomorphic to the orthogonal complement of W.

This result also allows us to determine the dimension of V/W as a function of that of V and of W in finite dimensions. In this case, $\dim(V) = \dim(W)+\dim(H)$ and $\dim(H) = \dim(V/W)$, then $\dim(V/W) = \dim(V)-\dim(W)$.

COROLLARY A1.1 (Dimension of V/W).– *Let V be a vector space of finite dimension and W a vector subspace of V, then:*

$$\boxed{\dim(V/W) = \dim(V) - \dim(W)}$$

Many problems in both pure and applied mathematics require us to consider situations where V and W are of infinite dimension, while V/W is of finite dimension. In this case, $\dim(V/W)$ is known as the codimension of W in V and written as $\mathrm{codim}(V/W)$.

Once the dimension of V/W and the linear isomorphism with H have been determined, Corollary A1.2 concerning the bases of V/W in finite dimensions is almost immediate.

COROLLARY A1.2 (Bases of V/W).– *Let V be a vector space of finite dimension n and W a vector subspace of V, then the linear subvarieties $(e_i+W)_{i=1}^n \subset V/W$ form a basis of the quotient vector space V/W if and only if the representatives $(e_i)_{i=1}^n \subset V$ constitute a basis for a supplementary subspace of W in V.*

Note that the zero of V/W is evidently the linear subvariety which contains the 0 of V, i.e. $0_V + W \equiv W$ is the zero of V/W.

We conclude our analysis of V/W by considering the natural projection of V onto V/W:

$$\begin{aligned} \pi : V &\longrightarrow V/W \\ v &\longmapsto \pi(v) = v + W \end{aligned}$$

The properties of π are as follows:

– *π is surjective*: this stems from the fact that each element in V/W is represented by a vector in V;

– the fibers of π, i.e. the counter-images of the elements in V/W through π, are the elements of V/W interpreted as a subvariety of V:

$$\pi^{-1}([v_0]) = \{v \in V \ : \ v + W = v_0 + W\}$$

but the equality between sets $v + W = v_0 + W$ is only verified for $v = v_0 + W$, thus:

$$\pi^{-1}([v_0]) = v_0 + W$$

where $[v_0]$ is interpreted, first, as the equivalence class corresponding to the element of V/W identified by $v_0 \in V$, then as $v_0 + W$, seen as a subset of V;

– π is a linear application, by the fact that V/W is well defined;

– the kernel of π is W: $\ker(\pi) = W$. By Lemma A1.1, $v_0 + W = W$ if and only if $v_0 \in W$.

Appendix 2

The Transpose (or Dual) of a Linear Operator

Any linear operator $A : V \to W$, where V and W are two finite-dimensional vector spaces, can be univocally associated with a linear operator A^t known as the *transpose* or *dual operator* of A, defined as:

$$\begin{aligned} A^t : W^* &\longrightarrow V^* \\ \varphi &\longmapsto A^t\varphi = \varphi \circ A \end{aligned}$$

that is:

$$\begin{aligned} A^t\varphi : V &\longrightarrow \mathbb{K} \\ v &\longmapsto A^t\varphi(v) = \varphi(Av) \end{aligned}$$

This definition is natural, as it only uses A and the elements supplied by the vector spaces themselves.

Using canonical notation to express the action of a linear functional, we can rewrite $A^t\varphi(v) = \varphi(Av)$ as:

$$\langle A^t\varphi, v\rangle = \langle \varphi, Av\rangle \qquad \text{[A2.1]}$$

The fact that this is well defined, that is, the linearity of the functional $A^t\varphi$, is guaranteed by the fact that for a fixed φ, the function $v \mapsto \varphi(Av)$ is linear, as it is a composition of linear applications. The uniqueness of this definition can also be easily proven. Let A_1^t and A_2^t be two transpose operators such that $A_1^t\varphi(v) = \varphi(Av) = A_2^t\varphi(v)$, that is, $(A_1^t - A_2^t)\varphi(v) = 0$. Taking an arbitrary fixed $\varphi \in V^*$ and leaving v free within V, it is evident from equation $(A_1^t - A_2^t)\varphi(v) = 0$ that $(A_1^t - A_2^t)\varphi$ is the identically zero functional. This holds for all $\varphi \in V^*$, implying that $A_1^t - A_2^t = 0$, that is, $A_1^t = A_2^t$.

Now, let V and W be two finite-dimensional Banach spaces. In this case, the definition remains valid as long as, for all $A \in \mathcal{B}(V, W)$, the transpose operator defined above is continuous, that is, $A^t \in \mathcal{B}(W^*, V^*)$, and if $A^t\varphi$ is a bounded linear functional on V whenever φ is a bounded linear functional on W.

Let us verify these properties.

– $A^t\varphi$ is a bounded linear functional on V $\forall\varphi \in W^*$: linearity is evident by definition, so we only need to prove that $A^t\varphi$ is bounded:

$$\begin{aligned}\|A^t\varphi\| &= \sup_{\|v\|=1} \|(A^t\varphi)v\| \underset{\text{def of } A^t}{=} \sup_{\|v\|=1} \|\varphi(Av)\| \underset{\varphi \text{ bounded}}{\leqslant} \sup_{\|v\|=1} \|\varphi\|\|Av\| \\ &\underset{A \text{ bounded}}{\leqslant} \sup_{\|v\|=1} \|\varphi\|\|A\|\|v\| = \|\varphi\|\|A\| < +\infty\end{aligned}$$

– $A^t \in \mathcal{B}(W^*, V^*)$:

$$\|A^t\| = \sup_{\|\varphi\|=1} \|A^t\varphi\| \underset{\text{def of } A^t}{=} \sup_{\|\varphi\|=1} \|\varphi \circ A\| \underset{\varphi \text{ bounded}}{\leqslant} \sup_{\|\varphi\|=1} \|\varphi\|\|A\| = \|A\| \underset{A \in \mathcal{B}(V,W)}{<} +\infty$$

If $V = W = \mathcal{H}$, where $\mathcal{H}$ is a Hilbert space, then the Riesz isomorphism T : $\mathcal{H} \to \mathcal{H}^*, \mathcal{H} \ni x \mapsto T(x) = T_x$, where $T_x(y) = \langle y, x\rangle$ $\forall y \in \mathcal{H}$ is associated with the adjoint operator defined in section 6.4 *via* the expression:

$$A^\dagger = T^{-1} A^t T.$$

Appendix 3

Uniform, Strong and Weak Convergence

Sequences of operators may be shown to converge with respect to different topologies than the one induced by the operator norm. The same can be said for sequences of elements in Banach or Hilbert spaces.

To take a concrete example, consider the following case. Let $(u_n)_{n\in\mathbb{N}}$ be an arbitrary Hilbert basis in a Hilbert space $\mathcal{H}$. For all $n \in \mathbb{N}$, we define the linear operator:

$$\begin{aligned} A_n : \mathcal{H} &\longrightarrow \mathcal{H} \\ x &\longmapsto A_n x = \sum_{m=0}^{n} \langle x, u_m \rangle u_m \end{aligned}$$

From the geometric characterization of projection operators (see Theorem 6.32), we know that A_n is the orthogonal projector on the vector subspace of $\mathcal{H}$ generated by $u_1, \ldots, u_n$: $S_n = \text{span}(u_1, \ldots, u_n)$.

Since any $x \in \mathcal{H}$ may be written as $x = \sum_{n=0}^{\infty} \langle x, u_n \rangle u_n$, it would seem that the sequence of projectors $(A_n)_{n\in\mathbb{N}}$ converges toward $id_{\mathcal{H}}$ when $n \to +\infty$.

Nevertheless, since $S_n \subset S_{n'}$ $\forall n < n'$, we know by Theorem 6.35 that $A_{n'} - A_n$ is the projector onto $S_{n'} \cap S_n^{\perp} = \text{span}(u_{n+1}, u_{n+2}, \ldots, u_{n'})$, thanks to the orthogonality of the vectors $(u_n)_{n\in\mathbb{N}}$. As we have seen, all orthogonal projectors onto non-trivial subspaces have a unitary norm, that is, $\|A_{n'} - A_n\| = 1$ $\forall n < n'$, and thus the sequence $(A_n)_{n\in\mathbb{N}}$ is not a Cauchy sequence in $\mathcal{B}(\mathcal{H})$ with respect to the operator norm; thus, it cannot be convergent because $\mathcal{B}(\mathcal{H})$ is complete and so convergent and Cachy sequences coincide.

The sequence $(A_n x)_{n\in\mathbb{N}}$ in $\mathcal{H}$, however, converges to x for all $x \in \mathcal{H}$, by the fact that $(u_n)_{n\in\mathbb{N}}$ is a Hilbert basis. This highlights the need to define an alternative form of convergence in order to assign a precise meaning to the intuitive notion that the sequence $(A_n)_{n\in\mathbb{N}}$ converges to $id_{\mathcal{H}}$.

Similar examples are encountered in Banach and Hilbert spaces; for this reason, we have organized our presentation of alternative forms of convergence into separate sections for different spaces.

A3.1. Strong and weak convergence in Banach spaces

Let $(V, \|\ \|)$ be a Banach space. By definition, a sequence $(x_n)_{n\in\mathbb{N}} \subset V$ converges toward $x \in V$ if $\|x_n - x\| \xrightarrow[n\to+\infty]{} 0$. A different type of convergence can be defined in V by using the continuous linear functionals of its dual space V^*.

DEFINITION A3.1 (Weak convergence in a Banach space).– *Let V be a Banach space. The sequence $(x_n)_{n\in\mathbb{N}} \subset V$* converges weakly *toward $x \in V$ if, for all $\varphi \in V^*$:*

$$\varphi(x_n) \xrightarrow[n\to+\infty]{} \varphi(x)$$

where the convergence in this case is that of sequences of scalars in $\mathbb{K}$. x is the weak limit *of the sequence $(x_n)_{n\in\mathbb{N}}$ and we write $x_n \xrightarrow[n\to+\infty]{w} x$, with w for* weak.

We note that, for all $\varphi \in V^*$ and $x \in V$, $\|\varphi(x)\| \leqslant \|\varphi\|\,\|x\|$, thus, if $\|x_n - x\| \xrightarrow[n\to+\infty]{} 0$, then:

$$\|\varphi(x_n) - \varphi(x)\| = \|\varphi(x_n - x)\| \leqslant \|\varphi\|\,\|x_n - x\| \xrightarrow[n\to+\infty]{} 0$$

that is, "standard" convergence implies weak convergence. For this reason, "standard" convergence in a Banach space is also referred to as *strong convergence*.

Counter-examples show that the inverse is not generally true. Thus, in a Banach space, the topology defined by weak convergence has fewer opens than the topology defined by strong convergence.

A3.2. Strong and weak convergence in a Hilbert space

A Hilbert space $\mathcal{H}$ is also a Banach space, thus the definition of strong and weak convergence given above also applies to Hilbert spaces. Nevertheless, by the Riesz representation theorem, we know that the action of any continuous linear functional on $\mathcal{H}$ can be identified with a scalar product. For this reason, an equivalent definition, which is more explicit for the purposes of calculation, can be used for weak convergence in a Hilbert space.

DEFINITION A3.2 (weak convergence in a Hilbert space).– *Let $\mathcal{H}$ be a Hilbert space. The sequence $(x_n)_{n\in\mathbb{N}} \subset \mathcal{H}$* converges weakly *toward $x \in \mathcal{H}$ if, for all $y \in \mathcal{H}$:*

$$\langle y, x_n \rangle \xrightarrow[n\to+\infty]{} \langle y, x \rangle$$

As in the case of Banach spaces, x is said to be the weak limit *of the sequence $(x_n)_{n\in\mathbb{N}}$ and we write $x_n \xrightarrow[n\to+\infty]{w} x$.*

A very simple counter-example can be used to show that weak convergence does not generally imply strong convergence in a Hilbert space.

Take any $y \in \mathcal{H}$ and $x_n = u_n \ \forall n \in \mathbb{N}$, where $(u_n)_{n\in\mathbb{N}}$ is an arbitrary orthonormal system in $\mathcal{H}$. By Bessel's inequality $\sum\limits_{n\in\mathbb{N}} |\langle y, u_n \rangle|^2 \leqslant \|y\|^2$, so the series is convergent and thus its general term tends toward 0.

Since $\mathcal{H}$ is complete, any series which is absolutely convergent is convergent, hence $\sum\limits_{n\in\mathbb{N}} \langle y, u_n \rangle^2$ is convergent and then $\langle y, u_n \rangle^2 \xrightarrow[n\to+\infty]{} 0$; however, this holds if and only if $\langle y, u_n \rangle \xrightarrow[n\to+\infty]{} 0$ for all $y \in \mathcal{H}$.

Hence, *any orthonormal system $(u_n)_{n\in\mathbb{N}}$ in a Hilbert space is weakly convergent toward 0.*

However, we know that the distance between any two elements of an orthonormal system is $\sqrt{2}$: $\|u_n - u_m\| = \sqrt{2} \ \forall n, m \in \mathbb{N}$, thus $(u_n)_{n\in\mathbb{N}}$ does not verify the Cauchy condition, and therefore it cannot be strongly convergent.

A3.3. Uniform, strong and weak convergence in the Banach algebra $\mathcal{B}(\mathcal{H})$

In the Banach algebra $(\mathcal{B}(\mathcal{H}), \| \ \|)$, where $\mathcal{H}$ is any Hilbert space and $\| \ \|$ is the operator norm, three different convergences can be defined for a sequence of operators $(A_n)_{n\in\mathbb{N}} \subset \mathcal{B}(\mathcal{H})$.

DEFINITION A3.3.– *We shall use u, s and w to denote* uniform, strong *and* weak. *Let $(A_n)_{n\in\mathbb{N}} \subset \mathcal{B}(\mathcal{H})$ be a sequence of bounded linear operators on the Hilbert space $\mathcal{H}$, and take $A \in \mathcal{B}(\mathcal{H})$.*

– Uniform convergence (*standard convergence, in operator norm*):

$$A_n \xrightarrow[n\to+\infty]{u} A \iff \|A_n - A\| \xrightarrow[n\to+\infty]{} 0$$

– Strong convergence:

$$A_n \xrightarrow[n\to+\infty]{s} A \iff A_n x \xrightarrow[n\to+\infty]{} Ax \iff \|A_n x - Ax\|_{\mathcal{H}} \xrightarrow[n\to+\infty]{} 0 \quad \forall x \in \mathcal{H}$$

– Weak convergence:

$$A_n \xrightarrow[n\to+\infty]{w} A \iff \langle y, A_n x\rangle \xrightarrow[n\to+\infty]{} \langle y, Ax\rangle \quad \forall x, y \in \mathcal{H}$$

As we saw at the beginning of this appendix, for any Hilbert basis $(u_m)_{m\in\mathbb{N}}$, the sequence:

$$\begin{aligned} A_n : \mathcal{H} &\longrightarrow \mathcal{H} \\ x &\longmapsto A_n x = \sum_{m=0}^{n} \langle x, u_m\rangle u_m \end{aligned}$$

does not converge uniformly $id_{\mathcal{H}}$. However, it converges strongly towards the identity operator, since, by the continuity of the norm, we have:

$$\lim_{n\to+\infty} \|A_n x - id_{\mathcal{H}}(x)\|_{\mathcal{H}} = \|\lim_{n\to+\infty} A_n x - x\|_{\mathcal{H}} = \|\sum_{m\in\mathbb{N}} \langle x, u_m\rangle u_m - x\|_{\mathcal{H}} = 0$$

having used the fact that $id_{\mathcal{H}}(x)$ is not dependent on n and the generalized Fourier expansion on the Hilbert basis $(u_n)_{n\in\mathbb{N}}$.

It is possible to show that, in $\mathcal{B}(\mathcal{H})$, uniform convergence implies strong convergence, which itself implies weak convergence. On the other hand, as we see from the example shown above, strong convergence does not imply uniform convergence. Other counter-examples can be used to show that weak convergence in $\mathcal{B}(\mathcal{H})$ does not imply strong convergence.

References

Abbati, M. and Cirelli, R. (1997). *Metodi matematici per la fisica – Operatori lineari negli spazi di Hilbert*. Città studi, Milan.

Bartle, R. (1966). *The Elements of Integration*. John Wiley & Sons, Hoboken.

Berberian, S. (1961). *Introduction to Hilbert Spaces*. Oxford University Press, Oxford.

Boggess, A. and Narcowich, F. (2015). *A First Course Wavelets with Fourier Analysis*. John Wiley & Sons, Hoboken.

Briane, M. and Pagè, G. (1998). *Théorie de l'intégration – cours et exercices*. Vuibert, Paris.

Debnath, L. and Mikusinski, P. (2005). *Introduction to Hilbert Spaces with Applications*. Academic Press, Cambridge.

Dunford, N. and Schwartz, J. (1958). *Linear Operators, Part 1*. Wiley Interscience, Hoboken.

El Hage Hassan, N. (2011). *Topologie générale et espaces normés : cours et exercices corrigés*. Dunod, Paris.

Frazier, M.W. (2001). *Introduction to Wavelets through Linear Algebra*. Springer, Berlin.

Gasquet, C. and Witomski, P. (2013). *Fourier Analysis and Applications: Filtering, Numerical Computation, Wavelets*, vol. 30. Springer Science & Business Media, Berlin.

Moretti, V. (2013). *Spectral Theory and Quantum Mechanics*, vol. 64. Springer, Berlin.

Saxe, K. (2000). *Beginning Functional Analysis*. Springer, Berlin.

Sondaz, D. (2010). *Bien maîtriser les mathématiques : limites, applications continues, espaces complets*. Cépaduès, Toulouse.

Vretblad, A. (2003). *Fourier Analysis and Its Applications*. Springer, Berlin.

Yosida, K. (1995). *Functional Analysis*. Springer-Verlag, Berlin-Heidelberg.

Index

G, H, I

K, L, M

N, O

P, R, S

T, U

2007

HARLAMOV Boris
Continuous Semi-Markov Processes

2006

CLERC Maurice
Particle Swarm Optimization